SI Conversion Factors

To Convert from	to	Multiply by
STRESS/PRESSURE		
pounds per square inch	pascal	6895
pascal	pounds per square inch	0.000 145
ENERGY		
foot·pound	joule	1.36
joule	foot·pound	0.738
Btu	joule	1055.
Btu	calorie	252.
calorie	joule	4.19
calorie	Btu	0.003 97
kilowatthour	kilocalorie	860.
kilowatthour	Btu	3413.
POWER		
horsepower	kilowatt	0.746
kilowatt	horsepower	1.34
heat rate (Btu/hr)	horsepower	0.000 393
MOMENT (TORQUE)		
pound·foot	newton·meter	1.36
pound·inch	newton·meter	0.113
VELOCITY		
feet per second	miles per hour	0.682
miles per hour	feet per second	1.47
feet per second	meters per second	0.305
meters per second	feet per second	3.28
VOLUME FLOW (WATER ONLY)		
cubic feet per minute	gallons per minute	7.48
cubic feet per minute	pounds per minute	62.4
gallons per minute	cubic feet per minute	0.1337

Note: 1 joule = 1 newton·meter

Introduction to Applied Physics

Introduction to Applied Physics

Second Edition

ABRAHAM MARCUS

JAMES R. THROWER

Massasoit Community College

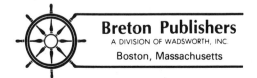

Breton Publishers
A DIVISION OF WADSWORTH, INC.
Boston, Massachusetts

Breton Publishers
A Division of Wadsworth, Inc.

Library of Congress Cataloging in Publication Data

Marcus, Abraham.
    Introduction to applied physics.

    Includes index.
    1. Physics. I. Thrower, James R. II. Title.
QC23.M338 1985   530   84–29207
ISBN 0–534–04746–7

Printed in the United States of America
1 2 3 4 5 6 7 8 9—89 88 87 86 85

ISBN 0-534-04746-7

Interior design by Ellie Connolly
Cover design by Trisha Hanlon
Cover photo courtesy of Automatix, Inc.

Contents

Preface

In recent years, the student population in career-related programs has become extremely diversified. Students with varied backgrounds, skills, and abilities are increasingly enrolled in technical career programs that require technical physics courses. The standard texts in "technical physics" often prove to be too much for many of these students.

This text was written with this broad range of readers in mind. Coverage is limited to the aspects of physics that prospective technicians must know to understand concepts taught in their other courses. The only mathematics prerequisite is elementary algebra.

By combining the principles of physics and the applications required by vocational technical students, the text lets the career-oriented student know why particular topics are important in the world of work. The examples and practical applications make it easier to grasp and understand the more difficult conceptual material.

The purpose in preparing this second edition has been to expand explanations in areas that cause students unnecessary difficulty, to add end-of-chapter problems that will aid the learning process, and to provide computer programs that will give students related computer operating experience. The text remains organized into "parts." Part I provides a general introduction to physics and measurement. The remaining parts are based on the type of energy discussed in the chapters. Part II covers mechanics; Part III is on heat; Part IV, sound and light; and Part V, electricity.

Each part begins with a discussion of physical principles and concepts. Each continues with descriptions of industrial applications and examples of typical problems encountered by practicing technicians. For example, in Part III, the concept of heat energy is introduced in Chapter 9. Solar heating and the solution of heat-transfer problems related to making buildings energy efficient are discussed in Chapter 10.

Most students should cover Parts I and II first because they present concepts basic to understanding any topic in physics. The topics covered in these two parts will be referred to throughout the text. Beyond Parts I and II, coverage and organization of topics is flexible, depending on the wishes of the instructor and the needs of the students. There is sufficient material for a two-semester course, but the text can be used for one semester by omitting selected topics.

The following learning aids are featured in this second edition:

> Each chapter begins with a set of objectives.
> Key terms and definitions are highlighted throughout the text.
> Key formulas are highlighted in the text and indexed on the back endpapers.
> Numerous sample problems, including metric problems, are presented with labeled, step-by-step solutions.
> A table of SI conversion factors is included on the front endpapers for easy reference.
> A summary is located at the end of each chapter.
> A list of key terms and formulas is provided at the end of each chapter.
> Practice problems appear at the end of each chapter and are grouped according to the type of application.
> Computer programs are presented at the ends of most chapters for use with the chapter's problems.

While the computer programs presented in this book are not vital to learning physics, they are intended to increase the user's appreciation for a computer's abilities and to encourage the use of computers in appropriate situations. The programs are written in BASIC, a very popular language that most small computers can use. In most cases, only minor variations in commands need be made in the programs to accommodate different brands of computers.

Many people assisted in the final preparation of the manuscript and bound book. I wish to thank the following professors who reviewed the manuscript for this second edition and who offered many valuable suggestions: Robert L. Petersen, College of Marin, CA; Eda Davis-Butts, Savannah Technical, GA; Doyle Davis, New Hampshire Vocational-Technical College-Berlin, NH; Paul Feldker, St. Louis Community College at Florissant Valley, MO.

I am also grateful to the editorial and production staff at Breton Publishers, particularly George Horesta, who provided support throughout the manuscript development phase, and Sylvia Dovner, who skillfully guided the book through the various stages of production. Finally, I wish to thank my family for their patience and support throughout the entire process.

James R. Thrower

Part I

Introduction to Physics

Molten iron from the blast furnace is charged into a basic oxygen furnace at a Bethlehem Steel plant. Once charged, the furnace will return to its upright position for the oxygen "blow." That is, oxygen is forced into the furnace and through the molten metal to burn off impurities. This process "refines" the iron into steel, which is then poured into ingot molds where the steel cools to a solid state. (Courtesy of Bethlehem Steel Corp.)

Chapter 1

When you finish this chapter, you will be able to:

☐ Discuss the organization of knowledge of natural events (science) and, in particular, how physics fits into that organization.

☐ Describe situations in which physics is applied in ways that involve technicians.

☐ Analyze word problems by identifying the question to be answered, selecting a likely equation that answers the question directly, substituting known data into the formula, and arithmetically working out the solution.

☐ List and compare the three states of matter.

☐ Sketch and describe the model of an atom showing location of neutrons, protons, and electrons. Indicate where electrical charge is to be found.

☐ Compare atoms and molecules.

What Is Physics?

This chapter will attempt to answer the questions: "What is science? What is physics? Why do technicians need to know physics?" We then begin the study of physics with a discussion of the atomic structure of matter. This information will be useful in later chapters because explanations of physical phenomena often must refer to the atoms and molecules of a substance. For example, after studying this chapter, we will have a clearer understanding of such discussions in this text as how a light bulb gives off light, what happens when ice melts, or why the pressure of a gas in a container increases when heated.

1.1 Science

At all times, human beings have tried to understand and explain the world we live in. To accomplish this task, we have developed an accurate system of collecting, arranging, and explaining important facts, like the change of seasons and the cause of lightning and thunder.

We call this system for organizing knowledge **science.** The knowledge is obtained by two means—observation and planned experimentation. To illustrate, primitive people saw lightning and explained it by saying that an angry god was hurling thunderbolts at an enemy. Early scientists, however, observed a relationship between lightning and the clouds in the sky. They noticed that lightning occurred most frequently in the summer, during hot weather. They smelled a peculiar odor when the lightning bolt struck nearby. These, and other, observations about lightning were carefully noted and collected.

---

* science

The organization of knowledge of natural events. This knowledge is obtained by observation and planned experimentation.

Having observed these facts, scientists were then able to go a step further—they experimented. For example, Benjamin Franklin sent a kite up during a thunderstorm and drew a spark from the end of its wet string. Thus, he had a clue about the relationship between lightning and electricity. Soon scientists were creating miniature lightning flashes by means of electric current. Today, we are able to produce considerable flashes using millions of volts of electricity. The information gathered from all

PHOTO 1.1   Experiments have shown that lightning flashes occur in two stages. The first stage is the formation of invisible electric currents in the clouds. The second stage is the contact of these currents with an object, such as a tree or the ground, creating the highly visible flash of light. (Courtesy of the National Oceanic and Atmospheric Administration)

these experiments was added to the facts already collected, so that now we have a fairly good idea of what causes lightning.

Our observations and experiments with lightning have not stopped. The process still goes on. Recently, photographs of lightning flashes gave us new facts. As a result, our explanations had to be altered slightly. In this way, we continue to increase our knowledge.

When new facts are discovered, they are carefully examined and placed in their proper classifications. When scientists are faced with a phenomenon that they have no explanation for, they examine all the facts known about it and try to develop an explanation that seems to describe these facts. This tentative explanation is called a *hypothesis* or *theory*.

Scientists are constantly devising new experiments to check the accuracy of existing theories. Starting with the theories, scientists make predictions concerning the behavior of substances under a given set of circumstances. Experiments are then conducted to check the accuracy of these predictions. In this way, evidence is gathered to confirm or disprove the theories.

When evidence is produced that tends to discredit a theory, the theory may be modified to include this new evidence. If modification is impossible, a new theory is formulated. Theories sometimes stand for many years before they are proved true or false. When enough evidence is gathered to prove the theory correct in all instances, the theory is said to be a *law of nature*. Newton's laws of motion, which we will discuss later, are examples of laws of nature.

1.2 Organization of Science and Scope of Physics

Simply collecting knowledge is not enough. We must arrange, or organize, our knowledge and apply names to the various groups or classes of knowledge.

Scientists have given a name to the class of knowledge that deals with living things. They call it *biology,* the science of life. Even this category is too large, so the science of biology has been broken down into two other groups: *zoology,* which deals with animals, and *botany,* which deals with plants. There are even subdivisions of these two groups.

Knowledge concerning nonliving things has also been arranged in several categories. *Astronomy* is the science of the heavenly bodies: the sun, moon, planets, stars, and galaxies. *Geology* is the science of the composition and structure of the earth and the history of its development. *Chemistry* examines the makeup of substances and their interactions. That is, chemistry is useful in answering the question "How are substances made?" In the field of chemistry, investigations are made into the interactions and transformations that occur when two or more substances are brought together. For example, what happens when oxygen and iron combine? A new material is formed. It is called iron oxide (commonly known as rust). Chemistry is closely related to another branch of science called physics. **Physics** is the science of matter and energy and of the interactions between the two.

---

* physics

The science that deals with matter (solids, liquids, and gases), especially its relationship to energy (excluding the fields that are covered by biology and chemistry).

What is energy? When you stand in front of a campfire and feel warmth, you are receiving energy. When gasoline is burned in an engine, energy is released to run the automobile. Energy will be defined in more detail in Chapter 6.

In this text, we will want to know what happens to a substance (whether a solid, liquid, or gas) when it is pushed, pulled, squeezed, moved, or heated. That is, what happens when some form of energy is applied to a substance? We will also want to know how energy is transmitted and how it affects matter. The characteristics (properties) of different types of energy will be discussed as well.

1.3 Organization of the Text and the Importance of Physics to Technicians

This text is divided into five parts. Except for this first introductory section, each part is determined by the type of energy involved. Part II is titled "Mechanics" and refers to mechanical energy. The other sections discuss heat, sound, light, and electricity.

The organization of the text is intended to aid future technicians' understanding of technology. **Technology** means "applied science," especially science as applied by industry. As a technician, you will be expected to use your head as well as your hands. You will have to apply your knowledge of physics to help solve industrial problems. You will be assisting engineers or working on your own. For example, suppose you are asked to specify what is required to raise a 4000 pound automobile on a hydraulic lift with a mere 20 pound push of a piston. To perform your job, you must understand the appropriate laws of physics and how to apply them.

∗ technology   The application of science to industrial operations.

Not the least of your problems will be communicating with people at the extremes of the technical spectrum. You must understand, and be able to talk with, engineers and scientists. They, like you, will be dealing with the laws of physics daily. It is also quite likely that you will have to explain the operation of a machine, or machine part, to someone with a very limited knowledge of physics.

You will be expected to know how to use data and formulas from handbooks. For instance, you may have to refer to a table listing the insulating properties of various materials and to use this data to calculate the heat loss through a wall. Or you may have to use a graph that relates the horsepower of a gear to the number of teeth and speed of rotation. All this information comes from the science of physics.

1.4 General Procedure for Solving Word Problems

The sample problems in this text are grouped according to the concept being discussed. Each concept area usually includes a formula that is to be used in solving a word problem. The problems are set up according to the following steps, which are helpful when solving any type of word problem:

Steps for Solving Word Problems

Step 1 Wanted: Identify the question being asked, and write down exactly what you need to find for the answer.

Step 2 Given: Write down all the given values. (Some of the values may have to be taken from tables in the chapter or in the appendix.) Draw a sketch of the setup, if appropriate.

Step 3 Formula: Write down on the left-hand side of your paper the formula you need to obtain your answer. This practice helps you to keep your work organized and aids the instructor in reviewing your work. Generally, you will have just one formula to deal with in each concept area. Remember that a formula can be used to find only one unknown value at a time; all other variables must be given, or be obtainable from tables of data.

Step 4 Calculation: Rearrange the formula if necessary, substitute the known information in the formula, and solve for the unknown value.

Let us follow these steps to solve Sample Problem 1.1.

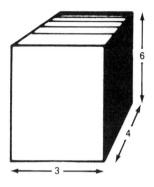

FIGURE 1.1 Block for Sample Problem 1.1.

SAMPLE PROBLEM 1.1	The block shown in Figure 1.1 is 3 in. wide, 4 in. deep, and 6 in. high. Find the volume (V).
SOLUTION	
Wanted	We wish to find the volume: $V = ?$
Given	We are told: $w = 3$ in., $d = 4$ in., and $h = 6$ in.
Formula	The formula for the volume of a block is:
	$$V = w \times d \times h$$
Calculation	$V = 3$ in. $\times$ 4 in. $\times$ 6 in. $= 72$ in.^{3}

1.5 What Is Matter?

We begin our study of physics with a general discussion of the atomic and molecular structure of matter. We have said that physics is the science that deals with matter. What do we mean by "matter"? All around us are a great many things: air, water, wood, iron, stone, plants, animals, and so forth. We call these things **matter**.

∗ matter	Any substance or material.

We know that matter appears in three forms: solid, liquid, and gas. (Note that some scientists consider plasmas, described later, as a fourth form.) *Solids,* such as a lump of iron, have a definite shape and volume. If the lump of iron were placed in a can, it would retain its original shape and volume, regardless of the shape or size of the can.

Liquids, such as water, have a definite volume, but their shape depends upon the vessel in which they are placed. If you pour a certain amount of water into a can, the volume of the water remains the same regardless of the size of the can. But the water assumes the shape of that portion of the can it occupies.

Gases, such as air, not only take the shape of their container, but their volume, too, depends upon that of the container. Thus, if air is placed in an empty can, it will fill the entire can regardless of its size. (By "empty," we mean a container without any gas or other substance in it—that is, a *vacuum.*)

The form, or state, of matter may vary according to conditions. For example, ice is a solid. If we heat ice, it melts and becomes water, a liquid. If we heat the water enough, it changes to steam, a gas. For another example, consider wax, which is a solid. If we apply pressure to the wax, we can make it flow like a liquid and take on the shape of the containing vessel. Again, carbon dioxide is a gas, but if we cool it enough, it may become a solid, called dry ice.

All matter, regardless of state, has two things in common: It takes up space, and it has weight. If air is blown into a balloon, the air will force the sides of the balloon out. If an empty steel tank is filled with a gas—even hydrogen, the lightest of all substances—the tank will weigh more with the gas than when it was empty.

1.6 The Composition of Matter

People have constantly sought to explain the composition of matter. At first, matter was thought to be continuous—that is, a solid mass rather than a lot of particles with no spaces within it. So far as we know, the ancient Greeks were the first people to suspect that matter was granular in nature—that is, made up of a large number of tiny, invisible particles tightly packed together.

The Greeks reasoned that if you were to take one of the countless substances in the world—water, for example—and divide it enough times, you eventually would reach a particle so small that it could no longer be divided. Nevertheless, this particle would exhibit all the properties and tendencies of the original substance. We call this ultimate particle of a substance a *molecule,* and we know that there are as many different kinds of molecules as there are substances.

Ideas about the composition of matter did not advance until about the beginning of the nineteenth century. At that time, the molecular theory had to be revised, chiefly as a result of the work of John Dalton (1766–1844), an English scientist. It was found that many substances could be broken down into two or more simpler substances that had no resemblance to the original. For example, an electric current that was passed through water would break down the water into two gases, oxygen and hydrogen. Thus, certain molecules could be broken down into smaller particles, and the molecule could no longer be considered the smallest particle of matter.

Since the time of Dalton, scientists have found that there are 92 naturally occurring substances that cannot be broken down into simpler substances. These substances are called **elements.** A substance that can be broken down into two or more distinct elements is called a **compound.** Water is a compound; oxygen and hydrogen are elements.

---

* **element** A substance that cannot be broken down into simpler substances.

* **compound** A substance that can be broken down into two or more distinct elements.

The smallest particle of an element is called an **atom,** and the smallest particle of a compound is a **molecule.** Although molecules are usually made up of a group of atoms bound together, some inert gases with only one atom qualify as molecules. There are as many kinds of molecules as there are kinds of matter, but there are only 92 kinds of atoms (plus more than 14 synthetic, or man-made, ones). Just as a few different types of bricks may be used to build a great many types of buildings, so these relatively few types of atoms may be combined to form the enormous number of known substances. The atom may be considered the building block of matter. A list of both natural and synthetic elements is presented in Table 1.1.

TABLE 1.1 The Elements and Their Symbols

Number	Element	Symbol	Number	Element	Symbol
1	hydrogen	H	56	barium	Ba
2	helium	He	57	lanthanum	La
3	lithium	Li	58	cerium	Ce
4	beryllium	Be	59	praseodymium	Pr
5	boron	B	60	neodymium	Nd
6	carbon	C	61	prometeum	Pm
7	nitrogen	N	62	samarium	Sa
8	oxygen	O	63	europium	Eu
9	fluorine	F	64	gadolinium	Gd
10	neon	Ne	65	terbium	Tb
11	sodium	Na	66	dysprosium	Dy
12	magnesium	Mg	67	holmium	Ho
13	aluminum	Al	68	erbium	Er
14	silicon	Si	69	thulium	Tm
15	phosphorus	P	70	ytterbium	Yb
16	sulfur	S	71	lutecium	Lu
17	chlorine	Cl	72	hafnium	Hf
18	argon	Ar	73	tantalum	Ta
19	potassium	K	74	tungsten	W
20	calcium	Ca	75	rhenium	Re
21	scandium	Sc	76	osmium	Os
22	titanium	Ti	77	iridium	Ir
23	vanadium	V	78	platinum	Pt
24	chromium	Cr	79	gold	Au
25	manganese	Mn	80	mercury	Hg
26	iron	Fe	81	thallium	Tl
27	cobalt	Co	82	lead	Pb
28	nickel	Ni	83	bismuth	Bi
29	copper	Cu	84	polonium	Po
30	zinc	Zn	85	astatine	At
31	gallium	Ga	86	radon	Rn
32	germanium	Ge	87	francium	Fr
33	arsenic	As	88	radium	Ra
34	selenium	Se	89	actinium	Ac
35	bromine	Br	90	thorium	Th
36	krypton	Kr	91	protactinium	Pa
37	rubidium	Rb	92	uranium	U
38	strontium	Sr	93	neptunium	Np
39	yttrium	Y	94	plutonium	Pu
40	zirconium	Zr	95	americium	Am
41	niobium	Nb	96	curium	Cm
42	molybdenum	Mo	97	berkelium	Bk
43	technetium	Tc	98	californium	Cf
44	ruthenium	Ru	99	einsteinium	Es
45	rhodium	Rh	100	fermium	Fm
46	palladium	Pd	101	medelevium	Md
47	silver	Ag	102	nobelium	No
48	cadmium	Cd	103	lawrencium	Lw
49	indium	In	104	unnilquadium	Unq
50	tin	Sn	105	unnilpentium	Unp
51	antimony	Sb	106	unnilhexium	Unh
52	tellurium	Te	107	unnamed	Un
53	iodine	I	108	none yet	
54	xenon	Xe	109	made in 1983 but unnamed	
55	cesium	Cs			

* atom	The smallest particle of an element.
* molecule	The smallest particle that exhibits all the properties and tendencies of the original substance.

1.7 The Electron Theory of Atomic Structure

For nearly one hundred years, the atom was considered the smallest particle of matter. During this period, however, many scientists all over the world were digging more deeply into the makeup of the atom. Finally, in 1897, Joseph J. Thomson, an English physicist, announced that he had definite proof that atoms, under certain conditions, shot out smaller particles of matter.

Each of these tiny particles, too small to be seen by the most powerful microscope, carries a charge of electricity. For that reason, they are called *electrons*. The amazing thing about electrons is that they are all alike, regardless of what substance emits them. Here, then, is a true building block of matter.

Once this discovery was made, scientists probed deeper into the structure of the atom. We shall not attempt here to review the various experiments and calculations that were performed in order to discover what an atom is made of. It is enough for our purposes to consider the conclusions that were reached. However, we must keep in mind that these conclusions are still in the theoretical stage and are subject to modification.

Today, most scientists believe in the **electron theory**, in which the atomic structure of matter is broken down into three types of smaller, or subatomic, particles. These subatomic particles are the **electron,** which carries a negative electrical charge; the **proton,** which carries a positive electrical charge; and the **neutron,** which carries no electrical charge.

* electron theory	A scientific theory stating that all matter is composed mainly of three types of building blocks: electrons, protons, and neutrons.
* electron	A subatomic particle that carries a negative electrical charge.
* proton	A subatomic particle that carries a positive electrical charge.

∗ neutron   A subatomic particle that carries no electrical charge.

All atoms are composed of these particles; the atoms differ from one another only in the number of particles they contain and in the arrangement of these particles. Niels Bohr (1882–1962), a Danish scientist, has given us the picture of atomic structure that is most widely accepted.

According to Bohr, the atom is composed of a central **nucleus**, which is surrounded by revolving electrons, somewhat as our sun is surrounded by revolving planets. In fact, the electrons that revolve around the nucleus are called **planetary electrons.** As shown in Figure 1.2, the nucleus contains all the protons and neutrons in the atom.

An atom of one element differs from an atom of any other element in the number of protons contained in the nucleus. The number of protons in the nucleus is called the **atomic number** of the element and varies from 1, for the element hydrogen, to 109, which has yet to be named. The atomic number of helium, whose structure is shown in Figure 1.2, is 2.

Since the proton is a particle carrying a positive charge, the nucleus contains a total positive charge that is equal to the number of protons present. In a normal atom, the positive charge of the nucleus is exactly neutralized by the negative charges of the planetary electrons revolving about it. Since the negative charge carried by an electron is equal and opposite to the positive charge carried by a proton, the normal atom has as many planetary electrons as it has protons in the nucleus. (An atom may temporarily have more protons than electrons, or vice versa, as we will see in later chapters.) Consequently, the number of planetary electrons revolving around the nucleus varies from 1 to 109. Note that in the helium atom shown in Figure 1.2, there are two protons and two planetary electrons.

Except for most hydrogen atoms, all atoms contain one or more neutrons in the nucleus. The helium atom (Figure 1.2) contains two neutrons; the uranium atom may contain 146 neutrons. The neutron, as we have said, carries no electrical charge.

Although the opposite electrical charges carried by an electron and a proton are equal in magnitude, the **mass** of a proton is about 1840 times as great as the mass of an electron. The mass of a neutron is about equal to that of a proton. We can readily see that almost the entire mass of the atom is contained in the nucleus.

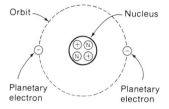

FIGURE 1.2 Theoretical Structure of an Atom. In this atom of the element helium, two planetary electrons revolve around a nucleus containing two protons and two neutrons.

---

* mass

The amount of matter that makes up any given substance or material in terms of its molecules or atoms or even its subatomic particles.

The number of protons and neutrons in the nucleus of an atom determines its **mass number.** The mass number of atoms varies from 1, for hydrogen, which is the lightest of the elements (one proton and no neutrons in its nucleus), to 238, for uranium, which is the heaviest naturally occurring element (92 protons and 146 neutrons in its nucleus). The mass number for the helium atom shown in Figure 1.2 is 4 (two protons and two neutrons).

All the atoms of the same element contain the same number of protons in their nuclei. The atoms of one element differ from those of any other element in this number of protons. For example, each hydrogen atom contains one proton in its nucleus, and each uranium atom has 92 protons. But the atoms of one element may differ in mass number because there may be different numbers of neutrons in their nuclei.

Figure 1.3 shows the theoretical structure of the hydrogen atom. In Figure 1.3A, we see an ordinary hydrogen atom with an atomic number of 1 (one proton) and a mass number of 1 (only one building block in the nucleus). In Figure 1.3B, we see the atomic structure of a rare form of hydrogen called deuterium. The deuterium atom has an atomic number of 1 (one proton) but a mass number of 2 (one proton and one neutron in the nucleus).

This atom has exactly the same properties as the atom shown in Figure 1.3A, except that it is twice as heavy.

We call these two different atoms of the same element **isotopes**. Most elements are known to have two or more isotopes. Natural uranium, for example, has three known isotopes. All the uranium isotopes have the same atomic number of 92 (92 protons in the nucleus). The most common isotope of uranium has the mass number 238 (92 protons and 146 neutrons). Over 99 percent of all natural uranium is of this type. Another isotope of uranium has mass number 235 (92 protons and 143 neutrons). About 0.7 percent of uranium in the natural state is of this type. This isotope is famous, since it is the first element that man was able to use for nuclear energy. A third type of uranium has mass number 234 (92 protons and 142 neutrons). This type is extremely rare, accounting for only about 0.006 percent of natural uranium.

∗ isotopes

Atoms of an element having the same number of protons but different numbers of neutrons.

It is interesting to note that scientists have been able to produce artificial isotopes by bombarding the nuclei of atoms with neutrons. In this way, an atom of uranium with a mass number of 238 sometimes captures a neutron, raising the mass number to 239 (92 protons and 147 neutrons).

So far, we have concentrated on the nucleus of the atom. Now let us turn our attention to the electrons revolving around

⊖ = Electron
⊕ = Proton
Ⓝ = Neutron

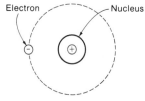

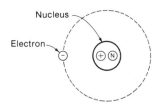

A. Hydrogen with a mass number of 1   B. Hydrogen with a mass number of 2

FIGURE 1.3   Isotopes of Hydrogen.

the nucleus of the Bohr model of the atom. As previously stated, the normal atom has one planetary electron for each proton in the nucleus. Thus, the number of such electrons will vary from 1, for hydrogen, to 92, for uranium (and higher for the synthetic elements).

These electrons do not revolve around the nucleus in a disorderly fashion. Their orbits, or paths, form concentric *shells,* or *layers,* somewhat like the layers of an onion. There is a certain maximum number of electrons that each shell can contain. If this number is exceeded, the excess electrons arrange themselves in the next outer shell.

The maximum number of electrons that each shell can contain follows a definite pattern. The shell nearest to the nucleus may contain up to two electrons ($2 \times 1^2 = 2$). If there are more than two electrons, the excess form a second shell around the first. This second shell may hold up to eight electrons ($2 \times 2^2 = 8$). Then a third shell will be formed that may hold up to 18 electrons ($2 \times 3^2 = 18$). The fourth shell may hold up to 32 electrons ($2 \times 4^2 = 32$), the fifth shell up to 50 electrons ($2 \times 5^2 = 50$), and the sixth shell up to 72 electrons ($2 \times 6^2 = 72$). In some complicated atoms, electrons may be found in outer shells even before some of the inner ones have been filled for reasons that require a rather involved explanation, which will be omitted here since it is not important to our understanding at this stage.

Figure 1.4A shows the theoretical structure of the helium atom. The nucleus contains two protons and two neutrons. The two planetary electrons revolve in a single shell around the nucleus. In Figure 1.4B, the structure of the carbon atom is shown. Six electrons revolve around a nucleus containing six protons and six neutrons. The first two electrons form the shell nearest to the nucleus. Since this shell can contain no more than two electrons, the remaining four electrons form a second shell. In Figure 1.4C, we see the structure of the lead atom. The 82 electrons are contained in six shells, or layers, around the nucleus. (Note that in Figure 1.4C the individual electrons are not illustrated as in Figure 1.4A and 1.4B, but are indicated with a number beside the circles.) The outer shell of the lead atom has four electrons, and the next-to-outer shell contains 18 electrons. We can see that in the lead atom, the sixth shell begins to fill before the fifth shell is complete.

Except for the electrons in the outermost shell, the particles of the atom are held together tightly, and relatively large forces are required to pry them apart. Most difficult to break apart is the nucleus of the atom: When we do succeed in exploding it, as occurs in nuclear reactors, an enormous amount of energy is

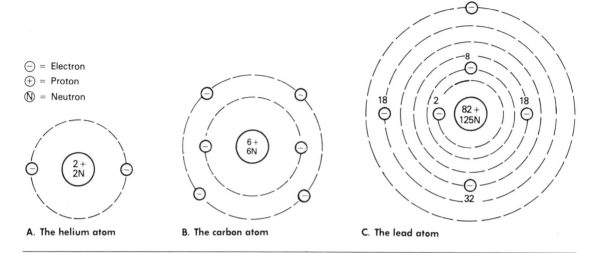

FIGURE 1.4   Theoretical Structure of Atoms with Varying Numbers of Electrons.

released. The next most difficult task is removal of the electrons in the shells closest to the nucleus. When these electrons are disturbed, we obtain energetic effects such as emission of X-rays. Least difficult to disturb are the electrons in the outermost shell. As a matter of fact, under normal conditions, electrons are constantly leaving and entering this outer shell.

When an atom gains or loses an electron, it is no longer electrically neutral; it has a negative or positive charge and is called an **ion.** A group of ions is called **plasma.** Associated with the electrons in the outer shell is the chemical and physical behavior of the atom. We will discuss this physical behavior later in the book.

＊ ion	A negatively or positively charged atom.
＊ plasma	A group of ions. Plasma is regarded by some scientists as the fourth state of matter.

We must pause at this point to consider the brilliant scientists who have pried into the secrets of atomic structure, who have weighed, measured, and explained these incredibly tiny and invisible particles. If 250 million hydrogen atoms were placed side by side, they would extend about one inch. And if 100,000 electrons were placed side by side, they would be as large across as a single hydrogen atom. How were these invisible particles weighed and measured?

It is impossible in an introductory book to explain the techniques and procedures of our greatest scientists. But to gain some idea of the methods used, assume that you wish to measure the weight of an object by means of a spring balance. You hang the object from the balance and look at the dial to find its weight. It makes no different whether the object is visible or invisible. You can tell its weight from the effect it has on the balance—by how far it stretches the spring. In other words, you weigh the object, not by seeing it, but by noting its effect on another object—in this case, the spring. So it was with these particles. They were measured and weighed by the effects they had on other particles—effects that could be seen and measured.

You may have noticed by now what is, perhaps, the most astonishing fact about the atom. The atom is made up almost entirely of empty space! The nucleus and the planetary electrons account for only a very tiny part of its volume, almost insignificant in size. Note this parallel to our solar system: The sun is at the center, and the planets move in their orbits around it. But between the sun and planets, and comprising most of the volume of the entire system, is empty space.

Summary

Science is the observation, description, experimental investigation, and explanation of natural phenomena. A theory, or hypothesis, is a tentative explanation of some natural phenomena. Theories become laws of nature when enough evidence is gathered to prove the theory correct in all cases. Physics is the science that deals with matter and energy and the connection between the two.

An object, or substance, or material, is said to be composed of matter. This matter may appear as a solid, liquid, or gas. The simplest substances are called elements. Oxygen, hydrogen, iron, and carbon are examples of elements.

When two or more elements combine, they form compounds. Water, sugar, alcohol, and salt are examples of compounds. The smallest particle of an element is an atom, and the smallest particle of a compound is a molecule.

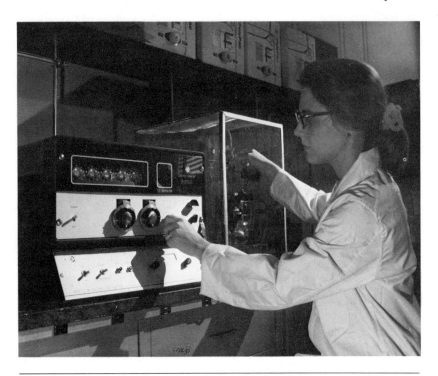

PHOTO 1.2   Modern electronic equipment can measure incredibly small quantities with high accuracy. This special counter can measure the size and number of nonmetallic impurities in a metal part. (Courtesy of Bethlehem Steel Corp.)

Atoms are composed of electrons, protons, and neutrons. The nucleus of the atom contains the neutrons and protons. The electrons orbit about the nucleus in shells, or layers. An atom that has gained or lost an electron is called an ion, and a group of ions is called plasma.

Key Terms

* science   * atom
* physics   * molecule   *2 or more Elements*
* technology   * electron theory
* matter   * electron
* element   * proton
* compound   * neutron

            * nucleus   * mass number
            * planetary electrons   * isotopes
            * atomic number   * ion
            * mass   * plasma

Questions

1. By what two means has the knowledge of science been obtained? *Observation & planned experiments*

2. A tentative explanation of some phenomenon is called *theory*.

3. A theory that has been proved correct in all instances is called *laws*.

4. The science that deals with matter in general, and its relationship to energy (excluding biology and chemistry), is called *physics*

5. State the three forms of matter. Give several examples to illustrate the fact that matter has weight and occupies space. *Solids liquid Gas.*

6. What do we mean by a compound? An element? Give three examples of each. *2 or more elements combine H2O, Alcohol, Salt. Simple substance O, H. Fe C*

7. Describe the theory of the structure of matter in terms of electrons, protons, and neutrons. *- Charge + Charge No charge*

PG 14   8. In terms of the Bohr model of the atom, how do the atoms of one element differ from those of another? How do the atoms of one isotope differ from the atoms of another isotope of the same element?

H 1 Proton   1 Neutron.   9. What do we mean by the atomic number of an element? The mass number? *#1 Protons in Nucleus. #1 " & Neutrons in Nucleus.*

10. Draw the hypothetical structure of an oxygen atom (atomic number = 8, mass number = 16).

11. If an atom has 4 electron shells, and each shell contains its maximum number of electrons, how many electrons are present? *36*

12. What element has 88 protons in its nucleus? 26 protons? *Radium   Iron*

13. How many electrons does the neon atom have? *10*

14. What do we mean by an isotope? Give an example. *C14 More Neutrons*

15. What is an ion? *+ or - Charge atom*

16. What is plasma? *4th form/matter   Group/ions.*

A thorough knowledge of measuring techniques and equipment is essential for technicians at all levels. Here, a mechanical technician uses an oversize micrometer. (Courtesy of the L.S. Starrett Company)

Chapter 2

Objectives

When you finish this chapter, you will be able to:

☐ Compare the SI metric and U.S. customary systems of measurement by listing the name and symbol for units of length (L), mass (M), time (T), and force (F) for both systems.
☐ Convert a numerical solution from one unit to another, as in re-expressing an answer in feet to one in miles.
☐ Distinguish between mass and weight (force) by considering a given mass and its weight if first placed on earth and then on the moon.
☐ Use conversion tables when changing expressions using one system of units to another expression using different units, like feet to meters.
☐ Use symbols for units, such as lb for pound.
☐ Use prefixes to shorten numerical notation, such as kilogram (kg) for 1000 grams, or picometer (pm) for 0.000 000 000 001 meter.

Measurements and Their Units

When we observe the physical world and want to describe events and objects, we must use physical terms that have specific meanings. In order to describe the amount or quantity of matter and the effect of energy on matter, we must use quantitative terms such as force, mass, length, and time. These quantities, and how they are measured, are defined in this chapter. Other quantitative terms such as density, stress, and heat are discussed in the appropriate sections of this text. This chapter also describes a prefix method of writing extremely large and extremely small numbers.

2.1   The Importance of Measurement

A small-town baker once brought a farmer to court. The baker complained that each day he purchased 1 pound of butter from the farmer, but the butter no longer weighed 1 pound. The judge, on learning that the farmer had a very good balancing scale but no balance weights, asked, "How can you weigh the butter with no balance weights?" "Well, your honor," the farmer replied, "each day I buy a 1 pound loaf of bread from the baker and use that."

Technicians use sophisticated instruments to measure physical quantities. So a technician not only needs to know how to make a measurement and what units to use, but he or she must also know something about the physical quantity being measured. For example, you have probably noticed that a pencil looks bent when placed in a glass of water. In the chapter on light, we will see that the light rays are actually bent, or refracted, when going from one medium (in this case, air) to another (water). For most people, this concept explanation would be sufficient. However, as a technician, you may be involved in building telescopes, cameras, microscopes, or other optical devices connected to electrical or mechanical equipment. In this situation, it is not enough to understand the concept; you must also be able to make appropriate and accurate measurements in order to know exactly "how much" bending occurs when light goes from one substance to another.

When making measurements, a technician must be familiar with the units of two different measuring systems. The **U.S. customary system,** which is familiar to most of us, includes such units of measure as ounces, pounds, inches, and feet. However, a conversion to the metric system, referred to as the **SI system,** is under way in the United States. The full name of the SI system is the International System of Measuring Units; the official name is in French and is "Le Système International d'Unités." As a technician, you will need to be able to solve problems in both systems and to convert units from one system to the other. For quick reference, a table of metric units with conversions to and from the U.S. system is presented inside the front cover of this book.

2.2 Defining and Measuring Force

Suppose you wished to remove a large boulder from your path. You could lift or push it out of the way, using your muscles. Or you might use a machine to do it for you. Another way might be to plant a charge of gunpowder beneath the boulder and blast it out of the way. In any event, a push or pull would be applied to the rock. This push or pull is called **force.**

---

* **force**

A push or pull applied to some object.

We say that force is the action of one body on another. Generally, the two bodies must be in contact with each other for the force to be applied, but not always. For example, if the boulder just mentioned were held up off the ground, the force of the gravitational attraction, or more simply, the force of **gravity,** of the earth would tend to pull it down.

---

* **gravity**

The attraction that any two objects, even separated by a distance, have for each other.

Assume that the boulder weighs 100 pounds. Thus, the gravitational attraction pulls it toward the center of the earth with a force of 100 pounds. If you tried to lift the boulder, you would have to overcome the pull of gravity by pulling up with an equal force—that is, 100 pounds. Thus, **weight** is the measure of the pull or force due to gravitational attraction. Of course, a force can be produced in many ways other than gravitationally, as we will discuss in later chapters.

---

* **weight**

A measure of the force of gravity.

The basic U.S. customary unit of force is the **pound.** The symbol is lb. One-sixteenth (1/16) of a pound is the ounce (oz). The basic unit of force in the SI metric system is the **newton,** whose symbol is N. Notice that the name is not capitalized but the symbol is. One-thousandth (0.001) of a newton is one milli-newton (mN), and a force of one thousand newtons is a kilonewton (kN). Notice that m in the symbol mN refers to the word milli, and the k in kN refers to the word kilo. In the SI system, milli means one-thousandth, and kilo means thousand. These terms, and others, will be described again later in this chapter.

Until the time comes when the United States has converted completely to the SI system, technicians must be adept at converting to and from the SI system units. To convert pounds to newtons, we multiply pounds by 4.45, since there are 4.45 newtons per pound. To convert newtons to pounds, we multiply newtons by 0.225. For example, to convert 12.0 pounds to newtons:

$$12.0 \text{ lb} \times 4.45 \text{ N/lb} = 53.4 \text{ N}$$

To convert 12.0 newtons to pounds:

$$12.0 \text{ N} \times 0.225 \text{ lb/N} = 2.70 \text{ lb}$$

In these examples, the conversion numbers (multipliers) are shown with their units of measure. However, handbook tables (and the tables in this text) leave the units off. So, when converting from one measuring system to another in this text, we will not bother stating the units of measure.

2.3 Defining and Measuring Mass

We saw in Chapter 1 that the **mass** of an atom depends upon the number of its protons, neutrons, and, to a lesser degree, electrons. Thus, to find the mass of an object, we would have to count each kind of atom in it and multiply by the mass of each kind of atom. Since this task is impossible, we must look for some other means to calculate mass.

The force of gravity can be used to compare the masses of two different objects. Suppose we have two rocks and a spring scale (a bathroom scale is a type of spring scale). One rock weighs 6 pounds, and the other weighs 12 pounds. Thus, the heavier rock is twice as massive as the lighter rock because the force of gravity increases directly as the mass of an object increases. Therefore,

PHOTO 2.1   At a certain distance between the earth and the moon, the gravitational attraction of the earth just balances the gravitation attraction of the moon. The astronaut shown here feels no gravitational attraction and thus feels weightless, but he still has mass. (NASA photo)

weight can be used to compare one mass with another. Making such comparisons can be useful in a lot of everyday situations. For instance, we know which box in the supermarket has more cereal (more mass) between the 15 ounce size and the 10 ounce size.

Because mass and weight are different, we must have a separate unit for mass; the force unit cannot be used. To illustrate the difference between mass units and force (in this case, weight) units, Figure 2.1 takes us on a trip to the moon. The moon is smaller than the earth, less massive, and therefore its gravitational attraction for other masses is less. In fact, the force of gravity on the moon is only one-sixth that on the earth. At home, you stand on your bathroom scale, and the scale might indicate 150 pounds. Then, you join the astronauts on a trip to the moon.

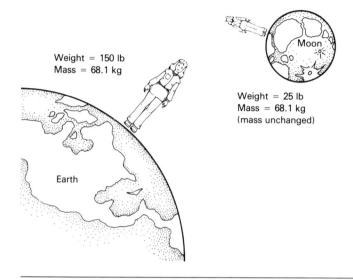

Weight = 150 lb
Mass = 68.1 kg

Earth

Moon

Weight = 25 lb
Mass = 68.1 kg
(mass unchanged)

FIGURE 2.1　Weight and Mass on Earth and on the Moon.

If you stand on your scale on the moon, it will read:

$$\frac{1}{6} \times 150 \text{ lb} = 25 \text{ lb}$$

What happened? You are still the same person as on earth. Your *mass* has not changed, but the force due to the gravitational attraction of the moon is less. Thus, the spring scale indicates a smaller *weight*. This example shows why we need to measure the mass of an object in such a way that this measure does not change when the object is subjected to various types of actions.

The U.S. customary unit of mass is the **slug.** An object with a mass of 1 slug has a weight (here on earth) of 32 pounds. In the SI system, the standard unit of mass is the **kilogram** (kg). One gram (g) is one-thousandth (0.001) of a kilogram. An object with a mass of 1 kilogram has a weight (here on earth) of 9.8 newtons, or in U.S. customary units, 2.2 pounds.

In spite of the fact that mass and force are different, we may want to convert mass units to force units, or vice versa. Why? Well, either set of units can be used for most everyday transactions where we want to *compare* the amounts, or masses, of competing products. In the United States, we are in the habit of using weight (force) units, but people in countries on the metric system are in the habit of using mass units. For example, packages in supermarkets have the contents listed in ounces (force units) as well as in grams or kilograms (mass units). And just about every physics laboratory has a spring scale with ounce graduations on one side and grams on the other. Thus, the conversion between mass units and force units is an important and useful procedure.

Table 2.1 shows how to convert between mass units and force units. Let's use it to convert 17 pounds to kilograms and slugs:

$$17 \text{ lb} \times 0.454 = 7.7 \text{ kg}$$
$$17 \text{ lb} \times 0.0311 = 0.53 \text{ slug}$$

Normally, it is not necessary to convert kilograms to slugs, even though both are mass units. Usually, we do not leave the answers in slugs, but in pounds.

2.4 Measuring Space and Time

The unit of length is, in reality, a unit of space. We measure the distance between two points by the number of units of length between them. An area depends on the number of units along the length of the surface and the number of units along the width. If volume is to be measured, we must know the number of units along the length, width, and height.

The standard unit of length in the SI system is the meter (m). (Note that "meter" is not the official spelling for the unit of length. The official spelling is "metre." However, until most handbooks are revised with this spelling, and to avoid unnec-

TABLE 2.1 Units of Force and Mass in the U.S. and SI Systems

To Convert from	To	Multiply by
Kilogram (mass)	Newton (force)	9.81
Newton (force)	Kilogram (mass)	0.102
Kilogram (mass)	Pound (force)	2.20
Pound (force)	Kilogram (mass)	0.454
Pound (force)	Slug (mass)	0.0311
Slug (mass)	Pound (force)	32.2

PHOTO 2.2   The atomic clock shown here, operated by the National Bureau of Standards, is one of the world's most accurate standards for time. If it were allowed to run without adjustment for one million years, the clock would still be accurate to better than 4 seconds. (Courtesy of the National Bureau of Standards)

essary confusion, we will continue with the spelling "meter.") One thousand meters is a kilometer (km), and one-thousandth of a meter is a millimeter (mm). The centimeter (one-hundredth of a meter) is not considered a standard in the SI system because the governing organization is trying to reduce the number of multiples and submultiples. Prefixes in steps of 1000 are recommended (Section 2.5 discusses prefixes). We are already familiar with the units of length in the U.S. customary system: the foot (ft) and, of course, one-twelfth (1/12) of a foot is the inch (in.). The yard, a U.S. customary unit equal to 3 feet, is rarely used in technical applications.

Table 2.2 lists conversions between units of length in the U.S. system and the SI system. These conversions, and others for areas and volumes, are listed in the Conversion Factors Table inside the front cover of this book. Let's try a sample problem.

TABLE 2.2   Units of Length in the U.S. and SI Systems

To Convert from	To	Multiply by
Foot	Meter	0.305
Inch	Meter	0.0254
Inch	Millimeter	25.4 (exactly)
Mile	Kilometer	1.61
Kilometer	Mile	0.621
Millimeter	Inch	0.0394
Meter	Foot	3.28

SAMPLE PROBLEM 2.1

As a small manufacturer, you are used to placing the following orders: 1/2 in. diameter steel rods in 12 ft lengths, and 150 ft² of insulation in rolls 5.0 ft wide by 30 ft long. However, to order this material from an overseas supplier, you need to use the SI system of units. Convert inches to millimeters, feet to meters, and square feet to square meters.

SOLUTION

For all the conversions, we simply look up the proper multipliers in a table in this chapter or at the front of the book, and apply them to the given dimensions. The rod diameter is small, so it must convert to millimeters; the other lengths will convert to meters.

Wanted

rod diameter in mm and other lengths in meters

Given

rod diameter = 1/2 in. and lengths = 12 ft, 5 ft, and 30 ft

Calculation

We convert inches to millimeters:

$$1/2 \text{ in.} \times 25.4 = 12.7 \text{ mm}$$

We convert feet to meters:

$$12 \text{ ft} \times 0.305 = 3.7 \text{ m}$$
$$5.0 \text{ ft} \times 0.305 = 1.5 \text{ m}$$
$$30 \text{ ft} \times 0.305 = 9.2 \text{ m}$$

We convert square feet to square meters:

$$150 \text{ ft}^2 \times 0.093 = 14 \text{ m}^2$$

Volumes for liquid measurements have different units. The standard for liquid measure in the SI system is the liter (L). One thousand liters is 1 kiloliter (kL), and one-thousandth of a liter is 1 milliliter (mL). In the U.S. customary system, the standard for liquid measure is the gallon (gal), and 4 quarts (qt) equal 1 gallon. Frequently, we need to convert from liquid measure units to and from regular volume units. The relationships in Table 2.3 will be helpful.

SAMPLE PROBLEM 2.2

Your company manufactures a centrifuge with a volume of 130 in.³ For use in sales literature going to different countries, determine the volume in quarts, liters, and cubic centimeters.

SOLUTION

Wanted

volume in quarts, liters, and cubic centimeters

Given

volume = 130 in.²

Calculation

We obtain the proper multiplier from the list in this chapter or from the Conversion Factors Table at the front, and we convert cubic inches to quarts:

130 in.³ × 0.0173 = 2.25 qt

TABLE 2.3 Units of Liquid Measure in the U.S. and SI Systems

To Convert from	To	Multiply by
Cubic centimeter	Cubic inch	0.0610
Cubic centimeter	Liter	0.001 00
Cubic inch	Cubic centimeter	16.39
Cubic inch	Liter	0.0164
Cubic inch	Quart	0.0173
Gallon	Liter	3.78
Liter	Cubic centimeter	1000.0 (exactly)
Liter	Cubic inch	61.0
Liter	Gallon	0.264
Liter	Quart	1.06
Quart	Cubic inch	57.8
Quart	Liter	0.946

We convert cubic inches to liters:

$$130 \text{ in.}^3 \times 0.0164 = 2.13 \text{ L}$$

We convert cubic inches to cubic centimeters:

$$130 \text{ in.}^3 \times 16.39 = 2131 \text{ cm}^3$$

In addition to space, time is an essential quantity in physics. Technicians must be able to measure such phenomena as velocity, acceleration, and vibration—events that require the measurement of time. The units of time—the second (s), minute (min), and hour (hr)—are the same in both the U.S. and SI systems.

2.5 How to Write Extremely Large and Extremely Small Numbers

It may be well at this point to consider the question of how to write very large and very small numbers. Occasionally in this book, we will encounter numbers so large that they become awkward to handle. Scientists and mathematicians have worked out several methods for dealing with such numbers.

One method is to use a series of prefixes that signify certain numerical quantities. (The word "prefix" comes from Latin and refers to letters that appear at the beginning of words or symbols to change their meanings.) Prefixes placed in front of SI system units, such as the meter and gram, indicate very large or very small quantities. Thus:

The prefix	tera (T)	means a trillion	(1 000 000 000 000)
The prefix	giga (G)	means a billion	(1 000 000 000)
The prefix	mega (M)	means a million	(1 000 000)
The prefix	kilo (k)	means a thousand	(1000)
The prefix	milli (m)	means a thousandth	(0.001)
The prefix	micro (μ)	means a millionth	(0.000 001)
The prefix	nano (n)	means a billionth	(0.000 000 001)
The prefix	pico (p)	means a trillionth	(0.000 000 000 001)

Note that the prefix "giga" is pronounced "jig-a," and the abbreviation for the prefix "micro" is the Greek letter mu, μ. Also notice that large numbers are not separated by commas but by

PHOTO 2.3   This picture of the earth taken by satellite shows South America in front and North America at the upper left. The diameter of the earth is 7918 miles, or 6 371 000 meters, or 6.371 megameters. (NASA photo)

spaces, in accordance with the SI metric system. A group of four digits may be written with or without the space (1000 or 1 000). With handwritten numbers, it is a good idea to use commas for clarity.

To see how the prefixes are used, let's suppose we know the mass of some object is 54 000 grams. We probably would need to leave the number as is if we were using it in a formula calling for grams. However, we would find it more convenient to use a prefix when writing a report. What prefix should we use? Note from the listing of the prefixes that they change in multiples of one thousand. How many thousand grams do we have? We have 54 thousand grams. Instead of using the word "thousand," we use the word "kilo" and tie it to the word "grams." We would write "54 kilograms."

Going down the scale below "one" may be more confusing, but again, the changes are in multiples of thousandths. For example, suppose the diameter of a metal rod is 0.008 meter (read: "eight-thousandths of a meter"). We go to the prefix list and note that the prefix "milli" can be substituted for "thousandths." Therefore, the diameter of the rod is 8 millimeters.

On the other hand, if we are given a prefix but must write the number without the prefix, we simply multiply the given number by the value represented by the prefix. To illustrate: The volume of an oil storage tank is 3500 kiloliters. We will write this number without the prefix. The prefix "kilo" tells us to multiply 3500 by 1000. So, $3500 \times 1000 = 3\,500\,000$ liters.

We can summarize the procedure for using prefixes as follows:

1. To convert a number in digits to a prefix number, determine the word describing the number and replace it with the SI prefix.
2. To convert a prefix number to a number in digits, multiply the given number by the value represented by the prefix.

SAMPLE PROBLEM 2.3	Rewrite the following numbers with proper prefixes: The force on a bridge support is 29 400 000 000 N, and a fatigue crack in the metal support is 0.000 125 m wide.
SOLUTION	We note that there are less than 30 billion newtons; in fact, the number can be stated as 29.4 billion newtons. The prefix corresponding to "billion" is "giga." So we write the number: 29.4 giganewtons (GN) The number 0.000 125 m can be written as 125 millionths of a meter. "Micro" is substituted for "millionths," and we write the number: 125 micrometers (μm)

SAMPLE PROBLEM 2.4

The moon is 240 000 miles from the earth. Convert this distance to meters and use the appropriate prefix.

SOLUTION

Wanted

distance in meters with the appropriate prefix

Given

distance in miles

Calculation

First, we want to convert the given distance from the U.S. system to the metric system. Using data from the Conversion Factors Table, we note that to convert from miles to kilometers, we must multiply by 1.61:

$$240\ 000 \times 1.61 = 386\ 000 \text{ km}$$

Now that we have the distance in kilometers, we wish to convert it to meters. The prefix "kilo" tells us to multiply 386 000 by 1000. Therefore:

$$386\ 000 \times 1000 = 386\ 000\ 000 \text{ m}$$

This number represents a value of 386 million meters, and the prefix for million is "mega." The answer as a prefix number is 386 megameters (Mm).

2.6 Significant Figures

The number of significant figures (or digits) has nothing to do with the position of the decimal point, but simply refers to the number of digits that are accurate, or true. All digits, except zero, are significant all the time. Zeros may or may not be significant. A zero is significant in the following circumstances:

When it is located between two nonzero digits—for example, in the number 105;

When it stands without a nonzero digit to the right of the decimal point—for example, in the numbers 12.0, 12.00, or 12.000 (all zeros are significant in each of these examples);

When it stands to the right of a decimal number—for example, in the numbers 0.120, 0.0120, or 0.001 20 (only the last zero in each of these examples is significant);
When it has an overbar—for example, in the number 8$\overline{0}$0.

A zero is not significant when it is used merely to locate the decimal point. For example, none of the zeros is significant in the following numbers: 80 000, 8000, 800, 80, 0.8, or 0.08.

It is common practice to "round off" answers when the last digits are in doubt or not needed. When the digit being dropped is a 5 or greater, we round the number up; if the digit being dropped is a 4 or less, we round the number down. For instance:

| | Rounded off to | | |
Number	4 figures	3 figures	2 figures
3.141 593	3.142	3.14	3.1
385 940	385 900	386 000	390 000
0.695 42	0.6954	0.695	0.70

Whether working with digits or zeros, the general procedure for multiplying and dividing is to round off the answer to the same number of significant digits as appears in the least accurate given data. Here is a concrete example. In U.S. customary units, the mile is defined as exactly 5280 feet. This number can be rounded off to 5300 feet (2 significant digits or figures). The two zeros are called place holders and represent the positions that can have digits varying from 0 feet to 99 feet. Therefore, if another person, who does not know the exact value, were shown the number 5300 feet, he or she would know that a more accurate value lies between 5300 + 50 (5350 ft) and 5300 − 50 (5250 ft). And so it does. Now we can use the 5300 value to calculate the circumference of a circle 1 mile in diameter. Pi is taken as 3.14 and is accurate to 3 significant figures in our calculations:

circumference = pi × diameter
circumference = 3.14 × 5300 feet = 16 642 ft

The calculator gives an answer with 5 digits, implying that we know the circumference accurately to 5 significant digits. Of course, this answer is inconsistent since we also know the answer can't be more accurate than the least accurate number being used—in this case, 2 significant digits. So, the answer is rounded to 17 000 feet. The true value is between 17 000 + 500 (17 500 ft) and 17 000 − 500 (16 500 ft).

When adding or subtracting, we perform the operation and then round off by eliminating any digits resulting from operations on broken columns on the right. For example:

$$
\begin{array}{r}
71.8 \\
5.66 \\
+ \ \ 0.333 \\
\hline
77.793
\end{array} \qquad \text{round to} \qquad 77.8
$$

Most engineering problems require an accuracy of only 3 or 4 significant figures. The data in this text will generally be given to three significant figures unless indicated otherwise.

Summary

Technicians are expected to take precise measurements and convert these measurements into useful information. They are required to use two systems of measurement: the U.S. customary system and the SI system (also called the metric system).

The measurements discussed in this chapter are force, mass, space, and time. The most confusing measurements are those of force and mass. A force is a "push" or "pull" of one object on another. Weight is a measure of the force of gravity between two objects—most commonly between the earth and the object we are weighing. The force units most useful to technicians are pounds and newtons.

Mass is the amount of matter that makes up an object, or substance. Mass units important to technicians are slugs and kilograms. For everyday comparisons of the mass of two or more objects, weight units are commonly used in the U.S. customary system. Here on earth, a mass of 1 kilogram has a weight of 9.81 newtons or 2.20 pounds. Also, a mass of 1 slug has a weight of 32.2 pounds or 143 newtons. However, when using formulas calling specifically for mass or force units, the proper units must be used; they cannot be interchanged.

To measure space, technicians generally have to convert miles, feet, and inches to and from kilometers, meters, and millimeters. Liquids are generally measured in quarts and gallons in the U.S. system, and in liters and milliliters in the SI system. Time is measured in seconds, minutes, and hours in both the SI system and the U.S. customary system.

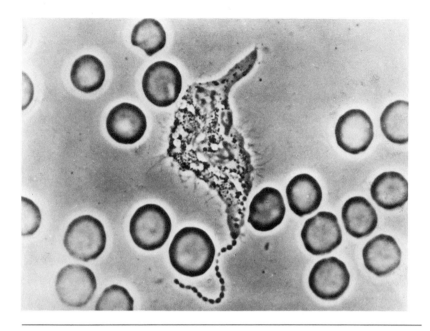

PHOTO 2.4   The irregularly shaped object in this photo is a white blood cell, devouring a chain of bacteria. The round disks surrounding the white blood cell are red blood cells. A red blood cell is about 0.0003 inch in diameter, or 0.000 008 meter, or 8 micrometers. (Courtesy of Pfizer, Inc.)

Key Terms

* U.S. customary system
* SI system
* force
* gravity
* weight

* pound
* newton
* mass
* slug
* kilogram

Questions

1. What two measuring systems must you be familiar with?
2. What is a force? What are its standard units in the U.S. customary and SI systems?
3. Define the term "weight."
4. What is the standard for mass in the SI system?

5. What is the difference between the mass and the weight of an object?

6. What is a slug?

7. What is the standard unit of length, and its symbol, in the SI system?

8. When used as a prefix, what do the following terms mean: mega-, micro-, kilo-, and milli-?

9. What is the procedure for writing prefixes?

Problems

Wherever possible, diagrams should be used to help clarify the answers to these problems. Although these diagrams need not be elaborate, they should be neatly drawn with the significant portions clearly labeled.

1. The force on a tool bit during a particular machining operation is 25 lb. In the SI system, this is _____ newtons.

2. When your product is packaged and ready for shipment, it weighs 38 lb. For overseas shipment, you must give this weight in equivalent SI mass units. What does your product weigh in SI units?

3. An automobile weighs $25\overline{0}0$ lb. Find its mass in kilograms.

4. For a particular engineering calculation, you must convert the weight of a $32\overline{0}$ lb satellite to U.S. customary mass units. Its mass is _____.

5. A man weighs 185 lb. What is his mass in (a) U.S. customary units? (b) SI metric units?

6. A carpenter must saw a floor joist 4.5 m long. How long is the joist in feet?

7. A welder is required to make a fillet weld 335 mm long. What is the length in inches?

8. A 3/4 in. diameter bolt has a diameter of _____ mm.

9. An industrial robot's forearm is 5.25 ft long. What is its length in meters?

10. How many kilometers is 25 mi equal to?

11. How many cubic feet are there in a cubic meter?

12. Your company manufactures 275 gal fuel oil tanks. Specify the capacity in liters for your overseas literature.

13. Susan owns a car with a $40\overline{0}$ in.³ engine. What is the size of the engine in liters?

14. Write 156 000 N with the proper prefix.

15. A balancing scale has a scale range of $0–15\overline{0}$ kg. This is equivalent to 0–_____ grams.

16. One year equals about 31 600 000 seconds. Write the number of seconds using the proper prefix.

17. A certain measuring instrument can measure lengths as short as 0.000 000 300 m. Write this length with the appropriate prefix.

18. A cross hair in a surveyor's transit is about $4\overline{0}$ μm in diameter. What is its measurement in meters?

19. An explosion may occur in one hundred-millionth of a second. Write this time interval with a prefix.

20. Write out with the proper prefix the number of seconds in 60 years.

Computer Program

Use this program to solve some of the chapter problems, or to place into the computer the weights of individuals so as to determine their masses. Also, use the program to answer the question "Does the mass double if the weight doubles?"

PROGRAM

```
10   REM   CH TWO PROGRAM
20   PRINT "PROGRAM TO CONVERT THE WEIGHT OF AN OBJECT OR PERSON TO "
21   PRINT "MASS IN SLUGS AND KILOGRAMS. TYPE WEIGHT IN POUNDS."
30   INPUT W
40   PRINT
50   PRINT "WEIGHT = ";W;" LB."
60   LET M = W * 0.0311
70   PRINT
80   PRINT "MASS = ";M;" SLUG."
90   LET K = W * 0.454
100  PRINT
110  PRINT "MASS = ";K;" KG."
120  END
```

NOTES

Lines 20 and 21 are used to control the length of each line printed so words won't be broken when listing.

Line 60 converts the weight to U.S. customary units—that is, slugs.

Line 90 converts the weight to SI units—that is, kilograms.

Part II

Mechanics

Some metals have the property of being plastic. The more plastic the metal, the more easily the force of hammer blows can deform and shape the metal part. (Courtesy of Department of Photography and Cinema, Ohio State University)

Chapter 3

Objectives

When you finish this chapter, you will be able to:

- [ ] Explain the kinetic theory of matter and demonstrate how it relates to gaseous, liquid, and solid states of matter.
- [ ] Explain that in solid objects, molecules attract most strongly because they are closer together, thus giving rise to the familiar characteristics of solids.
- [ ] Demonstrate how forces applied to solids may become either tensile or compressive forces.
- [ ] Recognize the stress equation $S = F/A$, showing the relation between stress (S), force (F), and area (A), and use this equation to solve for any one of the three symbols.
- [ ] Show how a stress applied to an object (like a spring) will cause that object to deform its shape (stretch or compress).
- [ ] Recognize the strain equation $e = \Delta L/L$, showing the relation between strain (e), stretch (ΔL), and original length (L), and use this equation to solve for any one of the three symbols.
- [ ] Recognize the elasticity equation $E = S/e$, showing the relation between elasticity (E), stress (S), and strain (e), and use this equation to solve for any one of the three symbols.
- [ ] Explain what is meant by elastic range, elastic limit, plastic range, and ultimate strength, and show how they are related.
- [ ] Demonstrate how various solids differ by comparing their ductility, malleability, brittleness, and hardness.

Mechanical Properties of Solids

Solids have numerous properties (or characteristics) classified as mechanical, electrical, chemical, thermal, and so on. What are properties? Properties help identify the various types of materials. Each material has its own set of properties that either affect our senses in a particular way or cause the material to behave differently from other materials when some form of energy is applied. For instance, sugar is sweet, glass is transparent, and grease is slippery. If mechanical energy is applied in the form of a hammer blow, rocks will crumble and copper will deform and flatten out. Thermal energy, in the form of heat, will melt lead at a lower temperature than steel.

In this chapter, we will cover some of the mechanical properties of solids. Mechanical properties are probably the most important for the technician to understand. Our discussion will cover primarily the strength of a solid (metals, in particular) and how metals deform under a load. Material selection for a metal product is an important task of technicians. For instance, in order to drill straight holes, we want a drill bit strong enough to resist bending. On the other hand, material for the roof of a car must bend enough under manufacturing operations so that the roof can be shaped in one operation of a press.

3.1 The Kinetic Theory of Matter

Considering the extremely small sizes of molecules and atoms, they are spaced relatively far apart. The distance between molecules depends on the state of matter. In solids, the molecules are close together. In liquids, they are farther apart, and in gases they are much farther apart. According to the **kinetic theory of matter** ("kinetic" comes from the Greek word meaning "to move"), the atoms and molecules of all substances are in constant motion. In gases and liquids, the movement is a random, zigzag motion; in solids, it is a vibrating, back-and-forth motion.

＊ kinetic theory of matter

A scientific theory stating that the atoms and molecules of all substances, regardless of whether the substance is a gas, a liquid, or a solid, are in constant motion.

Although the molecules of matter are relatively far apart and in a constant state of motion, there is, nevertheless, an attraction between them. This attraction depends on the nature of the molecules and the distances between them. The closer the molecules are, the greater is the attraction. Thus, the molecules of solids attract each other more strongly than do the molecules of liquids, and those of liquids attract each other more strongly than those of gases. We can see now why solids tend to retain their sizes and shapes; why liquids take on the shapes of their containing vessels, although retaining their own volumes; and why gases conform to the shapes and volumes of their containing vessels.

All matter, regardless of whether it's in the solid, liquid, or gaseous state, has two common properties: It has mass and it occupies space.

Since it is convenient to consider matter according to its state—solid, liquid, or gas—we will examine some of the properties peculiar to the solid state. These properties are determined by the nature of the substance and by the arrangement and motions of its atoms and molecules.

3.2 Structure of Solids

Each vibrating atom of a solid is confined to a small, definite space between its neighboring atoms. The atomic arrangement may take either of two forms. In **crystalline solids,** the atoms are

PHOTO 3.1   The atoms of crystalline solids are arranged in definite geometric patterns. Common table salt is an example of crystalline solids. (Courtesy of Morton Salt Division of MortonNorwich)

arranged in a geometric pattern, such as cubic or hexagonal. Common table salt, diamonds, quartz, ice, and metals are examples of crystalline solids. In **amorphous solids,** the atoms are arranged in no definite pattern. Amorphous solids are not numerous but do include glass, rubber, and some waxes.

---

* crystalline solids   Materials whose atoms and molecules are arranged in a geometric pattern.

* amorphous solids   Materials whose atoms and molecules are arranged in no definite pattern.

Figure 3.1A shows the theoretical structure of a single salt crystal (common table salt), and Figure 3.1B shows the structure of a single iron crystal. Salt is a combination of two atoms, sodium and chlorine. The pattern of its structure, which is called a *space lattice,* is cubic. The pattern for iron is called *body-centered cubic* because one iron atom is positioned in the center of the cube, or body. Not shown is the structure of aluminum, copper, gold, and some other metals that have a *face-centered cubic* structure. That is, there is an atom in the middle of each face of the cube as well as at the cube's corners.

As previously stated, atoms and molecules tend to attract each other more strongly in the solid state, when they are closer together. Although the attractive force between individual atoms

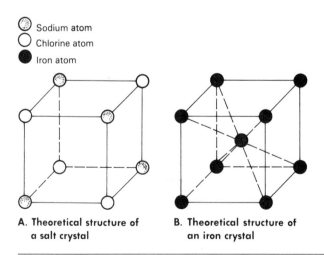

A. Theoretical structure of a salt crystal

B. Theoretical structure of an iron crystal

FIGURE 3.1   Examples of the Atomic Structure of Solids.

is small, the combined effects of the billions of atoms that make up a substance may be enormous. As a result, a steel cable 1 inch in diameter is able to support a weight of many tons without breaking.

3.3 The Property of Strength: Tensile and Compressive Forces and Stress

Let us consider the mechanical property of strength. When designing a part for a machine, the technician must make sure the part is strong enough to resist the forces that will be applied to it under normal use. When an external force—either a **tensile force** or a **compressive force**—is applied to a solid, the force is transmitted through the solid to its other end. Figure 3.2 shows a tensile force (F) being applied to a rod used to support a small crane. We can imagine that every cross section of the rod is pulling on every adjacent cross section.

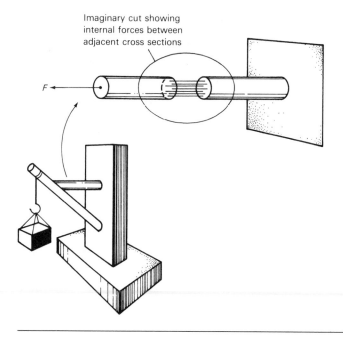

Imaginary cut showing
internal forces between
adjacent cross sections

F

FIGURE 3.2 Tensile Force on a Cross-Sectional Area.

* tensile force   An external force that creates a pulling or stretching action on
                         a material.

* compressive   An external force that creates a pushing or squeezing action on
   force   a material.

     We can also imagine the internal forces acting like a bunch
of rubber bands pulling on the adjacent cross sections. The internal force acting on each square inch of cross-sectional area is
called **stress**. Stress may be either tensile or compressive, depending on the type of applied force. The units of stress are pounds
per square inch (lb/in.^{2} or psi) in the U.S. customary system and,
in the SI system, newtons per square meter (N/m^2) or, more commonly, pascal (Pa). Note that 1 pascal is equal to 1 newton per
square meter (Pa = N/m^2). Handbooks usually indicate the ultimate (maximum) stress a given material can support without
rupturing.

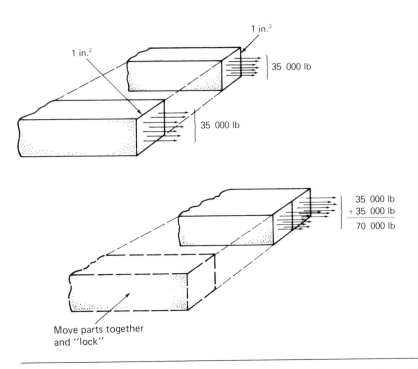

FIGURE 3.3   Tensile Stress of 35 000 Pounds per Square Inch on Cross-Sectional
Areas.

* stress

An internal force applied to each square inch (or square meter) of the cross-sectional area of a material.

To demonstrate the use of the term "stress," let's say that a particular material can take a tensile stress of 35 000 pounds per square inch. If a machine part made of this material has a 2 square-inch area, it can take twice the value of 35 000, or 70 000 pounds, as is shown in Figure 3.3. Conversely, if our part is only 1/2 square inch in area, it can take only 1/2 of 35 000 pounds, or 17 500 pounds.

We say that the force is equal to the stress multiplied by the area. To put what we have done in algebraic form:

$$F = S \times A$$

However, the formula is more commonly written:

* STRESS FORMULA

stress = force ÷ area

$$S = \frac{F}{A}$$

where S = the tensile or compressive stress in lb/in.^{2} or psi; in the SI system, in Pa
F = the tensile or compressive force in lb or N
A = the cross-sectional area in in.^{2} or m^2

Of course, all the atoms and molecules in a machine part are resisting the applied force, but only the area of the part that is perpendicular to the tensile or compressive force is used in calculations. Sample Problem 3.1 illustrates how to proceed with a typical stress problem.

SAMPLE PROBLEM 3.1

The chief engineer gives you a rod of aluminum alloy that is 15 in. long and has a cross-sectional area of 0.88 in.^{2} He wants to know the maximum (ultimate) tensile force (F) that can be applied along the axis of the rod. Note that the length of the rod is not needed for this problem. However, technicians frequently must select the data they need from a mass of information.

SOLUTION

Wanted

$F = ?$

Given

We first look up the ultimate stress for this material in the appendix of this text. Stress is listed in the table of Mechanical Properties of Materials. The aluminum alloy has an ultimate tensile stress (S) of 45 000 lb/in.² The cross-sectional area is given as 0.88 in.² Therefore, we are given $S = 45\,000$ lb/in.² and $A = 0.88$ in.²

Formula

We use the stress formula $S = F/A$, but rearrange it to solve for F:

$$F = S \times A$$

Calculation

$$F = 45\,000 \text{ lb/in.}^2 \times 0.88 \text{ in.}^2 = 39\,600 \text{ lb}$$
$$(\text{round to } 4\overline{0}\,000 \text{ lb})$$

What is the force in newtons? We check the Conversion Factors Table at the front of the book and see that we have to multiply pounds by 4.45:

$$39\,600\,\text{lb} \times 4.45 = 176\,000\,\text{N} = 176\,\text{kN} \,(\text{round to } 180\,\text{kN})$$

If a force is applied greater than that calculated, the part might rupture.

3.4 The Properties of Deformation and Strain

The English scientist Robert Hooke (1635–1703) noticed that most materials deform in direct proportion to the force applied. This concept is referred to as **Hooke's Law**:

∗ Hooke's Law

The deformation of an object is directly proportional to the applied force.

For instance, suppose, as is shown in Figure 3.4, that a single strand of steel wire, originally 100.0 inches long, has a $10\overline{0}0$

PHOTO 3.2 When forces in the earth's interior created by heat and pressure build up enough stress, the earth ruptures, causing a landslide. This landslide destroyed homes in Oakland, California, in the winter of 1969–70. (USDA, Soil Conservation Service; L.E. Welch, photographer)

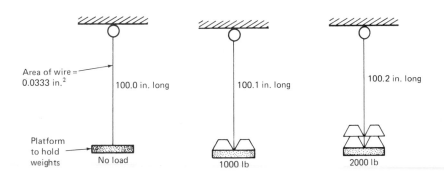

Area of wire = 0.0333 in.²

100.0 in. long

100.1 in. long

100.2 in. long

Platform to hold weights

No load

1000 lb

2000 lb

FIGURE 3.4 Deformation, Strain, and the Modulus of Elasticity.

pound weight placed on it. The wire is measured with the load on it, and the new length is determined to be 100.1 inches. Then the 10$\overline{0}$0 pound load is removed and a 20$\overline{0}$0 pound load is placed on the wire. The wire is now stretched to a length of 100.2 inches. Accordingly, we judge that for every 1000 pounds placed on the wire, the wire will stretch 0.1 inch. Thus, the deformation (change in length) is directly proportional to the load (up to some limit).

In order to continue our discussion of this relationship, L will be used to represent the original length. A deformation, or change in length, is represented by ΔL. The symbol "Δ" is the capital Greek letter "delta," representing "change." (In later chapters, delta will be used to indicate changes in other properties, such as temperature, area, time, or velocity.)

Going back to our example, when a 2000 pound tensile force is placed on the wire, $\Delta L = 0.2$ inch. This total change in length is referred to as deformation, and we say "the deformation is 0.2 inch." The technician must also determine the *deformation for each inch* of the wire's length. This value is called **strain.** The symbol for strain is e.

---

* strain

The change in length for each inch (meter) of a part.

STRAIN = CHANGE IN length / Initial length

The formula for strain is:

Δ = CHANGE
Initial
Final
ΔL = F - I

* STRAIN FORMULA

strain = deformation ÷ original length

$$e = \frac{\Delta L}{L}$$

where e = the strain in in./in. (m/m)
ΔL = the deformation in in. (m)
L = the total original length in in. (m) (If the original length is in inches, then ΔL must also be in inches and not in feet or some other unit.)

The 1000 pound load produced a strain of

$$e = \frac{\Delta L}{L} = \frac{0.1 \text{ in.}}{100 \text{ in.}} = 0.001 \text{ in./in.}$$

The 2000 pound load produced a strain of

$$e = \frac{0.2 \text{ in.}}{100 \text{ in.}} = 0.002 \text{ in./in.}$$

Although we will continue to express strain in units of in./in. (m/m) in this text, the current practice is to discontinue use of the units because, as we see, the inches (or meters) cancel each other. Since the units cancel, strain is called a *dimensionless number,* just as the value pi (π) is dimensionless. That is, pi is used in circular measurements and calculations, as in the formula for the circumference of a circle.

circumference $(C) = \pi \times$ diameter (d)

SAMPLE PROBLEM 3.2	The length (L) of a short metal tension member in a machine is 2.0 in., and it is 0.50 in. in diameter. It has an 8230 lb tensile force applied, and the total deformation (ΔL) is measured as 0.0028 in. Find the strain (e).
SOLUTION	
Wanted	$e = ?$
Given	We actually have more information than we need, so we just write down what we need to find the strain: the original length $L = 2$ in. and the total deformation $\Delta L = 0.0028$ in.
Formula	$e = \dfrac{\Delta L}{L}$
Calculation	$e = \dfrac{0.0028 \text{ in.}}{2.0 \text{ in.}} = 0.0014 \text{ in./in.}$

To determine the strain in SI units, we look up the conversion factor in the table at the front of the book. To convert from inches to meters, multiply by 0.0254:

$$\Delta L = 0.0028 \text{ in.} \times 0.0254 = 0.000\,071 \text{ m}$$
$$L = 2.0 \text{ in.} \times 0.0254 = 0.051 \text{ m}$$
$$e = \frac{\Delta L}{L} = \frac{0.000\,071 \text{ in.}}{0.051 \text{ in.}} = 0.0014 \text{ m/m}$$

Since we multiplied the numerator and denominator of the strain formula by the same number (0.0254), the answer stays the same. So without bothering with the conversion, whenever you have strain in one system of units, you immediately have the strain in the other system of units.

Another English physicist, Thomas Young (1773–1829) discovered there was a relationship between stress and strain. Remember, stress is the force per square inch of cross-sectional area, and strain is the deformation per inch of length of a given material. Since the total deformation is proportional to the total force applied, strain has to be proportional to stress. In addition, Young found that for any given material (say, steel), any value of strain can be multiplied by some constant to give the appropriate stress. Referring back to Figure 3.4, we can use the given cross-sectional area of 0.0333 square inch to calculate the stress for a 1000 pound load:

$$S = \frac{F}{A} = \frac{10\bar{0}0 \text{ lb}}{0.0333 \text{ in.}^2} = 3\bar{0} \text{ 000 lb/in.}^2$$

For the 2000 pound load:

$$S = \frac{20\bar{0}0 \text{ lb}}{0.0333 \text{ in.}^2} = 6\bar{0} \text{ 000 lb/in.}^2$$

Young noticed an interesting fact. If the strain produced by the first load (0.001) is multiplied by 30 000 000, the stress for the first load is obtained:

$$0.001 \times 30\,000\,000 = 30\,000 \text{ lb/in.}^2$$

Also, if the strain produced by the second load is multiplied by 30 000 000, the stress for the second load is obtained:

$$0.002 \times 30\,000\,000 = 60\,000 \text{ lb/in.}^2$$

Young investigated further and discovered that almost every material has some particular number by which the strain can be multiplied to obtain the related stress. Thirty million is correct for most steels. This value is called Young's modulus or, as listed in the Mechanical Properties table in the appendix, the **modulus of elasticity**. The symbol for the modulus of elasticity is E. Mathematically speaking, this modulus is the ratio of stress to strain. This ratio remains constant for most metals (and other engineering materials) over a range of applied forces from zero to some maximum value. This maximum value varies depending on the type of material.

---

* modulus of elasticity
A constant value for each material that relates stress to strain.

One thing that hasn't been mentioned yet is the unit for the modulus of elasticity. Notice that in the above calculations, we have strain (which is dimensionless) on one side of the equal sign and stress (in lb/in.²) on the other. Of course, in a formula, the units on one side of the equal sign must equal the units on the other side. We have:

strain (dimensionless) $\times E$ = stress (lb/in.²)

We can see that the modulus of elasticity must have lb/in.² for units, and so it does. In the SI system, the units are pascals. Writing the formula in algebraic symbols, we have:

$e \times E = S$

Rearranging the formula in its most common form:

---

* STRESS–STRAIN FORMULA
modulus of elasticity = stress ÷ strain

$$E = \frac{S}{e}$$

where E = the modulus of elasticity in lb/in.² (Pa)
S = the stress in lb/in.² (Pa)
e = the strain (Although strain is dimensionless, this text will use the units in./in. or m/m to emphasize the concept of strain.)

SAMPLE PROBLEM 3.3	Suppose the stress (S) on the rod in Sample Problem 3.2 is 42 000 lb/in.² The strain (e) was found to be 0.0014 in./in. A. Find the modulus of elasticity (E). B. What material does this appear to be?

SOLUTION A

Wanted

$E\ =\ ?$

Given

The information needed to find E is stress $S\ =\ 42\,000$ lb/in.² and strain $e\ =\ 0.0014$ in./in.

Formula

Use the stress–strain formula:

$$E\ =\ \frac{S}{e}$$

Calculation

$$E\ =\ \frac{42\,000\ \text{lb/in.}^2}{0.0014\ \text{in./in.}}\ =\ 3\bar{0}\,000\,000\ \text{lb/in.}^2$$

SOLUTION B

Turning to the appendix, we note that structural steel and ultra strength steel both have a modulus of elasticity of $3\bar{0}\,000\,000$ lb/in.²

SAMPLE PROBLEM 3.4	If the rod in Figure 3.2 has a cross-sectional area (A) of 0.000 07 m² and an applied force (F) of 3500 newtons, what is the stress (S) in pascals?

SOLUTION

Wanted

$S\ =\ ?$

Given

Force $F\ =\ 3500$ N and area $A\ =\ 0.000\,07$ m²

Formula

Use the stress formula:

$$S\ =\ \frac{F}{A}$$

Calculation

$$S\ =\ \frac{3500\ \text{N}}{0.000\,07\ \text{m}^2}\ =\ 50\,000\,000\ \text{Pa}$$
$$=\ 50\ \text{megapascals (MPa)}$$

3.5 Other Properties

It may be surprising to find that most metals are elastic. Because we are accustomed to calling materials like rubber elastic, we may not easily think of metal as elastic. Rubber can be stretched several times its own length and then return to its original shape. However, physicists define an elastic material as a substance that will stretch under a tensile load, or force (or compress under a compressive load) and then return to its original dimensions when the load is removed. No mention is made of how much change must take place.

PHOTO 3.3   The mechanical properties of ductility, malleability, brittleness, and hardness determine how, and on what machines, the material will be shaped. In this picture, a hard cutting tool is being ground to size and shape. (Courtesy of Union Pacific Railroad)

We are, of course, aware that if too much force is applied in stretching a rubber band, it will break. Metals, too, will rupture if too great a tensile load is applied; however, another interesting phenomenon occurs to many metals. Above a certain load, but before rupture, the metal ceases to be elastic and becomes plastic.

The graph of stress plotted against strain in Figure 3.5 shows the "picture" obtained when a sample part made of structural steel is placed in a testing machine and the tensile force is gradually increased until the part ruptures. Stress is proportional to strain only up to the **proportional limit**. This straight-line part of the curve is called the **elastic range**. A point on the stress–strain curve very near the proportional limit is the **elastic limit**. For practical purposes, we may consider the two limits at the same point on the curve. The steel will return to its original dimensions when the stress (any value up to the elastic limit) is removed. Above the proportional (or elastic) limit, the curve is said to be in the **plastic range**. That is, if a stress above the elastic limit is placed on the material and then removed, the steel will not return to its original dimension. It will remain in a stretched condition. Note that in the plastic range, the curve is not a straight line. Thus, the stress–strain formula ($E = S/e$) does not apply to the plastic range; it applies only to the elastic range.

✶ elastic range   The range of stress values for which a material remains elastic.
                             No damage to material.

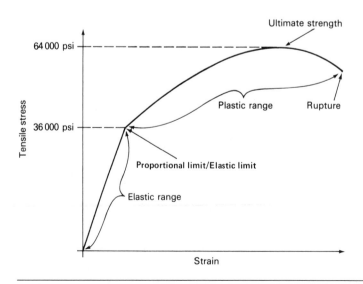

FIGURE 3.5   Typical Stress–Strain Curve for Structural Steel.

* **elastic limit/ proportional limit**	Limits that are very near each other and that represent the maximum stress that can be applied to a material and still have the material retain its elastic properties.
* **plastic range**	The range of stress values above the elastic limit. If applied, the material may take a permanent deformation.

PERMANENT DAMAGE to MATERIAL

The stress–strain curve also shows us that the maximum (or ultimate) stress this type of steel can take is 64 000 pounds per square inch. As the steel is being stretched, its cross-sectional area reduces slightly. But if we try to load the steel to a value greater than 64 000 pounds per square inch, the cross-sectional area begins to reduce rapidly. In fact, the area reduces so rapidly that we can no longer obtain the ultimate stress. The strength of the part weakens as the point of the steel's **ultimate strength** is passed, and the area decreases until the part ruptures.

* **ultimate strength**	The maximum stress value that a material can withstand. An attempt to go beyond this value can result in rupture.

The plastic property, or **ductility**, of steel is very important, particularly in manufacturing operations. For instance, some steels can be drawn out to form very thin wires. These steels are referred to as ductile steels. A ductile material is one that elongates a considerable amount before rupture. For example, some structural steels will stretch about 20% more than their original length, and some brasses may double in length before rupturing. Gold, silver, copper, and some steels are considered ductile metals.

* **ductility**	A measure of the plasticity of metals. The more a material will stretch under a tensile force, the more ductile it is.

Let's place a compressive stress on the steel represented in Figure 3.5. The elastic range will be the same, and we can use the stress–strain formula for compressive loads. However, as the stress gets up into the plastic range, the material "squashes" out. This presents a larger cross-sectional area to the load, so, in turn, a larger load can be supported. The mateial may not rupture at all. Gold, for example, can be hammered into a sheet about 1/300 000 of an inch thick. This plastic property is called **malleability.** The term "malleable" comes from the Latin for "to hammer" and refers to metals that can be hammered or squeezed into a different shape. Both malleability and ductility refer to the same property—plasticity. The terms are interchangeable but originated in different industries. Malleability (or ductility, if you prefer) is required when stamping out coins and when bending sheet metal to form air ducts or enclosures for equipment.

---

* malleability

A measure of the plastic property of metals. Like ductility, the term refers to how much a material will deform when squeezed or hammered by a compressive force.

A material that has a very short plastic range is called *brittle*. That is, it will deform only very slightly before rupture. Most cast irons, high-strength steels, concrete, clay products, and glass are brittle. Usually, **brittleness** is not a desirable property in a product, and this has to be evaluated with the other properties or characteristics. For instance, the most important characteristic of glass is that we can see through it; what's more, in most window applications, brittleness is not important. On the other hand, it would be dangerous to build bridges out of cast iron, which is brittle. The reason is that vibrations caused by traffic on the bridge can result in high localized stresses around the rivets joining the parts of the bridge. A ductile material will "stretch" a small amount and thus even out the stresses, but a brittle material will develop cracks that progressively become larger and could destroy the bridge.

---

* brittleness

The inability of a material to deform very much before rupture. The shorter the deformation required for rupture, the more brittle the material. Brittle materials have very short plastic ranges.

Hardness is yet another property of materials. When a solid is scratched or dented, some of its molecules are torn away or pushed out of their structural pattern. The ability to resist scratching or denting is called **hardness.** Diamonds, high-carbon steel, and ceramic materials are extremely hard. Lead and copper are considered to be soft metals.

---

* hardness

The ability of a material to resist scratching or denting.

If we are asked to specify a metal for a saw blade, the metal must be harder than the material to be sawed, in order to perform its task. On the other hand, a hardened hammer that is used to loosen machine assemblies could concentrate the blow in too small an area and possibly damage the machine. A lead hammer would then be required because it is soft and will deform easily, thus distributing the force of the blow.

Summary

The atoms and molecules of a substance are in a constant state of motion, even though they may be locked into a definite structure. The motion of each atom of a solid is a vibrating motion about some specific point in a space lattice structure. The attractive forces between the atoms and molecules of many solids are very strong, thus allowing structural parts to support large loads without rupturing. Handbooks list the strengths of various materials in terms of stress.

When a force is applied to a material, it deforms, or changes its dimensions. If a tensile or compressive force is applied to a material, it will change to some new dimension. The change in length, per inch of length, is called strain. Strain is related to stress by a value known as the modulus of elasticity.

Solids and, in particular, metals have elastic and plastic properties. Almost all metals have elastic properties—that is, the ability to return to initial dimensions once an applied force has been removed. Some metals become plastic if a force is applied that exceeds the elastic limit of the material. Those metals that have long plastic ranges are called ductile or malleable.

Other mechanical properties of materials include brittleness and hardness.

Key Terms

* kinetic theory of matter
* crystalline solids
* amorphous solids
* tensile force (load)
* compressive force
* stress

* stress formula
* Hooke's Law
* strain
* strain formula
* modulus of elasticity
* stress–strain formula

* elastic range

* elastic limit (proportional
  limit)

* plastic range

* ultimate strength

* ductility

* malleability

* brittleness

* hardness

Questions

1. Name four classifications of the properties of solids.

2. What is meant by the kinetic theory of matter?

3. The atomic arrangement of solids may take two forms. Name them.

4. What is the difference between a tensile and a compressive force?

5. Define stress and give its units in the U.S. customary system and in the SI system.

6. What is the difference between strain and deformation?

7. Write the stress–strain formula.

8. What property or properties do you think are important for (a) handsaws, (b) a hoisting chain, (c) gears, (d) springs on a car, (e) chisels?

Problems

1. Refer to Sample Problem 3.1 and Figure 3.2. If the rod has a cross-sectional area of 1.35 in.², what maximum tensile force can be applied along the axis?

$S = \frac{F}{A}$ $A = .442 \text{ in}^2$

$F = 10,000 \#$

$S = \frac{10,000 \#}{.442 \text{ in}^2} = 22,624 \, \#/\text{in}^2$

2. Suppose a rod is part of a brake assembly for a large roll in a paper-making machine. If the diameter of the rod is 0.750 in. (area = 0.442 in.²), what stress is developed when a tensile force of 10 $\overline{0}$00 lb is applied along the axis?

3. Refer to Figure 3.6. Two steel plates, each 1/2 in. deep by 3.00 in. wide, are butt welded together. Assume the weld has the same dimensions as the plates. If the weld metal has a tensile strength of 20 500 psi, what tensile load can be placed on the plates?

$A = .005 \text{ m}^2$ $\frac{200}{.005} = 40,000 \, Pa$

$F = 200$

4. A metal part with a cross-sectional area of 0.005 m² has a compressive force of 20$\overline{0}$ N applied to it. Find the stress in pascals.

$A = 3 \times .5 = 1.5 \, in^2$
$S = 20,500 \, \#/in^2$
$F = S \times A$
$1.5 \times 20,500 = 30,750 \#$

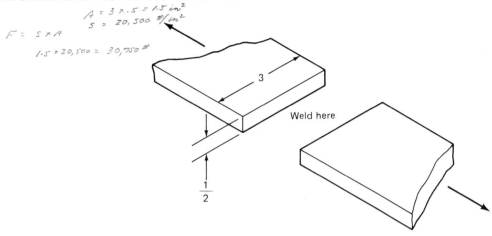

FIGURE 3.6 Steel Plates for Problem 3.

$S = \frac{F}{A} \quad \frac{530\, \#}{.196\, in^2} = 2729 \, \#/in^2$

$F = 530\#$

$DIA = \frac{1}{2}\, in$

$A = .196 \, in^2$

5. A 1/2 in. diameter manila rope used to hoist a sail can support a safe load of $53\bar{0}$ lb. If we assume the rope fibers completely fill a 1/2 in. diameter circle, with an area of 0.196 in.², what will be the stress on the cross section of the rope?

$S = \frac{F}{A} = \quad S = 1400 \, \#/in^2$
$A = 16 \, in^2$
$F = S \times A = 22,400\#$
$S = 1400 \, \#/in^2$
$A = 12.25 \, in^2$
$F = S \times A = 17,150\#$

6. Refer to Figure 3.7. A 4 in. by 4 in. by 6.00 in. high timber block is used to support a load. If the compressive strength is 1400 psi:
   a. Find the total load that can be supported if the dimensions are 4.00 in. by 4.00 in.
   b. A standard 4 in. by 4 in. block has actual dimensions of 3.50 in. by 3.50 in. What load can be supported by this block?

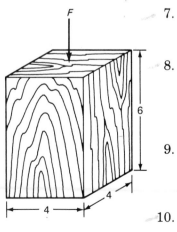

7. Given the data in Problem 2, and the fact that the rod is 15.0 in. long and made of stainless steel, find (a) the strain and (b) the deformation.

8. Refer to Figure 3.4. Assume the steel wire is 6.00 m long and has a cross-sectional area of 0.000 300 m². A $35\bar{0}0$ kg mass is placed on the platform. Determine (a) the stress in pascals and (b) the strain. (*Remember:* convert the kilograms to newtons.)

9. If the $10\bar{0}$ in. long steel wire in Figure 3.4 has a cross-sectional area of 1/30 in.², find (a) the stress, (b) the strain, and (c) the total deformation, when a $15\bar{0}0$ lb load is placed on the platform.

FIGURE 3.7 Timber Block
for Problem 6a.

10. If the wire in Problem 9 is made of copper and the diameter and length are kept the same, calculate (a) the stress, (b) the strain, and (c) the total deformation, when a $10\bar{0}0$ lb load is placed on the platform.

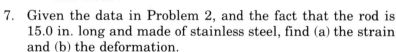

$9.A \quad S = \frac{F}{A} \quad \frac{1500}{\frac{1}{30}} = 45,000 \, \#/in^2 \qquad 9b. \quad e = \frac{S}{E} = \frac{45,000}{30,000,000} = .0015\# \quad 9c \quad \Delta L = e \times l$
$.0015 \times 100 = 0.15 \, in$

11. A steel block has a cross-sectional area of 8.00 in.² and is 5.00 in. high. It supports a 175 000 lb vertical load.
    a. What is the stress on the horizontal cross-sectional area?
    b. What is the strain?
    c. What is the total deformation of the block?
    d. If the stress on the block cannot exceed 15 $\overline{0}$00 lb/in.², how large an area is required to support the given load?

Computer Program

Use this program to solve some of the chapter problems. Also, using a constant area and load, compare the strains for the various materials.

PROGRAM

```
10   REM   CH THREE PROGRAM
20   PRINT "PROGRAM FOR FINDING THE TENSILE STRESS OR COMPRESSIVE."
21   PRINT "STRESS AND STRAIN. TYPE THE APPLIED FORCE IN POUNDS,"
22   PRINT "AND THEN THE CROSS SECTIONAL AREA IN SQ.IN."
30   INPUT F,A
40   PRINT "FORCE = ";F;" LB."
50   PRINT "AREA = ";A;" SQ.IN."
60   PRINT
70   PRINT "IF MATERIAL IS CAST IRON-TYPE 1, IF STEEL-TYPE 2,"
71   PRINT "AND IF ALUMINUM-TYPE 3."
80   INPUT T
90   PRINT "YOU HAVE SELECTED --- ";T
100   IF T = 1 THEN   GOTO 200
110   IF T = 2 THEN   GOTO 300
120   IF T = 3 THEN   GOTO 400
130   LET S = F / A
140   LET D = S / E
150   PRINT
160   PRINT "STRESS = ";S;" PSI."
170   PRINT "STRAIN = ";D;" IN./IN."
180   END
200   LET E = 15000000
210   GOTO 130
300   LET E = 30000000
310   GOTO 130
400   LET E = 10000000
410   GOTO 130
```

Lines 20, 21, and 22 help control the length of each line.

Lines 100, 110, and 120 select the proper modulus of elasticity, E.

Lines 130 and 140 contain the stress and strain formulas.

Large liquid storage tanks are necessary in our modern society. An understanding of mechanical properties of fluids is essential for designing tanks such as these. (Courtesy of W.R. Grace & Company)

Chapter 4

Objectives

When you finish this chapter, you will be able to:

- [ ] Explain that as you work with fluids, the ratio F/A becomes known as pressure, rather than stress as in work with solids.
- [ ] Recognize the pressure–force equation $p = F/A$, showing the relation between pressure (p), force (F), and area (A).
- [ ] State Pascal's hydraulic principle: Pressure applied to an enclosed fluid is transmitted undiminished to every portion of the walls of the container.
- [ ] Determine pressure under a column of liquid if given the height and weight density—that is, apply the pressure–height formula $p = (H)(D)$.
- [ ] Distinguish between substances in regard to their densities.
- [ ] Determine weight density for a substance if given its weight and volume.
- [ ] Compare various substances with water as to their specific gravities.
- [ ] Distinguish between gage pressure and absolute pressure.
- [ ] Recognize Boyle's Law, $p_1 V_1 = p_2 V_2$, showing how the product of initial states of pressure (absolute) and volume ($p_1 V_1$) equals the product of a later state ($p_2 V_2$).

Mechanical Properties of Fluids

The word *fluid* comes from the Latin meaning "to flow." Just about everyone recognizes that liquids are fluids, but few of us realize that gases are also fluids. This chapter will discuss such properties of fluids as ability to flow, pressure, density, specific gravity, and the compressibility or incompressibility of a fluid. We will discover how pressure is developed in a fluid and how this pressure can be used to produce a tremendous force. We will talk about the pressure developed due to the weight of a column of liquid. The section on gases will cover the relationship between pressure and volume. We will use this knowledge of fluid properties to help solve technical problems and to investigate the operation of some common liquid and gas pumps.

Engineers and technicians must know the information in this chapter in order to properly design and maintain storage tanks, pipes, dams, and machines that use fluid pressure to perform work. Fluid applications are all around us: We ride on a bed of air maintained in automobile tires; fluid under pressure transforms a light tap of a person's foot into a force sufficient to stop a 4000 pound automobile; the heavy plow on a bulldozer can be lifted easily by the movement of a small lever; and large industrial presses, operated by fluid power, squeeze sheet metal into complete car roofs in one operation. And what about the water coming from your household faucet? How is the pressure produced? What is required to get water to the 30th floor of a skyscraper? Technicians must work with these applications.

4.1   Properties of Liquids

In liquids, as in solids, the molecules of matter attract each other and are in a state of constant motion. The molecules of liquids, however, are farther apart than those of solids, and as a result, their attraction is less. Because of this smaller degree of attraction, they are able to travel greater distances. Thus, the molecules do not arrange themselves in a rigid pattern as do those of solids; instead, they adapt to the shape of the container. That is, they have the ability to flow, a mechanical property of fluids. This section also discusses the properties of pressure, density, and specific gravity.

THE PROPERTY OF PRESSURE

Pressure is similar to stress, which was discussed in Chapter 3. Recall that stress describes forces acting on internal areas of solids and that it may be tensile or compressive. Pressure, on the other hand, is only compressive and is normally used in connection with fluids. (Sometimes we may talk about the pressure applied to the surface of a solid, as we will in our later discussion of Figure 4.1.) Pressure can be applied externally to a fluid or transmitted internally through the fluid.

*** pressure**

The force applied to an area. Normally, it is considered to be applied by or to a fluid.

Pressure is defined as the force applied to 1 square inch (or 1 square meter) of area. The unit of pressure in the U.S. customary system is pounds per square inch (lb/in.^{2}). In the SI system, the unit of pressure is the pascal (Pa). The pascal is equal to 1 newton per square meter.

On occasion, people confuse pressure with force, partly because of our tendency to take short cuts in our speech and writing. For example, when you want to put air in your automobile tire, you may use an air pump that simply states "32 pounds" (which we know to be units of force). Actually, the pump supplies a *pressure* of 32 pounds per square inch of tire surface to your tire. Figure 4.1 provides a specific example that illustrates the use of the word "pressure." A gallon of water is poured into the tall flask A, and another gallon is poured into the long, low container B. In each case, the water adapts to the shape of the container. Since a gallon of water weighs 8.34 pounds, each container is exerting a force on the table of 8.34 pounds (we will neglect the

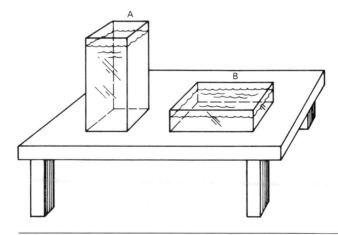

FIGURE 4.1 Illustration of Pressure Applied by Liquid.

weights of the containers). However, the area of the table covered by the tall flask is smaller than that covered by the low container. Therefore, the force (or weight) on each square inch of the table under the tall flask is greater. This force per square inch is called pressure, which can be stated as a formula:

---

* PRESSURE–FORCE FORMULA

pressure = force ÷ area

$$p = \frac{F}{A}$$

where p = the pressure in lb/in.² or Pa
F = the force (or weight) in lb or N
A = the area in in.² or m²

Suppose the tall container in Figure 4.1 has a bottom area of 3.00 square inches. The pressure on the table due to the weight of the water then is:

$$p = \frac{F}{A} = \frac{8.34 \text{ lb}}{3.00 \text{ in.}^2} = 2.78 \text{ lb/in.}^2$$

If the low container has a bottom area of 12 square inches, the pressure on the table is:

$$p = \frac{F}{A} = \frac{8.34 \text{ lb}}{12 \text{ in.}^2} = 0.70 \text{ lb/in.}^2$$

Now, let's move a second column of water next to the first in Figure 4.1, such that the second column has the same dimensions as the first. The combined weight is pressing down on 6.00 square inches. But, the pressure on the table is still 2.78 pounds per square inch, as we can see:

$$\frac{16.68 \text{ lb}}{6.00 \text{ in.}^2} = 2.78 \text{ lb/in.}^2$$

PASCAL'S PRINCIPLE

The French scientist Blaise Pascal (1632–1662) studied pressure in fluids and gave us **Pascal's Principle:**

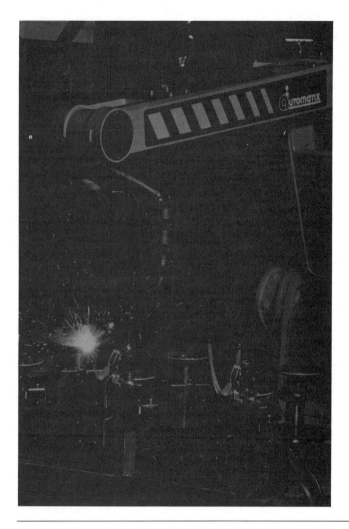

PHOTO 4.1 This high technology welding robot is operated by fluid power. (Courtesy of Automatix)

∗ Pascal's Principle

Pressure applied to an enclosed fluid is transmitted undiminished in all directions to every portion of the fluid and to the walls of the container.

Fluids have the property of **incompressibility**—that is, a quantity of fluid cannot be squeezed into a smaller space, regardless of the pressure applied. (Actually, fluids can be squeezed slightly, but for most technical problems, the change is so small that it can be disregarded.) When the fluid is incompressible, the pressure change is transmitted instantly.

* **incompressibility**	The inability of a fluid to decrease in volume, regardless of the pressure applied.

Figure 4.2 illustrates Pascal's Principle. The figure shows a body of water in a closed container with two pistons. (The container is shown cut away for viewing.) If a weight is placed on the left-hand piston and no weight is placed on the right-hand piston, the weight will slowly fall and force the right-hand piston up above the container walls, and the water will flow out. To prevent this action, we must place a weight on the right-hand piston. But how much weight do we need for balance? If the left-hand weight is 100 pounds, do we need more or less than 100 pounds for the right-hand weight? And how much? Sample Problem 4.1 demonstrates the procedure for answering this question.

SAMPLE PROBLEM 4.1

Figure 4.2 shows a test set-up to demonstrate Pascal's Principle:

1. Let the left-hand weight be designated $F_1 = 10\overline{0}$ lb.
2. Let the area of the left-hand piston be designated $A_1 = 2.0$ in.^{2}
3. Let the right-hand weight be designated $F_2 = ?$.

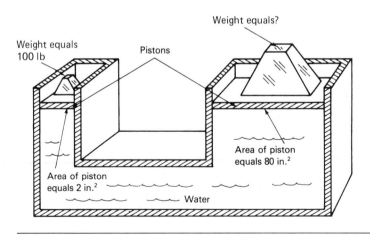

FIGURE 4.2 Test Demonstration of Pascal's Principle.

$$P = \frac{F_1}{A_1} = \frac{F_2}{A_2}$$

4. Let the area of the right-hand piston be designated $A_2 = 8\bar{0}$ in.^{2}

Find the weight, in pounds and in newtons, required to balance the 100 lb weight.

SOLUTION

Wanted

$F_2 = ?$

Given

$F_1 = 10\bar{0}$ lb, $A_1 = 2.0$ in.^{2}, and $A_2 = 8\bar{0}$ in.^{2}

Formula

Use the pressure–force formula:

$$p = \frac{F}{A}$$

Calculation

In this type of problem, we must use the pressure–force formula twice. The first time, the formula is used to find the pressure transmitted to the fluid by the left-hand piston:

$$p = \frac{F_1}{A_1} = \frac{100 \text{ lb}}{2 \text{ in.}^2} = 50 \text{ lb/in.}^2$$

Pascal's Principle says this pressure is transmitted to every portion of the fluid. Therefore, a pressure of 50 lb/in.^{2} must be pressing up on the right-hand piston. The formula must be rearranged from:

$$p = \frac{F_2}{A_2} \quad \text{to} \quad F_2 = p \times A_2$$

$$F_2 = 50 \text{ lb/in.}^2 \times 80 \text{ in.}^2 = 4\bar{0}00 \text{ lb}$$

The Conversion Factors Table tells us to multiply pounds by 4.45 to get newtons:

4000 lb $\times$ 4.45 = 17 800 N (or round to 18 kN)

This answer is the weight that must be placed on the right-hand piston for balance. If any extra weight should be added to the 100 lb weight, the added pressure will raise the 4000 lb weight.

Very practical uses resulted from Pascal's studies. For instance, Figure 4.3 represents a hydraulic lift. An air pressure of about 150 pounds per square inch is placed on a small piston, which transmits this pressure through the oil to a large piston that raises the lift. Sample Problem 4.2 represents a technical problem requiring the use of the pressure–force formula and Pascal's Principle.

SAMPLE PROBLEM 4.2

Figure 4.4 is a simplified representation of a piston pump and hydraulic cylinder on a bulldozer. Piston A must develop a force (F_a) of 10 $\overline{0}$00 lb to raise the blade on the bulldozer. The area of piston A is A_a = 70.0 in.² The area of piston B is A_b = 0.500 in.² Find the pressure (p) in the liquid and the force (F_b) required to develop the pressure. Give the answers in SI units.

SOLUTION

Wanted

p = ?, F_b = ?, and conversion to SI units

Given

A_a = 70.0 in.², F_a = 10 $\overline{0}$00 lb, and A_b = 0.500 in.²

Formula

$$p = \frac{F}{A}$$

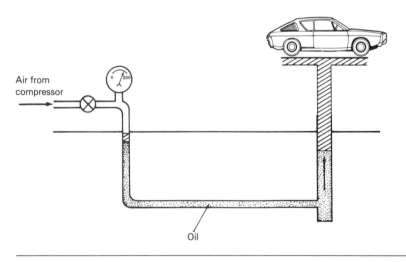

Air from compressor

Oil

FIGURE 4.3   Hydraulic Lift.

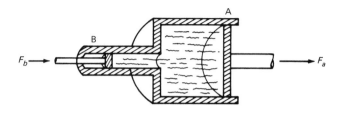

FIGURE 4.4   Piston Pump and Hydraulic Cylinder.

Calculation

As in Sample Problem 4.1, two steps are required. In the first step, we must use the pressure–force formula:

$$p = \frac{F_a}{A_a} = \frac{10\ 000\ \text{lb}}{70.0\ \text{in.}^2} = 143\ \text{lb/in.}^2$$

Since this pressure of 143 lb/in.² is transmitted through the fluid, the pressure on piston B must also be 143 lb/in.² Therefore, we must rearrange the formula to find F_b:

$$p = \frac{F_b}{A_b} \quad \text{to} \quad F_b = p \times A_b$$

$$F_b = 143\ \text{lb/in.}^2 \times 0.500\ \text{in.}^2 = 71.5\ \text{lb}$$

The Conversion Factors Table tells us to multiply lb/in.² by 6895 to get pascals:

143 lb/in.² × 6895 = 986 000 = 986 kPa

To convert from pounds to newtons, we multiply by 4.45:

71.5 lb × 4.45 = 318 N

With this information, an engineer can proceed to design the mechanism that will develop 71.5 lb, can specify the thickness of cylinder walls, and can specify wall thickness for any related piping.

For another application, let's consider the hydraulic brake system of an automobile. Look at Figure 4.5. The brake pedal is

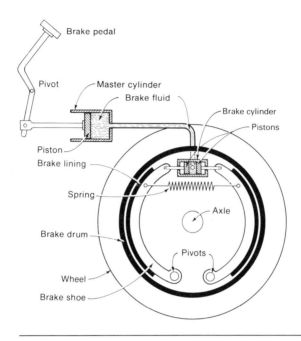

FIGURE 4.5 Hydraulic Brake.

located on the floor of the car near the driver's seat. When the driver presses it down, a piston on the master cylinder pushes against the enclosed brake fluid, usually oil. The pressure of the piston on the liquid in the master cylinder is transmitted, without loss, to the two pistons in the brake cylinder. As a result, these two pistons move outward and force the brake shoes against the brake drum, thereby stopping the vehicle. When the brake pedal is released, the pressure on the pistons in the brake cylinder is also released. The spring pulls the brake shoes away from the brake drum, releasing the wheel. In a four-wheel brake system, the liquid is carried in pipes from the master cylinder to each of the four brake cylinders. When the brake pedal is depressed, all four wheels are braked simultaneously.

PRESSURE DUE TO THE HEIGHT OF A LIQUID

Figure 4.6 shows a scene from a typical western. The good guys, firing at stagecoach robbers, are crouched behind a large water barrel. As the bandits' bullets hit the barrel, water flows from the holes. The lower the hole, the farther out the stream of water

FIGURE 4.6 Liquid Pressure Due to Height. The water pressure here varies according to the height of the water in the barrel.

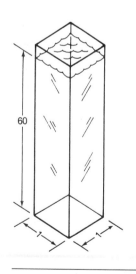

FIGURE 4.7 Method for Determining Pressure Due to the Height of a Fluid.

shoots. The mass of water over the lowest hole is greater than the mass of water over the top holes. The greater weight of water over the lowest hole creates a greater pressure, which pushes the stream farthest from the barrel. Note that the water starts out of the holes in a horizontal direction, indicating that the pressure due to the weight of water is exerted horizontally! In fact, below the surface of the water, the pressure on a given portion of water is *exerted in all directions*.

Technicians must be able to determine the pressure at the bottom of a column of liquid of any given height. Let us examine the method we will use. Figure 4.7 shows a column of water one inch on a side and 60 inches high. A cubic inch of water weighs 0.0361 pound. Since there are 60.0 cubic inches in the column, it weighs:

$$60.0 \text{ in.}^3 \times 0.0361 \text{ lb/in.}^3 = 2.17 \text{ lb}$$

This 2.17 pound force is exerted on the bottom of the container due to the weight of the water. Since the bottom is one square inch in area, we can say that the pressure at the bottom of the container is 2.17 pounds per square inch. Therefore, we have an easy method of finding the pressure at any level of a liquid:

PHOTO 4.2 Water falling from a great height does not build up the same pressure as water held back by a large dam, but it does generate great amounts of energy. The total height of Yosemite Falls in California is more than 2400 feet. Shown is Yosemite Upper Falls. (Courtesy American Airlines)

Simply multiply the height (H) of liquid in inches by the weight per cubic inch:

$$p = H \text{ in.} \times 0.0361 \text{ lb/in.}^3$$

This particular figure of 0.0361 is only for water. Other liquids have different weights, as will be discussed shortly.

It has already been stated that the pressure at any point in a fluid is exerted in all directions. Therefore, if a small hole is drilled in the wall of the container in Figure 4.7, just at the edge of the container bottom, the pressure forcing the water out of the hole would be 2.17 pounds per square inch. If a hole is drilled farther up the container wall, the pressure would be less, since the height from the hole to the surface of the water would be less.

Suppose the container were 10 inches on a side and 60 inches high. Would the pressure be any different? The answer is no, the pressure would still be 2.17 pounds per square inch at the bottom. We can check this answer by calculating the total weight of the water (the force on the bottom of the container) and dividing by the area. Thus, if you are building a dam across a river, it doesn't matter whether the water is backed up 10 feet or 10 miles, the dam has to withstand the pressure due to the *depth* of water.

You may have noticed the pressure due to the depth of water while swimming with a snorkel. When swimming on the surface, it is easy to breathe, but if you adjust your position in the water so that your body is vertical (of course the end of the snorkel is above the water), your chest is now one or two feet below the surface and breathing is harder. Your lungs must expand against the water pressure. For this reason, divers, whether using scuba equipment or a hose from the surface, must have air delivered to them at a pressure as great as the surrounding water pressure.

THE PROPERTIES OF DENSITY AND SPECIFIC GRAVITY

Before we can develop a formula for determining the pressure due to any height of any liquid, we need to look at two terms in common use by scientists, engineers, and technicians: density and specific gravity.

When we make the statement "Lead is heavier than aluminum," we are talking about **density.** We are comparing the weights of equal volumes of materials—that is, in this case, the weight of lead is heavier than an equal volume of aluminum. In other words, lead is denser than aluminum.

---

* density   The weight per volume (cubic inch) of a given material.

The formula for determining the density of a material is:

---

* DENSITY FORMULA   density = weight ÷ volume

$$D = \frac{w}{V}$$

PHOTO 4.3   A dam must be strong enough to withstand the pressure due to the depth of water, regardless of how much water is being held back. This dam being built on the Caroni River in Venezuela will be 106 meters tall. (United Nations photo)

Table 4.1 lists the weight densities (simply listed as "Density") for a number of substances in units of pounds per cubic inch and pounds per cubic foot. Currently, these U.S. customary units are more useful to technicians and engineers than the SI system of "mass density" using units of kilograms per cubic meter.

Mass density is the mass (in kilograms) per volume (cubic meter) of a given material. The formula for mass density is:

$$\text{mass density} = \frac{\text{mass}}{\text{volume}}$$

Since we will not be working with mass density, its symbol will not be presented here.

TABLE 4.1 Densities and Specific Gravities of Some Common Substances

Substance	Density		Mass Density kg/m³	Specific Gravity
	lb/in.³	lb/ft³		
Liquids				
Alcohol (ethyl)	0.0285	49.0	789	0.789
Gasoline*	0.026	45.0	720	0.72
Oil* (lubricating)	0.033	57.0	910	0.91
Mercury	0.489	845.0	13 550	13.55
Water (at 4°C, 39.2°F)	0.0361	62.4	1 000	1.00
Seawater	0.0372	64.3	1 030	1.03
Solids				
Aluminum	0.0971	168	2 690	2.69
Copper	0.323	558	8 940	8.94
Lead	0.409	707	11 340	11.34
Pine* (white)	0.016	27	430	0.43
Steel (machine)	0.282	487	7 800	7.80

*Average values.

SAMPLE PROBLEM 4.3

In order to design the support brackets for a gasoline tank on a truck, the engineer needs to know the total weight of the gasoline when the tank is full. The tank has a volume (V) of 6.00 ft³ (10 400 in.³). Find the weight (w) of the gasoline.

SOLUTION

Wanted

$w = ?$

Given

$V = 10\ 400$ in.³ From Table 4.1, we find the density of gasoline, $D = 0.026$ lb/in.³

Formula

Rearrange the density formula:

$$D = \frac{w}{V} \quad \text{to} \quad w = D \times V$$

Calculation

$w = 0.026$ lb/in.³ $\times$ 10 400 in.³ $= 270$ lb

The engineer can now decide how the tank will be supported.

The density of water has already been given: 0.0361 pounds per cubic inch. The previous formula for the pressure of water at any height can now be generalized to include all liquids. The formula is:

*** PRESSURE–
HEIGHT
FORMULA**

pressure = height × density

$$p = H \times D$$

where p = pressure in lb/in.^{2}
H = the height (or depth) in in.
D = the density in lb/in.^{3}

**SAMPLE
PROBLEM
4.4**

Company XYZ seals lubricating oil in 1 qt cans to be sold by retail stores. The company would like to store the incoming oil in a 40.0 ft high (H) storage tank. Find the pressure (p) that develops at the bottom of the tank. Give the answer in SI units also.

SOLUTION

Wanted

p = ?

Given

H = 40.0 ft, and from Table 4.1, we find the density of oil, D = 0.033.

Formula

$$p = H \times D$$

Calculation

The pressure–height formula requires the height to be in inches. Therefore:

$$H = 40.0 \text{ ft} \times 12 \text{ in./ft} = 480 \text{ in.}$$

We are now ready to use the formula:

$$p = H \times D = 480 \text{ in.} \times 0.033 \text{ lb/in.}^3 = 16 \text{ lb/in.}^2$$

The Conversion Factors Table tells us to multiply the pressure in pounds per square inch by 6895 to get pascals.

16 lb/in.^{2} × 6895 = 110 000 Pa = 110 kPa

An engineer can now use this information to determine the required wall thickness of the tank and pipes.

Sometimes, it is more convenient to compare the density of a material to the density of water. This ratio is called **specific gravity** and can be expressed as:

$$\text{specific gravity (S.G.)} = \frac{D_{\text{substance}}}{D_{\text{water}}}$$

For instance, aluminum has a specific gravity of 2.69. Thus, a cubic inch of aluminum weighs 2.69 times the weight of a cubic inch of water (2.69 × 0.036 lb/in.^{3} = 0.097 lb/in.^{3}).

---

* specific gravity   The density of a substance divided by the density of water.

It is important to note here that we are not investigating all the mechanical properties of fluids, just the ones most important to us. Also, both liquids and gases have many of the same properties, including pressure and density, already mentioned, and diffusion, which will be discussed below.

4.2 Properties of Gases

In this section we consider the properties of diffusion, compressibility, and pressure, as well as the relationship of pressure to the volume of a gas. As is true of matter in solid and liquid states, the molecules of a gas are also in a constant state of motion. However, because the molecules of a gas are farther apart, they can travel longer distances and attain greater speeds. The molecules of a gas are constantly rushing around, bumping their neighbors, and darting off in new directions. It is this motion of the gas molecules that causes the gas to spread itself quickly and uniformly throughout a container, regardless of its size and shape. This property is called *diffusion*.

PHOTO 4.4   Research balloons filled with hydrogen gas, such as this one, can lift scientific instruments to more than 80 000 feet above the earth, carrying 3 tons of apparatus. Shown here is a launch balloon; the main balloon inflates at 10 000 feet and contains 5.5 million cubic feet of gas. (NASA photo)

The motion of the molecules will not only cause the gas to diffuse throughout a container but will cause a constant bombardment against the walls of the vessel. This effect is known as *gas pressure*. Although the impact of an individual molecule is small, the combined effect of billions of molecules may be quite great. It is gas pressure, for example, that forces out the walls of a balloon when helium or some other gas is pumped in.

Like liquids, gases can be made to flow and transmit an applied pressure in accordance with Pascal's Principle. Gases also have densities, but because their densities vary with pressure and temperature, this property is not very useful to technicians. Gases (air is a gas, of course) exhibit another mechanical property that we will now mention—compressibility.

Since there is more space between the molecules of a gas than between the molecules of liquids or solids, it is easier to push these molecules closer together. This property of a gas is called **compressibility.** Remember, liquids and solids are considered incompressible. Of course, to compress a gas you must exert a pressure.

---

* compressibility

The capability of a gas to decrease in volume as the pressure on a given mass of gas is increased.

BOYLE'S LAW

The compressibility of a gas follows certain rules, or laws. English chemist Robert Boyle (1627–1691) discovered the relationship between the pressure and volume of a gas and gave us **Boyle's Law:**

* Boyle's Law

Provided the temperature of a gas remains constant, the absolute pressure on a given mass of gas changes inversely as the volume changes.

Boyle's Law says two things. First, if we decrease the volume on an enclosed gas, the pressure will increase. Second, if we know the amount of pressure change, we can determine the amount of volume change, and vice versa. For example, look at Figure 4.8. A gas is enclosed in a cylinder with a piston at the top. If we force the piston down so that the volume is one-half the original, we must exert twice the original pressure.

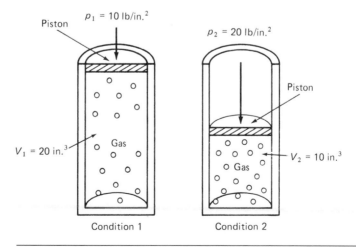

FIGURE 4.8 Boyle's Law. Pressure varies inversely as the volume changes.

To express Boyle's Law mathematically, we must identify which pressures and volumes go with the initial condition (condition 1) and which go with the changed condition (condition 2). So, we use subscripts to identify the conditions. p_1 and V_1 represent pressure and volume at condition 1; p_2 and V_2 represent pressure and volume at condition 2. Stated mathematically, then:

✱ FORMULA FOR BOYLE'S LAW

$$\frac{p_2}{p_1} = \frac{V_1}{V_2} \quad \text{or} \quad p_1 V_1 = p_2 V_2$$

where p_1 and p_2 = the pressures at conditions 1 and 2 in lb/in.^{2} or Pa

V_1 and V_2 = the volumes at conditions 1 and 2 in in.^{3} or m^3

If we apply the numbers in Figure 4.8 to Boyle's Law, we see that both sides of the formula are equal, as they must be:

$$\frac{20 \text{ lb/in.}^2}{10 \text{ lb/in.}^2} = \frac{20 \text{ in.}^3}{10 \text{ in.}^3}$$

Notice carefully that Boyle used the words **absolute pressure**, which is different from **gage pressure**. For a clear understanding of the difference, let's refer to Figure 4.9A. It shows a simple pressure gage attached to a gas cylinder. The gage is called a manometer. The U part of the tube is filled with a liquid, usually water. The higher the liquid goes in the tube, the higher the pressure in the cylinder. The liquid rises in the tube until the weight of the liquid (represented by height *H*) plus the atmospheric pressure at the open end of the tube creates a pressure equal to the pressure of the gas. Notice that the tube is open to the atmosphere (taken as 14.7 pounds per square inch) at the right-hand end. Therefore, the height *H* gives us the pressure of the gas in the cylinder over and above the air pressure in the room. If the gas in the cylinder were at atmospheric pressure, the liquid levels would be even: *H* would be zero, but 14.7 pounds per square inch pressure would be acting on both surfaces of the liquid. The readings from this type of manometer are called gage pressures because most pressure gages (including automobile tire gages) are set to read zero when the pressure in the container is

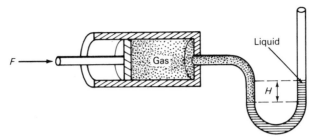

A. Manometer indicating gage pressure

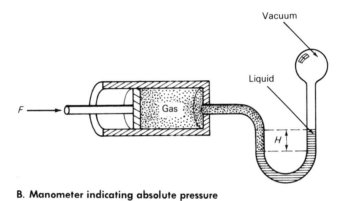

B. Manometer indicating absolute pressure

FIGURE 4.9   *Manometer Gages.*

the same as the outside atmosphere. The manometer in Figure 4.9B provides an absolute pressure reading because the gas only has to push against the weight of the water (represented by height *H*). There is no air pressure acting on the water surface on the right side of the manometer.

* absolute pressure	Pressure measured from a vacuum as a base. The pressure in a vacuum is indicated as zero pounds per square inch.
* gage pressure	Pressure measured from atmospheric pressure as a base. The pressure of a gas at atmospheric pressure is indicated by a gage as zero pounds per square inch.

If a problem requires absolute pressure and the given data is gage, simply add 14.7 pounds per square inch (101 kilopascals) to the gage pressure:

$$p_{\text{absolute}} = p_{\text{gage}} + 14.7 \text{ lb/in.}^2$$

In the SI metric system:

$$p_{\text{absolute}} = p_{\text{gage}} + 101 \text{ kPa}$$

SAMPLE PROBLEM 4.5	The cylinder in an automobile engine has a volume (V_1) of 40.0 in.³ when the piston is at the bottom of its stroke and a volume (V_2) of 5.00 in.³ when the piston is at the top of its stroke. Assuming the air drawn into the cylinder is at atmospheric pressure: A. What is the absolute pressure (p_{abs}) when the piston is at the top of its stroke? B. What is the gage pressure (p_{ga})?
SOLUTION A	
Wanted	If p_1 = pressure in the cylinder when the piston is at the bottom of its stroke and p_2 = pressure in the cylinder when the piston is at the top of its stroke, p_2 = ?
Given	Atmospheric pressure (absolute) is p_1 = 14.7 lb/in.² The air is drawn into the cylinder when the piston is at the bottom of its stroke; therefore, V_1 = 40.0 in.³ and V_2 = 5.00 in.³
Formula	$$\frac{p_2}{p_1} = \frac{V_1}{V_2}$$
Calculation	$$\frac{p_2}{14.7 \text{ lb/in.}^2} = \frac{40.0 \text{ in.}^3}{5.00 \text{ in.}^3}$$ $$p_2 = \frac{14.7 \text{ lb/in.}^2 \times 40.0 \text{ in.}^3}{5.00 \text{ in.}^3} = 118 \text{ lb/in.}^2 \text{ (absolute)}$$ For conversion to metric units, multiply pressure by 6895: $$p_2 = 118 \text{ lb/in.}^2 \times 6895 = 814\,000 \text{ Pa}$$ $$= 814 \text{ kPa (absolute)}$$

SOLUTION B

Wanted

$p_{ga} = ?$

Given

$p_{abs} = 118$ lb/in.^{2}

Formula

The formula to convert gage pressure to absolute is:

$$p_{abs} = p_{ga} + 14.7 \text{ lb/in.}^2$$

The formula must be rearranged:

$$p_{ga} = p_{abs} - 14.7 \text{ lbs/in.}^2$$

Calculation

$$p_{ga} = 118 \text{ lb/in.}^2 - 14.7 \text{ lb/in.}^2 = 103 \text{ lb/in.}^2 \text{ (gage)}$$

Rearranging the SI system formula, we have:

$$p_{ga} = 814 \text{ kPa} - 101 \text{ kPa} = 713 \text{ kPa (gage)}$$

This information can now be used by scientists and engineers to help determine the horsepower of the engine.

AIR (GAS) PUMPS

The air pump is illustrated in Figure 4.10. It consists of a metal cylinder in which an airtight piston slides. This piston generally consists of an oiled leather disk supported by iron washers. A piston rod and handle move this piston up and down in the cylinder. Set in the base of the cylinder are two nozzles. One, labeled vacuum nozzle, is attached to the chamber from which we wish to draw the air. If we desire to pump air into the chamber, we attach it to the pressure nozzle.

Inside each nozzle is a valve, which consists of a conical opening and a cone-shaped piece of metal that fits into the opening. When the piston is raised (Figure 4.10A), the air in the cylinder below the piston is expanded, and thus its pressure is reduced. The outside air pressure, being greater, opens the valve in the vacuum nozzle and the air rushes into the cylinder. But as the outside air tries to get into the cylinder through the pressure nozzle, the cone-shaped metal in its valve is forced into the opening, thus closing it.

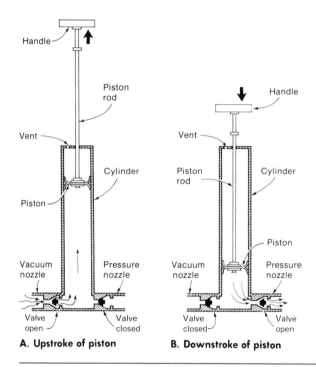

FIGURE 4.10   Operation of an Air Pump.

On the downstroke of the piston (Figure 4.10B), the air in the cylinder is compressed. This pressure causes the valve in the vacuum nozzle to close. But the valve in the pressure nozzle now is forced open and the air in the cylinder rushes out. Thus, as the piston is raised and lowered, air is drawn in through the vacuum nozzle and is forced out through the pressure nozzle. To enable the piston to move up and down freely, a vent is provided in the top of the cylinder so that the air above the piston may flow in or out.

WATER PUMPS OPERATED BY ATMOSPHERIC PRESSURE

There are a number of other devices that operate by atmospheric pressure. For example, consider the water pump used for drawing water from a well. If you were to put the end of a pipe into the water, as in Figure 4.11A, the air, pressing down on the water

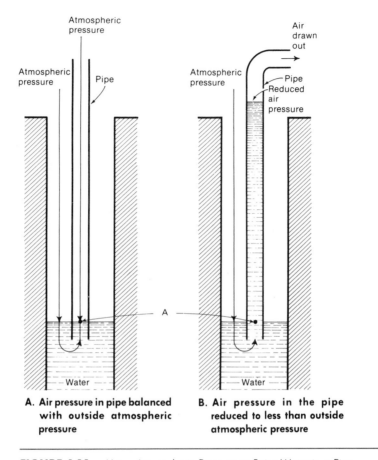

A. Air pressure in pipe balanced with outside atmospheric pressure

B. Air pressure in the pipe reduced to less than outside atmospheric pressure

FIGURE 4.11   Using Atmospheric Pressure to Raise Water in a Pipe.

of the well with a pressure of 14.7 pounds per square inch, would tend to force the water up the pipe. However, the air in the pipe, also pressing down with a pressure of 14.7 pounds per square inch, prevents the water in the pipe from rising above the level of the water in the well.

If we could reduce the pressure in the pipe by drawing out some of the air (as shown in Figure 4.11B), the greater outside pressure would then force the water up in the pipe. Why is this? Well, as we saw earlier in this chapter, the formula for determining the pressure at any level of water is:

$$p = H \times 0.0361$$

This formula applies in a unique way to our water pump. In Figure 4.11 we have shown a mark on the inside of the pipe and labeled it "A." It is at the same level as the surface of the water when the pressure on the surface of the water (both in and out of the pipe) is 14.7 pounds per square inch absolute—that is, atmospheric pressure. Now, suppose we remove all the air to create a perfect vacuum in the pipe. The absolute pressure on the surface of the water in the pipe would then be 0 pounds per square inch. However, the pressure at point A must be the same as the pressure at any other point in the same horizontal plane, both in and out of the pipe. Since the pressure outside the pipe is 14.7 pounds per square inch, the water must rise in the pipe until its weight creates a pressure of 14.7 pounds per square inch at point A. In this case, the height of water in the pipe (measured from A) can be found by rearranging the pressure–height formula to:

$$H = \frac{p}{0.0361} = \frac{14.7}{0.0361} = 407 \text{ in. or } 34 \text{ ft}$$

So the problem becomes one of reducing the air pressure in the pipe. Suppose that an airtight piston is placed in the pipe, as shown in Figure 4.12A. The air pressure in the pipe still balances the pressure outside the pipe, and the water does not rise. But if the piston is moved up (Figure 4.12B), the volume of the air in the pipe below the piston is increased. This increase of volume, according to Boyle's Law, causes a corresponding decrease in pressure. Therefore, the water will rise in the pipe.

While we now have the basic principle of the suction lift water pump, a number of details must be worked out before this device becomes practical. First of all, a valve must be placed in the pipe to prevent the water that has risen from falling back (see Figure 4.12C). This valve consists of a constriction in the pipe. A flap of metal or leather is placed over the opening in such a way that the rising water, pushing against the underside of the flap, forces it up and thus can pass up through the opening. On the other hand, when the water above the valve attempts to fall back down the pipe, its pressure on the top of the flap causes the flap to close the opening. The water is trapped above the valve.

The next problem is to get the water above the piston. To this end, a second valve, similar to the first, is inserted in the piston itself (Figure 4.12D). As the piston is forced down, the pressure on the top of the flap of the first valve closes that valve.

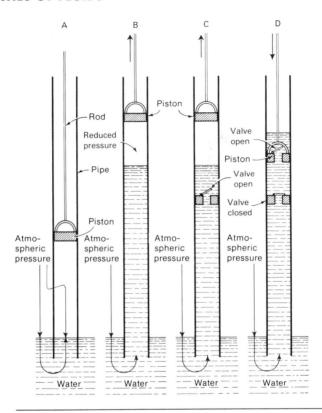

FIGURE 4.12   Development of a Suction Lift Water Pump.

The same pressure exerted on the underside of the flap of the valve in the piston, however, opens that valve and the water rises above the piston.

Of course, there are other types of liquid pumps. Some are designed like the air pump described previously. For industrial purposes, the pistons on either pump are simply linked to a motor. Controls are added to start and stop the motor at the desired pressures. Also, the pump casing must be made thick enough to contain the maximum pressure.

Summary

This chapter on properties of fluids covers both liquids and gases. The mechanical properties of liquids mentioned here are ability to flow, transmission of a pressure, density, and incompressibility. Pressure can be applied to a fluid with a piston. The amount of pressure is determined by the force on the piston and the area of the piston. This pressure is transmitted undiminished to all portions of the fluid. Liquids can transmit pressure changes instantly because they are incompressible. Pressure can also be created in a liquid as a result of the density and height of the liquid.

Gases have the properties listed above for liquids except that gases are compressible. If a piston is used to decrease the volume of a gas, the pressure will increase. This relationship between pressure and volume is known as Boyle's Law. Pressure is measured in terms of absolute pressure or gage pressure.

The air and water pumps described in this chapter rely on atmospheric pressure to aid in their operation. In both types of pumps, a piston creates suction in a chamber (the air pressure is reduced to below atmospheric pressure), and atmospheric air pressure forces air or water into the chamber, depending on the type of pump. Valves control the direction of flow. As the piston is pushed down, the air or water is forced out of the chamber through the proper valve.

Key Terms

* pressure
* pressure–force formula
* Pascal's Principle
* incompressibility
* density
* density formula

* pressure–height formula
* specific gravity
* compressibility
* Boyle's Law
* absolute pressure
* gage pressure

Questions

1. Name two properties of liquids.
2. Define the term "pressure."
3. State Pascal's Principle.
4. Explain the difference between "density" and "specific gravity."

5. Name two properties of gases.
6. What property do gases have that liquids do not have?
7. State Boyle's Law.
8. Explain the principle of the air pump.
9. Explain the operation of the suction lift water pump.

Problems

LIQUIDS

1. The pressure on a hydraulic piston that has an area of 2.56 in.² creates a force of 650 lb. What is the pressure?

2. The piston in a hydraulic pump similar to the master cylinder in Figure 4.5 has a cross-sectional area of 2.28 in.² A pressure of 35.0 lb/in.² is desired. How much force must be exerted by the piston?

3. What is the required area of a piston if a $25\bar{0}$ lb/in.² pressure is to exert a force of $150\bar{0}$ lb?

4. A force of 826 N is applied to an area of 2.00 m². What is the pressure in pascals?

5. A force of $80\bar{0}$ kN is exerted on a piston with an area of 0.125 m².
   a. What is the pressure in pascals?
   b. Convert this figure to pounds per square inch.

6. A hydraulic jack, with a cylinder similar to the one in Figure 4.4, has a pump piston with a 1.5 in.² cross-sectional area. If the maximum force that can be applied is $7\bar{0}$ lb, what weight can the lift piston raise if its cross-sectional area is 16 in.²?

7. Refer to Problem 6 above. The manufacturer of the pressure chamber states that the maximum safe pressure to place on the fluid in the cylinder is 175 lb/in.² Find the force on each piston at this pressure.

8. An aquarium has a large fish tank that is 20.0 ft high. What is the pressure on the bottom of the tank?

9. What is the pressure in the ocean a mile below the surface?

10. A submarine is $70\overline{0}$ ft below the surface of the ocean. What air pressure is necessary to blow the water out of the ballast tanks?

11. You have 5.00 ft³ of portland cement and you find it weighs 980 lb. What is its density in pounds per cubic inch?

12. Consider the following dimensions accurate to three significant figures. One cm³ of water has a mass of 1 g.
    a. One m³ of water has a mass of _____ kg.
    b. One m³ of water has a weight of _____ N.
    c. Water fills a cubic container 1 m on a side. The pressure at the bottom of the cube is _____ Pa.
    (*Note*: To convert from cubic meters to cubic centimeters, multiply by 1 000 000.)

13. What does 1.00 ft³ of water weigh?

14. What pressure must a pump supply to pump water up to the 31st floor of a skyscraper if there are 15.0 ft between floors? (Assume the reservoir of water is on the first floor, so no pressure is needed to supply the first floor.)

15. You are designing a small car with a $30\overline{0}0$ in.³ gas tank. When the tank is full, what will the gasoline weigh?

16. What is the density of ice if its specific gravity is 0.90?

17. Refer to Problem 11. What is the specific gravity of the cement?

18. The liquid in a fully charged automobile battery has a specific gravity of 1.30. What is its density in pounds per cubic inch?

19. A certain fluid is found to weigh 7.32 lb when it occupies a 15 in.³ volume. What does the fluid appear to be?

GASES

20. The pressure in a racing bicycle tire is 70.0 lb/in.² (gage). The absolute pressure is _____.

21. Refer to Problem 20. What is the gage pressure in SI units?

22. A boiler has a pressure reading of $14\overline{0}0$ kPa (gage). The absolute pressure is _____.

23. If the boiler in Problem 22 has a gage reading of $20\overline{0}0$ Pa, the absolute pressure is _____.

24. Manometers similar to that shown in Figure 4.9A are used in industry to indicate smokestack pressures. In our ex-

ample, the manometer gives a height reading of 4.00 inches (of water).

   a.   The stack pressure is _____ lb/in.^{2} (gage).
   b.   The stack pressure is _____ lb/in.^{2} (absolute).
   c.   The stack pressure is _____ pascals (gage).

25. If the manometer in Problem 24 gives a reading of 0.750 in. of mercury:

   a.   The stack pressure is _____ lb/in.^{2} (gage).
   b.   The stack pressure is _____ lb/in.^{2} (absolute).
   c.   The stack pressure is _____ pascals (gage).

26. Refer to Figure 4.8. If the initial conditions are $p_1 = 10.0$ (absolute) and $V_1 = 20.0$ in.^{3}, find the pressure necessary to compress the gas to a volume of 6.30 in.^{3}

27. An automobile tire has a volume of $30\overline{0}0$ in.^{3} and a pressure of 30.0 lb/in.^{2} (gage). What volume would the air occupy at atmospheric pressure?

28. An industrial plant pumps water into a large closed tank in which air is trapped. The air pressure is then used to force the water through the distribution pipes. The tank capacity is 40.0 ft^3 (69 100 in.^{3}). If the air pressure in the tank is at atmospheric pressure when the tank is empty of water, what is the air pressure (absolute) when 33.0 ft^3 (57 $\overline{0}$00 in.^{3}) of water are forced into the tank?

29. The pressure in a 1.5 m^3 tank is 1500 kPa (gage). What will be the gage pressure if the gas is allowed to expand and fill a tank with a volume of 7.50 m^3?

30. Refer to Figure 4.11. If a complete vacuum is drawn in the tube and the liquid is mercury instead of water, how high will the mercury rise? Give the answer in (a) inches and (b) millimeters.

31. Answer Problem 30 if the liquid is alcohol. Give the answer in (a) inches and (b) feet.

Computer Program

This program relates pressure to the height of a liquid. The program can be used to check the answers to some of the chapter problems. Also, the program can aid in plotting curves of pressure versus height for various liquids. Put the pressure scale on the Y-axis and the height scale on the X-axis.

PROGRAM

```
10   REM   CH FOUR PROGRAM
20   PRINT "PRESSURE-HEIGHT FORMULA. IF YOU WANT TO FIND PRESSURE - "
21   PRINT "THEN TYPE 1, TO FIND HEIGHT - TYPE 2."
30   INPUT A
40   PRINT "YOUR SELECTION IS # ";A
50   IF A = 1 THEN  GOTO 100
60   IF A = 2 THEN  GOTO 200
70   END
100   PRINT "TYPE IN HEIGHT OF LIQUID IN INCHES AND THEN TYPE IN THE"
101   PRINT "DENSITY IN LB/CU.IN."
110   INPUT H,D
120   PRINT "HEIGHT = ";H;" IN."
130   PRINT "DENSITY = ";D;" LB/CU.IN."
140   PRINT
150   PRINT "PRESSURE = ";H * D;" LB/SQ.IN."
160   GOTO 70
200   PRINT "TYPE IN PRESSURE IN LB/SQ.IN. AND THEN "
201   PRINT "THE DENSITY IN LB/CU.IN."
210   INPUT P,D
220   PRINT "PRESSURE = ";P;" LB/SQ.IN."
230   PRINT "DENSITY = ";D;" LB/CU.IN."
240   PRINT
250   PRINT "HEIGHT = ";P / D;" IN. OR ";P / (12 * D);" FT."
260   GOTO 70
```

NOTES

Lines 150 and 250 contain variations of the pressure = height times density formula.

Line 250 also converts the height from inches to feet.

This dramatic picture of a hydrofoil boat illustrates Newton's Laws of Motion. Balanced forces on the foils keep the boat above the water. An unbalanced force (applied by the propeller) accelerates the boat through the water. (Courtesy of Boeing Marine Systems, a division of The Boeing Company)

Chapter 5

Objectives

When you finish this chapter, you will be able to:

☐ State Newton's First Law of Motion.
☐ Explain that equilibrium is the term used to describe an object when all the forces affecting it balance one another.
☐ Identify situations that have constant velocity (including $V = 0$), the essential condition for equilibrium.
☐ Give examples of vectors: forces, displacements, velocities, acceleration. Illustrate how a vector has both a size (amount) and a direction.
☐ Given several forces acting at a common point and in various directions, show how they can be combined to determine the net effect using the graphic or parallelogram method.
☐ Given an object in motion, explain what is meant by its velocity and acceleration.
☐ Recognize the velocity and acceleration equations, $V = \Delta D/t$ and $a = \Delta V/t$ (D is distance, t is time).
☐ Express Newton's Second Law of Motion. Emphasize that changing velocity (acceleration) occurs only if an unbalanced force affects the accelerating object. Recognize the second law equation, $F = ma$, relating an unbalanced force, the mass of the object, and the resulting acceleration.
☐ State Newton's Third Law of Motion and give examples of its application.

Newton's Laws of Motion

Sir Isaac Newton (1642–1727) was an Englishman who made important contributions in the fields of optics, math, astronomy, and mechanics (his laws of motion). A knowledge of Newton's three laws of motion is a "must" in order to understand the operation of machines and the design of structures. If we are to use Newton's laws, we must learn the meanings of such terms as vectors, vector addition, concurrent forces, truss structures, velocity, speed, and acceleration. Newton's laws of motion are needed to help solve problems in almost every technical occupation. Any structure, bridge, or machine depends on Newton's laws. Of course, some bridges and large buildings were built before Newton, but these were built by trial and error without complete knowledge of the forces involved.

Some questions we will consider in this chapter are: How do we determine the forces in the members of a truss? How are velocity and acceleration calculated? What force is required to accelerate an object? What force is required to stop a moving object? Let's begin with Newton's First Law of Motion.

5.1   Newton's First Law

Newton's First Law of Motion states:

∗ Newton's First Law of Motion

> A body at rest tends to remain at rest, and a body in motion tends to continue in motion with constant speed in a straight line, unless acted upon by an unbalanced force.

When either of the above conditions exist, the body is said to be in **equilibrium**.

∗ equilibrium

A state that occurs when all the forces acting on a body (or object) balance each other—that is, the sum of all the forces equals zero.

There are two parts to Newton's First Law of Motion. The first part says a body at rest tends to remain at rest, unless acted on by an unbalanced force. To illustrate this point, imagine you are walking with a group of other students and you come across a rope lying on the ground. The rope is not moving about on the ground (as indeed, we would not expect it to) because there are no horizontal forces acting on it. The sum of the horizontal forces is zero. Suppose four of your friends grab the rope and two get

PHOTO 5.1   The force of gravity kept the Skylab space station circling the earth and prevented it from flying off toward the stars for several years without requiring additional force to keep it in constant motion. Eventually, the slight friction of the air and the earth's gravitational attraction did cause Skylab to fall in 1979. (NASA photo)

on each end for a tug-of-war. If we assume that everyone can exert the same force, then neither side can pull the rope across a dividing line on the ground—because the total force exerted by your two friends at one end of the rope exactly balances the force being exerted by the students at the other end. So even though forces are being applied, the rope remains at rest because the forces are **balanced**. Now another student joins the game and grabs one end of the rope. The total force exerted by the three students together is now greater than the force of the other two students. This difference between the forces is an **unbalanced force**, and the rope begins to move across the dividing line. The rope is no longer at rest (in equilibrium) because, as Newton says, the forces are unbalanced.

* balanced force	A force applied to some object that has an equal and opposite force applied to the same object.
* unbalanced force	The difference between the forces (the net force) if some force acting on an object is greater than any opposing force.

The more surprising part of Newton's first law is that an object in motion will continue in motion at constant speed and in a straight line unless acted on by an unbalanced force. This condition, too, is a state of equilibrium, although it is not as obvious as the first condition. For example, you know your car engine must provide the force necessary to move your car at a constant speed along a level stretch of road; however, Newton says an object doesn't need a force once it is moving. Why is there an apparent contradiction? Actually, your automobile engine must supply sufficient force to overcome friction and wind resistance. If we were in space, we could see Newton's point of view more easily because there would be no wind or friction to slow an object down. Once satellites have been fired into orbit, they spin around the earth at a constant speed for years with no force needed to keep them in motion.

When engineers or technicians design bridges, buildings, or the chair you are sitting in, their first objective is to make sure the structure won't collapse when loads (forces) are placed on them. This requires a knowledge of Newton's First Law of Motion (our discussion will be limited to objects at rest). Also, we must re-examine the definition of a force. Earlier, a force was defined as a push or pull applied to some object. For the more advanced problems in this chapter, the *direction* of a force is very important and must be considered. The direction of a force is indicated by an arrow, called a *vector*.

VECTOR QUANTITIES

Forces are **vector quantities.** Other vector quantities are: (1) wind velocity (for example, the weatherman states the direction from which the wind is blowing and the rate of wind velocity) and (2) the displacement of an object from one position to another (for example, the table must be moved from the east corner of the room to the west corner).

Figure 5.1 demonstrates that a force is a vector quantity. Figure 5.1A shows a book resting on a table. Since the book is

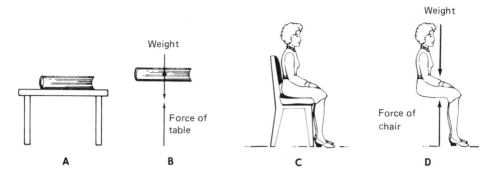

FIGURE 5.1   Examples of Equilibrium.

in equilibrium, the forces acting on it must be balanced. There are no sideways forces acting on the book. But we do know that gravity is acting in a downward direction, and that it pulls on the book with a force represented by the book's weight. Since the book doesn't fall to the floor, Newton's first law tells us that the table must be pushing up on the book with a force exactly equal to the book's weight. These forces acting on the book are shown as vectors in Figure 5.1B.

---

* vector quantities   Any physical quantity that has a direction as well as an amount.

Figure 5.1C is a picture of a student sitting in a chair, and Figure 5.1D shows the forces acting on the student. If the student weighs 163 pounds, Newton's first law tells us that the chair must push up on the student with a force that exactly balances her weight—163 pounds. If the chair pushed up with a force greater than 163 pounds, the student would fly up in the air. If the chair couldn't support the load, the student would fall. The chair will exert an upward force (called a **reaction force**) exactly equal to the weight of whoever sits on it. But how does the chair adjust to the different weights? As we discussed in Chapter 3, solids have elastic properties. So, the chair will deform an appropriate amount to match a particular load. Normally, we cannot see this deformation, but it can be measured with sensitive instruments.

Figure 5.2A shows a person holding a rope tied to a 20 pound bucket of sand. We know gravity is pulling the bucket of sand

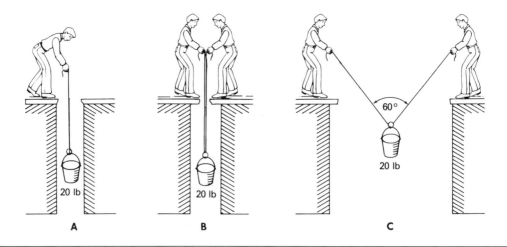

FIGURE 5.2   Applications of Force.

downward with a force of 20 pounds. Newton's first law tells us that the rope tied to the bucket must be pulling up with a force of 20 pounds and, in turn, the person holding the rope must be pulling on the rope with a 20 pound force. In Figure 5.2B, each rope and, in turn, each person must be pulling up with a force of 10 pounds to balance the weight of the 20 pound bucket of sand. The people must exert forces equal and opposite to the bucket's weight, for according to Newton's law, if the bucket is not to fall, the forces acting on it must be balanced.

Now what about Figure 5.2C? The ropes are inclined to the vertical, and therefore the people are not pulling straight up. Does each person still pull with a force of 10 pounds, and, in turn, does each rope pull on the bucket with a 10 pound force? The answer is no, because the rope forces on the bucket are not directly opposite the weight pulling down. However, the bucket is not in motion and, therefore, must be in equilibrium. The ropes must supply some vertical balance force to the weight.

The force systems in this chapter are all *concurrent* force systems. Concurrent means "meeting at the same point." Therefore, if all the forces acting on some object are concurrent, they meet at a single point, called the **point of concurrence**. The point of concurrence in each sketch of Figure 5.2 is located at the ring attached to the handle of the bucket.

PHOTO 5.2   These cranes loading cargo onto a ship use strong wire ropes, as well as the main boom of the crane, to help support the load. (United Nations photo)

∗ point of concurrence   A single point in space where concurrent forces meet.

VECTOR ADDITION

In this section, we will discuss the method of determining the forces in the ropes shown in Figure 5.2. Figure 5.3 shows the vectors that correspond to the forces shown in Figure 5.2. In Figures 5.3A and 5.3B, the forces are all vertical, so simple addition or subtraction can be used. However, the vectors in Figure 5.3C are not all vertical, so we must perform what is called **vector addition**.

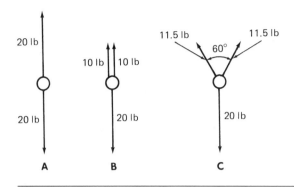

FIGURE 5.3   Vector Diagrams for Figure 5.2.

* **vector addition**

A method for adding vector quantities in which the angles they make with each other are taken into account by forming a parallelogram construction.

The procedure to follow in solving any equilibrium problem is to first draw a sketch of the setup, as in Figure 5.2C. The next step is to draw a vector diagram showing the forces acting at the point of concurrence, as in Figure 5.3C. The third diagram to draw, shown in Figure 5.4, is called a *parallelogram construction,* which is drawn to scale. The length of each vector represents the amount of force, and the direction of each vector is the same as the direction of the force. The 20 pound bucket weight is not drawn, since we wish to find the rope forces equivalent to the balance force to this weight.

We can graphically solve the problem of finding the rope forces by using the parallelogram. We need to find the lengths of the vectors representing the upward rope forces on the bucket. We use the parallelogram to show how long the vectors must be to be completely equivalent to the balance force. We know that the two rope-force vectors should be equivalent to the balance force because the bucket is in equilibrium. We also know the amount and direction of the upward balance force because it must be equal in amount, but opposite in direction, to the downward force on the bucket (its weight).

In Figure 5.5, we go through the step-by-step procedure for drawing the parallelogram in Figure 5.4. We will use the information given in Figure 5.2C—that is, the bucket weight is 20 pounds, and the ropes are 60 degrees apart.

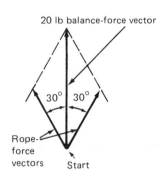

FIGURE 5.4   Force Vector Parallelogram for Figure 5.2C.

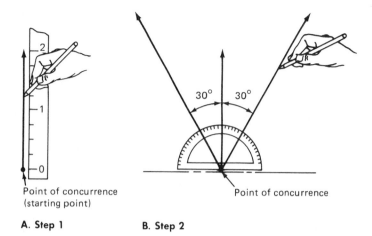

Point of concurrence
(starting point)

A. Step 1   **B. Step 2**

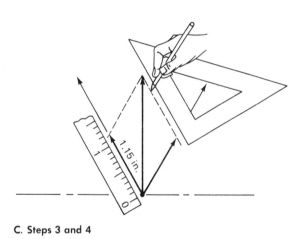

C. Steps 3 and 4

FIGURE 5.5   Step-by-Step Construction of the Force Vector Parallelogram Shown in Figure 5.4.

Steps for Drawing a Parallelogram Construction

Step 1   Locate the point of concurrence (the ring on the handle of the bucket in Figure 5.2C) at a convenient spot on the paper, and draw the balance-force vector upward and equal to 20 pounds. The *tail* of the vector must be at the point of concurrence. If the scale is 1 inch = 10 pounds, the vector will be 2 inches long (2 inches × 10 pounds per inch = 20 pounds).

Step 2   Starting at the point of concurrence, draw the rope-force vectors at the same angles as the ropes. These vectors must be at the proper angle, which, in this case, is a 30 degree angle to the balance-force vector, and their tails must be placed at the point of concurrence. At this point, we have no way of knowing how long the vectors should be.

Step 3   Starting at the top of the balance force, draw dotted lines that are parallel to the rope vectors. We now have a parallelogram, in which the lengths of the rope-force vectors—from the point of concurrence to the intersections of the dotted lines—represent the rope forces.

Step 4   Measure each rope vector from the starting point to the intersection with the dotted line. If the parallelogram has been drawn to the scale of 1 inch = 10 pounds, the rope-force vectors should measure 1.15 inches long. We can convert this dimension to pounds: 1.15 inches × 10 pounds per inch = 11.5 pounds.

This general procedure can be used for all the concurrent force problems in this chapter. Of course, when forces are compressive instead of tensile, an extra step may be necessary, as we will see when we get to compressive forces.

SAMPLE PROBLEM 5.1

Refer to Figure 5.2C. If the angle between the ropes is changed to 90.0° and the bucket of sand has a mass of 88.0 kg, find the force (F) in each rope. Note that in the SI system, it is customary to give loads in mass units (kg), but the forces in the supports must be in force units (N).

SOLUTION

Wanted

$F = ?$

Given

First, we must convert the given mass of 88 kg to force units by multiplying by 9.81: 88 kg × 9.81 = 863 N; thus, the bucket weight and the balance force = 863 N. We are also given an angle of 90° between the ropes.

Construction

Figure 5.6A is the vector diagram, and Figure 5.6B is the parallelogram with the balance force drawn vertically upward, opposite to the load. Let a scale of 1 in. = 400 N be selected for drawing the parallelogram. The balance force of 863 N is

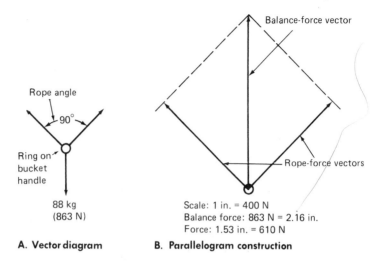

A. Vector diagram   **B. Parallelogram construction**

FIGURE 5.6   Diagrams for Sample Problem 5.1.

drawn 2.16 in. long. This length is obtained by dividing 863 by 400 N/in.:

$$\frac{863 \text{ N}}{400 \text{ N/in.}} = 2.16 \text{ in.}$$

The parallelogram is completed as outlined in the steps given previously. The force vectors are measured and found to be 1.53 in. long. We multiply to find the force in each rope:

$$1.53 \times 400 = 610 \text{ N}$$

With this information, required rope sizes can be determined.

In Chapter 3, we discussed tensile forces (forces pulling on an object) and compressive forces (forces pushing on an object). Applying our knowledge of these types of forces to equilibrium problems, we realize that the rope forces are tensile forces because, as in our examples, ropes are used to pull on objects. These tensile forces are applied along the longitudinal axis of each rope. Longitudinal means "running lengthwise." It would make no difference if the ropes in Figure 5.2 were replaced with rigid materials, such as steel or wooden poles. The important point to remember is that *the forces must be directed along the longitudinal axis of the member.*

Rigid materials may be used to apply either tensile or compressive forces, as illustrated in Figure 5.7A. The wire supporting the bucket is tied to pin A. The wire applies a tensile force to pin A. However, the members of the structure must apply compressive forces to pin A to support it in the position shown. The type of structure shown in Figure 5.7A is called a **truss structure**.

---

* truss structure

A structure made up of parts or members that transmit tensile or compressive forces only in a direction along their longitudinal axes.

The members of a truss must be connected only at their ends. Frequently, the connections are pinned connections, as in Figure 5.7A, and forces must be applied at the pins. All the trusses discussed in this chapter will use pin connections or (as we saw in Figure 5.2) ring connections, even though the members of large bridge trusses are generally riveted or welded together at their ends.

There are other structures whose members may have additional connections at points other than at their ends. In this case, the forces supplied by the members are not necessarily along the axis of the member. These structures are not trusses and are discussed in courses offering more detailed information on structures.

Now we must get back to our truss, whose members are in compression (Figure 5.7A). In order to solve for the forces in the truss members, we must follow the steps already outlined for

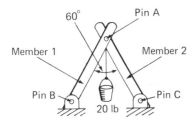

A. Sketch of truss

B. Vector diagram

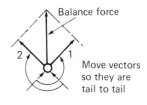

C. Parallelogram construction

FIGURE 5.7   Sketch and Diagrams of a Truss Structure.

parallelogram construction. There is, however, an extra step involved in drawing the parallelogram. Notice that in Figures 5.3 and 5.4, the vectors are positioned tail to tail, but in Figure 5.7B, two of the vectors are head to head. To avoid confusion, we will draw all our vector parallelograms tail to tail. We begin, as in step 1 earlier, by locating the point of concurrence (pin A) and then drawing the balance force to a scale of 1 inch = 10 pounds. Next, we simply move vectors 1 and 2 in Figure 5.7B along the directions in which they are heading so that their tails are at the point of concurrence. We can now proceed to complete the parallelogram (Figure 5.7C) by drawing dotted parallel lines from the head of the balance force until they cross the force vectors. When we measure the force vectors and convert this measurement into units of force, we get a value of 11.5 pounds, just as in the problem represented by Figure 5.2C.

Before beginning Sample Problem 5.2, let's go over the steps required for vector addition of forces in equilibrium—that is, the steps needed to complete the force parallelograms.

Steps for Vector Addition of Forces in Equilibrium

Step 1 Draw a sketch of the setup if none is shown.

Step 2 Draw a vector diagram showing the forces acting at the point of concurrence.

Step 3 Start the parallelogram construction, drawn to scale:
   a.   First draw the balance force equal and opposite to the weight of the load.
   b.   Next, draw the other force vectors tail to tail from the tail of the balance force. It may be necessary to reposition one or more of the forces so that the tail of its vector is located at the point of concurrence (as in Figure 5.7C).
   c.   Start at the head of the balance force and draw dotted lines parallel to the force vectors.
   d.   Measure the force vectors and convert to units of force.

Sample Problem 5.2 presents an interesting variation where one member of the truss is in tension and the other is in compression.

SAMPLE PROBLEM 5.2   Figure 5.8A illustrates a truss type of crane. Our first concern in designing this crane, so it will not fail, is to determine the forces in the wire and boom for the given load of 70.0 lb.

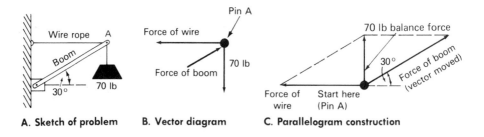

A. Sketch of problem   B. Vector diagram   C. Parallelogram construction

FIGURE 5.8   Sketch and Diagrams for Sample Problem 5.2.

SOLUTION

Wanted   force in wire = ? and force in boom = ?

Given   load weight = 70.0 lb

Construction   As we follow the steps for vector addition, we note that the first step—sketch the setup—has already been done for us in Figure 5.7A. (Most of the problems in this chapter will have the initial sketch already given to you.) The second step is to draw the vector diagram, as in Figure 5.8B. The third step is to start the parallelogram by drawing the balance force to scale (Figure 5.8C). A scale of 1 in. = 40 lb (40 lb/in.) will be satisfactory. Next, draw the wire and boom vectors, starting at the tail of the balance force. Notice that the boom vector had to be moved (in the direction in which it was acting) so its tail could be placed at the point of concurrence. Now, draw in the parallel lines from the head of the balance force. If the scale mentioned above is used, the boom vector should be 3.5 in. long, which represents a value of

3.5 in. × 40 lb/in. = 140 lb (force supplied by the boom)

The wire rope vector is 3.0 in. long, which represents a value of

3.0 in. × 40 lb/in. = 120 lb (force supplied by the wire rope)

This information allows the technician to obtain a wire rope and a boom strong enough to support the applied forces, but not so strong as to be uneconomical.

By describing the conditions of equilibrium, Newton's first law has aided us in determining the forces in the members of a truss. His first law will be used again to help solve problems in Chapter 9 on machines. Of course, the design of any type of structure or machine must start with Newton's first law, but the more complicated applications will be left for further courses in your particular technical area. Now, let us advance to Newton's Second Law of Motion.

5.2　Newton's Second Law

You have 25 seconds to extract a 500 pound red-hot steel billet from a furnace and get it to a forging hammer 50 feet away. What speed is required, and how much force is needed to attain this speed? What force is required to get a rocket up to 3600 miles per hour in 60 seconds? What force is needed to brake a car to a complete stop from a speed of 30 miles per hour? These questions represent the type of problems that Newton's second law addresses. **Newton's Second Law of Motion** states:

* Newton's Second Law of Motion

> When an unbalanced force is applied to a body, the body will accelerate.

Newton's second law is short enough, but a new term has been used: *accelerate*. Therefore, we must define the closely related terms of speed and velocity.

SPEED AND VELOCITY

Speed is defined as the time rate of change of distance. For example, a car can be traveling at a speed of 30 miles per hour (mi/hr). The cutting tool on a lathe may be moving at the rate of 0.08 inch per second (in./s). **Velocity** is speed in a given direction (along some straight line); thus, it is a vector quantity. Recall part of Newton's first law: ". . . and a body in motion tends to continue in motion with constant speed in a straight line. . . ." Scientists have realized the need for a single word to describe "speed in a straight line," and that word is velocity. Most technical problems and formulas require the use of the term velocity because direction may be important.

* speed

The time rate of change of distance—that is, a distance traveled in some time interval.

* velocity

The time rate of change of distance in a given direction. Velocity is a vector quantity, measured in standard units of feet per second or meters per second.

The symbol for velocity is V. It is the change in distance (ΔD) during a given time interval (t). The symbol delta (Δ) in front of D indicates a change in distance. The formula for velocity is:

* VELOCITY FORMULA 1

velocity = change in distance ÷ time

$$V = \frac{\Delta D}{t}$$

Formula 1 gives either the **average velocity** or the **constant velocity** of an object, depending on the circumstances.

* average velocity

The total change in distance during a total time interval, without regard to possible changes in velocity within that time interval.

* constant velocity

A time rate of motion that does not change over a given time interval.

Suppose you and a friend are each driving your cars to a town some 20 miles away. He drives at a constant speed (we can use velocity formula 1 for speed also) of 40 miles per hour. How long does he take to get there? We can arrive at the time by rearranging velocity formula 1:

$$V = \frac{\Delta D}{t} \quad \text{to} \quad t = \frac{\Delta D}{V}$$

and substituting:

$$t = \frac{\Delta D}{V} = \frac{20 \text{ mi}}{40 \text{ mi/hr}} = \frac{1}{2} \text{ hr}$$

You drive at speeds that vary from 10 miles per hour to 55 miles per hour and also arrive at the town one-half hour later. The formula indicates that your average speed was 40 miles per hour:

$$V_{avg} = \frac{\Delta D}{t} = \frac{20 \text{ mi}}{1/2 \text{ hr}} = 40 \text{ mi/hr}$$

Note that in some previous formulas, V indicated volume and D indicated density. We must be aware of the type of problem and use the correct formula.

SAMPLE PROBLEM 5.3	An airplane flew from New York City to Boston, a distance (ΔD) of 240 mi, in three quarters of an hour (t). What was its average velocity (V_{avg})?
SOLUTION	
Wanted	$V_{avg} = ?$
Given	$\Delta D = 240 \text{ mi}$ and $t = 0.75 \text{ hr}$
Formula	$V_{avg} = \dfrac{\Delta D}{t}$
Calculation	$V_{avg} = \dfrac{240 \text{ mi}}{0.75 \text{ hr}} = 320 \text{ mi/hr}$

This information can help pilots determine the amount of fuel needed.

Also, a statistical method may be used for determining the average velocity of an object. We must be given **instantaneous velocities,** and the object must be changing its velocity smoothly (at a uniform rate). An automobile speedometer, for example, gives instantaneous readings. Suppose you are in a car that is increasing its velocity smoothly along a straight stretch of highway. You record the car's velocity each second for, say, 10 seconds. We can simply add up these instantaneous velocities, divide by the number of seconds (10 in this case), and obtain the average velocity. This procedure is the same one you use when you want to determine your test grade average. For example, on four tests you receive grades of 100, 90, 80, and 85. Your average is

$$\frac{100 + 90 + 80 + 85}{4} = 88 \ 3/4$$

---

* instantaneous velocity

The time rate of change of distance during a very brief period of time.

For problems in which an object increases (or decreases) its velocity at a uniform rate, only two instantaneous velocities are needed: V_1 for the original, or initial, velocity, and V_2 for the final, or second, velocity. When these factors are known, the formula for average velocity is:

---

* VELOCITY FORMULA 2

$$V_{avg} = \frac{V_1 + V_2}{2}$$

In a number of problems, you will need to convert to other velocity units.

To convert ft/s to mi/hr, multiply by 0.682.
To convert mi/hr to ft/s, multiply by 1.47.
To convert ft/s to m/s, multiply by 0.305.
To convert m/s to ft/s, multiply by 3.28.

SAMPLE PROBLEM 5.4

You are in a car that is increasing its velocity smoothly. At a given instant, you record the velocity (V_1) as 44.0 ft/s, and a second later, you record the velocity (V_2) as 50.0 ft/s. For the 1 s interval (t):

A. Find the average velocity (V_{avg}).
B. Find the distance traveled (ΔD).

SOLUTION A

Wanted

$V_{avg} = ?$

Given

$V_1 = 44.0$ ft/s and $V_2 = 50.0$ ft/s

Formula

No distance is given; therefore, we must use velocity formula 2:

$$V_{avg} = \frac{V_1 + V_2}{2}$$

Calculation

$$V_{avg} = \frac{44.0 \text{ ft/s} + 50.0 \text{ ft/s}}{2} = 47.0 \text{ ft/s}$$

For meters per second, multiply by 0.305:

$$V_{avg} = 47.0 \text{ ft/s} \times 0.305 = 14.3 \text{ m/s}$$

For miles per hour, multiply by 0.682:

$$V_{avg} = 47.0 \text{ ft/s} \times 0.682 = 32.1 \text{ mi/hr}$$

SOLUTION B

Wanted

$\Delta D = ?$

Given

$t = 1$ s and $V_{avg} = 47.0$ ft/s

Formula

The distance traveled during the above 1 s time interval must be obtained by using the average velocity and rearranging velocity formula 1:

$$V_{avg} = \frac{\Delta D}{t} \quad \text{to} \quad \Delta D = V_{avg} \times t$$

Calculation

$$\Delta D = 47.0 \text{ ft/s} \times 1 \text{ s} = 47.0 \text{ ft}$$

For the distance in meters, multiply by 0.305:

$$\Delta D = 47 \text{ ft} \times 0.305 = 14.3 \text{ m}$$

This information could be useful in determining the wind resistance and friction acting on the car. It could also be useful in calibrating the car's speedometer.

ACCELERATION

Many people like to talk about their car's "pick-up"—in other words, how fast a car can change velocity. Assume a car starts from a stopped condition and increases its velocity smoothly. At the end of the first second, it is traveling at 10 feet per second; at the end of two seconds, it is traveling at 20 feet per second, and at the end of three seconds, it is traveling at 30 feet per second. These values are instantaneous velocities because the velocity is changing continuously. In fact, the velocity is changing at the rate of 10 feet per second for every second. The rate of change of velocity is called **acceleration**, and we would say the acceleration is 10 feet per second per second, or 10 feet per second squared (10 ft/s²).

---

* acceleration

The change in velocity of some object during some interval of time, divided by that time interval.

Just as velocity is a change in distance during a given time interval ($V = \Delta D/t$), acceleration (a) is a change in velocity (ΔV) during a given time period. The formula for acceleration is written:

---

* ACCELERATION FORMULA

$$a = \frac{\Delta V}{t}$$

where ΔV = the change in velocity: $\Delta V = V_2 - V_1$
        a = the acceleration, usually in units of ft/s² or m/s²

Engineers and technicians rarely convert from one system of units to another, so we will not bother with SI conversions of acceleration. Sample Problem 5.5 illustrates the use of the acceleration formula.

SAMPLE PROBLEM 5.5	A racing car increases its speed smoothly from a standing start (V_1) to a velocity (V_2) of 200.0 mi/hr in 15.0 s (t).

    A.   Find the acceleration (a) in ft/s².
    B.   Find the average velocity (V_{avg}) for the 15.0 s.
    C.   Find the distance traveled (ΔD) in ft/s.

SOLUTION A

Wanted

$a = ?$

Given

time $t = 15.0$ s, $V_1 = 0$, and $V_2 = 200.0$ mi/hr

Formula

$$a = \frac{\Delta V}{t} = \frac{V_2 - V_1}{t}$$

Calculation

We must first change V_2 from 200.0 mi/hr to ft/s by multiplying by 1.47:

$$V_2 = 200.0 \text{ mi/hr} \times 1.47 = 294 \text{ ft/s}$$

We can now use the acceleration formula:

$$a = \frac{V_2 - V_1}{t} = \frac{294 \text{ ft/s} - 0 \text{ ft/s}}{15.0 \text{ s}} = 19.6 \text{ ft/s}^2$$

SOLUTION B

Wanted

$V_{avg} = ?$

Given

$V_1 = 0$ and $V_2 = 294$ ft/s

Formula

Since the distance isn't known, we must use velocity formula 2:

$$V_{avg} = \frac{V_1 + V_2}{2}$$

Calculation

$$V_{avg} = \frac{0 \text{ ft/s} + 294 \text{ ft/s}}{2} = 147 \text{ ft/s}$$

SOLUTION C

Wanted

ΔD = ?

Given

V_{avg} = 147 ft/s and t = 15.0 s

Formula

To find the distance, we must rearrange velocity formula 1:

$$V_{avg} = \frac{\Delta D}{t} \quad \text{to} \quad \Delta D = V_{avg} \times t$$

Calculation

ΔD = 147 ft/s × 15.0 s = 2200 ft

This information can be used to determine the horsepower output of the engine. It can also be used as a check on other measurements.

SAMPLE PROBLEM 5.6

A bread packaging machine has an arm that pushes the bread into a wrapper at a constant speed for a distance of 0.750 m in 3.00 s (t). On the return, to pick up another loaf, the arm accelerates smoothly for a distance (ΔD) of 0.750 m in 0.600 s (t). In order to design the mechanism that will drive the arm:
  A. Find the constant speed (V) of the forward motion.
  B. Find the acceleration (a) on the return trip.

SOLUTION A

Wanted

V = ?

Given

ΔD = 0.750 m and t = 3.00 s

Formula

$$V = \frac{\Delta D}{t}$$

Calculation

$$V = \frac{0.750 \text{ m}}{3.00 \text{ s}} = 0.250 \text{ m/s}$$

SOLUTION B

Wanted

a = ?

Given

ΔD = 0.750 m and t = 0.600 s

Formula

This solution is not as direct as previous ones. We look at our three formulas and find that the only one we can use is $V = \Delta D/t$. Since we know the arm is accelerating, V must be the average velocity. This formula cannot give us the acceleration directly, but it can give us additional information we can use to calculate the acceleration. Therefore:

$$V_{avg} = \frac{\Delta D}{t} = \frac{0.750 \text{ m}}{0.600 \text{ s}} = 1.25 \text{ m/s}$$

The acceleration formula still can't be used, but velocity formula 2 can:

$$V_{avg} = \frac{V_1 + V_2}{2} = 1.25 \text{ m/s} = \frac{0 + V_2}{2}$$

$$V_2 = 2 \times 1.25 \text{ m/s} = 2.50 \text{ m/s}$$

Calculation

V_2 is the maximum velocity, and now we can find the acceleration:

$$a = \frac{\Delta V}{t} = \frac{V_2 - V_1}{t} = \frac{2.50 \text{ m/s} - 0 \text{ m/s}}{0.600 \text{ s}} = 4.17 \text{ m/s}^2$$

Notice that in this solution, we had to use three different equations to find the answer we wanted. To find the acceleration, we needed to find the change in velocity; to find the change in velocity, we had to calculate the average velocity; and we used the given information to calculate the average velocity. This procedure should be used whenever you cannot obtain the answer you want directly from the given data—that is, find out what data you do need to obtain your answer, and then see how you can use the given values to calculate this data.

This chapter has been dealing with constant, or smooth, acceleration, and we will continue to do so in this introductory text. But be aware that not all accelerations are constant. For instance, the acceleration of a drag racer may not be constant because the force transmitted from the engine may vary; also, the force of wind resistance is not constant, but changes with speed. Constant acceleration may be too expensive and not required for some machinery operations using levers and cranks. On the other hand, the high operating speeds of valves in gasoline engines may require constant acceleration motion for smooth performance.

A good everyday example of an object having constant acceleration is a falling object. The acceleration of a falling body due to the force of gravity has been calculated to be 32 feet per second for each second of fall. This is true for all objects on earth, regardless of weight and neglecting air resistance. Thus, if a body were falling at the rate of 100 feet per second at one instant, it would be traveling 132 feet per second one second later. We say that the acceleration due to gravity is 32 feet per second squared (ft/s^2). In the SI metric system, the acceleration is 9.80 meters per second per second, or 9.80 meters per second squared (m/s^2).

Sample Problem 5.7 illustrates the information that can be obtained from the velocity and acceleration formulas for a falling object.

SAMPLE PROBLEM 5.7

Suppose a rock is dropped from a high cliff. Determine its velocity at the end of the third (V_2) second.

SOLUTION

Wanted

$V_2 = ?$

Given

Since we are simply letting the rock drop, not throwing it down, we know: initial velocity $V_1 = 0$ and $a = 32$ ft/s^2

Calculation

Simply add 32 ft/s for each second. At the end of the first second, the velocity is 32 ft/s. At the end of 2 seconds, the velocity is:

32 ft/s + 32 ft/s = 64 ft/s

and at the end of 3 seconds, the velocity is:

64 ft/s + 32 ft/s = 96 ft/s

Using the same data, let's check this answer with the acceleration formula:

$a = \Delta V / t$

The formula is rearranged to:

$$\Delta V = a \times t$$

Then we replace ΔV with $V_2 - V_1$:

$$V_2 - V_1 = a \times t$$
$$V_2 - 0 = 32 \text{ ft/s}^2 \times 3 \text{ s} = 96 \text{ ft/s}$$

Technicians are often required to use the techniques in this problem to determine the velocities of a machine part at different points along its line of travel.

Let's see the difference between constant velocity and constant acceleration. In Figure 5.9A, a car is moving at a constant velocity; in Figure 5.9B, another car is moving at a constant acceleration. Notice that the distances traveled each second are constant for constant velocity, but continue to increase each second for the car moving at constant acceleration.

So far, we have talked about acceleration increasing the speed or velocity of an object. If an object already was moving at some velocity and an unbalanced force tended to slow it down, as in braking a car to a stop, the result would be negative ac-

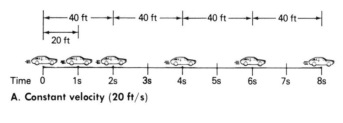

A. Constant velocity (20 ft/s)

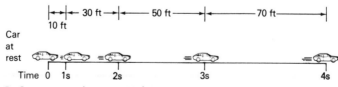

B. Constant acceleration (20 ft/s²)

FIGURE 5.9   Comparison of Constant Velocity and Constant Acceleration.

celeration, or *deceleration*. To illustrate this, let's look back at Sample Problem 5.5 again.

Suppose the car is braking to a stop in 15 seconds from a velocity of 294 feet per second. What is the deceleration? We are given $V_1 = 294$ feet per second and $V_2 = 0$. The equation is:

$$a = \frac{\Delta V}{t} = \frac{V_2 - V_1}{t} = \frac{0 - 294 \text{ ft/s}}{15 \text{ s}} = -19.6 \text{ ft/s}^2$$

Deceleration (negative acceleration) is indicated by the minus sign in the answer.

Now that we have discussed velocity and acceleration, let's get back to Newton's Second Law of Motion. The statement of Newton's second law is that an unbalanced force acting on a body causes that body to accelerate, not just move at some constant velocity. Remember that Newton's first law states that if an object is motionless or moving with constant velocity, no unbalanced force can be acting on it. Newton's second law means that if an object (or body) is accelerating, an unbalanced force *must* be acting on it. The relationship between the force (F), the mass of the object (m), and the acceleration can be expressed by the formula:

---

* FORMULA FOR NEWTON'S SECOND LAW

$F = m \times a$ or $F = ma$

where F = the unbalanced force in lb or N
m = the mass of the object being accelerated in slugs or kg
a = the acceleration in ft/s² or m/s²

Notice that the italic symbol m represents "mass," and the upright symbol m represents "meters." Mass appears in the formula and meters in the units, so you must be careful.

SAMPLE PROBLEM 5.8

An overhead crane, whose weight (w) is 15 000 lb with its load, must be able to reach a speed of 7.00 ft/s within 5.00 s after it starts. Before the senior technician can determine the size

of motor to do the job, she must know the acceleration and the force required to produce the acceleration.
- A. Find the acceleration (a).
- B. Find the force (F).

SOLUTION A

Wanted

$a = ?$

Given

$w = 15\ 000$ lb, $V_1 = 0$, $V_2 = 7.00$ ft/s, and $t = 5.00$ s

Formula

$$a = \frac{\Delta V}{t} = \frac{V_2 - V_1}{t}$$

Calculation

$$a = \frac{7.00 \text{ ft/s} - 0 \text{ ft/s}}{5.00 \text{ s}} = 1.40 \text{ ft/s}^2$$

SOLUTION B

Wanted

$F = ?$

Given

$w = 15\ 000$ lb and $a = 1.40$ ft/s^2

Formula

$F = ma$

Calculation

In order to use Newton's second law, the weight of the crane and load must be converted to mass units (slugs). The Conversion Factors Table tells us to multiply pounds by 0.0311:

$15\ 000$ lb $\times$ $0.0311 = 466$ slugs

Newton's second law gives the force required as:

$F = ma = 466$ slugs $\times$ 1.40 ft/s$^2 = 650$ lb

Technicians can use this information to determine the size of electric motor required and the size and type of gears needed for a particular job.

Let's review for a minute to put these first two laws in perspective. Suppose you are trying to push your car down the street. You push with a force of 50 pounds, but the car doesn't move. Where is the acceleration? Well, since the car hasn't accelerated up to some velocity, the second law doesn't apply, so the first law must. The force you exerted was not sufficient to overcome some opposing force—maybe a brick was under the wheel, or the brakes were on, or friction occurred in the moving parts. Any or all of these things could prevent motion. Now a friend gives you a hand and the car starts rolling. Let's say it takes 50 pounds of force to overcome friction, and the two of you exert a total of 100 pounds. As long as you both push, an unbalanced force of 50 pounds will cause the car to accelerate. If you both push until the car reaches, say, 3 miles per hour, your friend can stop and you can maintain a constant 3 miles per hour simply by overcoming the 50 pounds of friction. (Remember, the first law says forces are balanced when an object is moving at a constant velocity.) Of course, the car must be on a level stretch of road. If you have to push it uphill, more force is needed.

We will now move on to Newton's Third Law of Motion. Newton's third law doesn't give us a new formula, nor will there be problems to solve, but his third law will increase our understanding of the first two laws.

5.3 Newton's Third Law

Newton's **Third Law of Motion** states:

＊Newton's Third Law of Motion

> When an object exerts a force on a second object, that second object exerts an equal and opposite force (called a reaction) on the first object.

This law applies to objects at rest or in motion. We will discuss objects at rest first.

Figure 5.1 shows a person sitting in a chair. The weight of that person's body acts (applies a force) down on the chair. The chair reacts up on the person with an equal force. Also, the chair acts down on the floor and the floor reacts up on the chair. A similar situation exists in Figure 5.10. In Figure 5.10A, the person is pulling on the rope with a force of 100 pounds, as is indicated on the spring scale. Therefore, the rope (and also the wall attached to it) must be pulling on the person with exactly 100 pounds in the opposite direction. This force in the opposite direction, known as a *reaction force,* can be demonstrated by replacing the wall with another person, as is shown in Figure 5.10B. To remain in equilibrium, the second person must pull on the

PHOTO 5.3 Launching a rocket to the moon is an example of Newton's third law. To send the huge rocket aloft with enough acceleration to overcome the earth's gravitational attraction, the rocket engines must emit gases with an equal force in the opposite direction. The motion of the rocket is a reaction to the thrust of the rocket engines. (NASA photo)

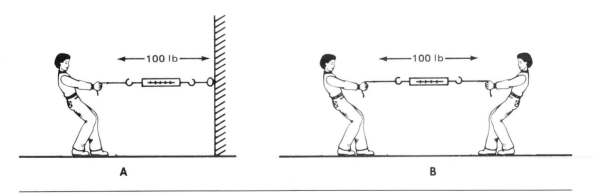

FIGURE 5.10 Action and Reaction Forces When Objects Are at Rest.

rope (just as the wall did) with 100 pounds of force. By the way, the tension in the rope is *not* 200 pounds (the sum of the forces) but simply the single force of 100 pounds, as is indicated by the scale.

Note carefully that the reaction force described by Newton's third law does not act on the same body originally acted on. For instance, the person pulling on the rope in Figure 5.10 causes a reaction force to arise in the rope, but the reaction force acts on the person, not on the rope itself. The person pulling on the rope is the applied force; the rope pulling on the person is the reaction force. Both forces are equal, act in opposite directions, and act on different objects. The rope remains in equilibrium not because it pulls on the person with an equal and opposite reaction force, but because it is tied to an immovable wall. The person pulls on one end of the rope, and the wall reacts with equal force on the other end of the rope. Thus, the rope experiences balanced forces and doesn't move.

Most of us can get a mental picture of action and reaction when objects are at rest, but objects in motion are sometimes more difficult to understand. How can an unbalanced force exist to provide acceleration if there is always an equal and opposite reaction? Yet, we know that objects can accelerate.

Look at Figure 5.11A. A balloon has been blown up with air by a student, stoppered with a cork, and placed on a table. In this case, the balloon and contained air weigh more than the displaced air in the room, so the balloon tries to sink down on the table. The reaction force of the table on the balloon prevents the balloon from falling to the floor. However, the balloon has

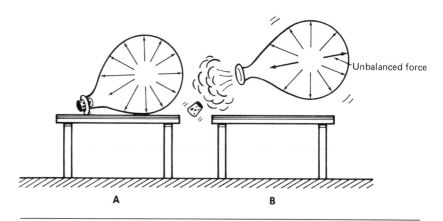

FIGURE 5.11   Action and Reaction When Motion Is Involved.

no tendency to move about on the table. Indeed, you would not expect it to. But why not? Although pressure in the balloon is creating forces in all directions on the balloon, all the forces acting on the balloon are balanced. The horizontal left force is balanced by the horizontal right force, and so on. Now, if someone pulls the cork, the balloon flies around the room. We have acceleration, and, of course, an unbalanced force (Figure 5.11B). The primary active force here is the air inside the balloon rushing out to release the pressure in the balloon. However, the force acting on the balloon itself is the unbalanced force, or reaction force to the force of the escaping air. Again, as with equilibrium situations, the action and reaction forces are equal, opposite in direction, and act on different objects. However, in this case, the forces acting on any one object are not balanced, and an acceleration is the result. The escaping air moves in one direction and the balloon moves in the opposite direction. By Newton's second law, the unbalanced force 1 is equal to the mass of the balloon multiplied by its acceleration: $F = ma$.

You can demonstrate this to yourself. Obtain a small cart from the physics laboratory, and pull or push it across a table with a 1 pound force. If you use a spring scale to make sure that you are exerting a 1 pound force, you will find that the cart accelerates. Also, if you are pulling on the spring scale with a force of 1 pound, the cart must be holding back (even as it accelerates) with an equal force (reaction). Otherwise, the scale could not give a reading.

Summary

Newton's three laws of motion, which relate forces applied to an object and the motion of that object, are needed to help solve problems in almost every technical occupation. Newton's first law basically states that if the forces applied to an object are balanced, the object is in equilibrium (constant or zero velocity).

Forces are vector quantities (have both amount and direction), and vector addition must be used when the forces are at angles to each other. This is especially true when determining the forces supplied by the members of a truss structure.

A knowledge of speed, velocity, and acceleration is needed to apply Newton's second law, which basically states that if the forces applied to an object are unbalanced, the object undergoes acceleration (changing velocity). Although speed and velocity are both "rates of motion," velocity is a vector quantity. Acceleration

PHOTO 5.4   The John F. Kennedy Bridge in Kentucky is an example of a large truss structure. The members of the truss are riveted together at their ends. (Courtesy of United States Steel Corp.)

is also a vector quantity. A good example of constant acceleration is given by a falling object. Since the unbalanced force (its weight) is constant, the acceleration is constant.

Newton's third law states that every applied force has a reaction. In the case of balanced forces, the reaction to an applied force is also a force, but it acts on the object applying the initial force. In the case of an unbalanced force, the reaction is supplied by the object's mass times its acceleration.

Key Terms

* Newton's First Law of Motion

* equilibrium

* balanced force

* unbalanced force

* vector quantities

* reaction force
* point of concurrence
* vector addition
* truss structure
* Newton's Second Law of Motion
* speed
* velocity
* velocity formula 1
* average velocity
* constant velocity
* instantaneous velocity
* velocity formula 2
* acceleration
* acceleration formula
* formula for Newton's second law
* Newton's Third Law of Motion

Questions

1. Give an illustration to explain Newton's First Law of Motion.
2. What is meant when we say a body is in equilibrium?
3. How do we know that a body is in equilibrium?
4. What is meant by the term "concurrent forces"?
5. Explain what is meant by a truss structure.
6. Give an illustration to explain Newton's Second Law of Motion.
7. Explain the difference between speed and velocity.
8. What is acceleration?
9. How do we know when to apply Newton's second law?
10. Explain Newton's Third Law of Motion when an object is at rest and when it is accelerating.
11. A man wants to jump from a rowboat over to a dock a few feet away. What is your advice to him?

Problems

For each problem, sketch a vector diagram and then draw a parallelogram to scale. Refer to the Conversion Factors Table on the front endpapers for conversion of units.

NEWTON'S FIRST LAW

1. Refer to Figure 5.12 and indicate which of the round objects are in equilibrium.

2. Refer to Figure 5.2C.
   a. If the bucket weighs 60.0 lb, what is the force in each rope?
   b. If the bucket weighs 60.0 N, what is the force in each rope (in newtons)?

3. Refer to Figure 5.2C. What is the force in each rope if the angle between the ropes is (a) 90.0°, (b) $12\overline{0}°$?

4. If the angle between the ropes in Figure 5.2C is 45° and the force in each rope is 40.0 lb, what is the total weight being supported?

5. The stoplight in Figure 5.13 is supported by two wires. The light weighs 75.0 lb, and the wires make an angle of 10.0° with the horizontal. What is the force in each wire?

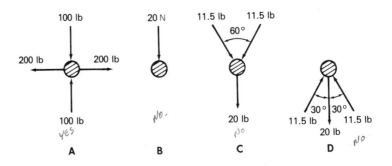

FIGURE 5.12 Diagrams for Problem 1. Only those forces shown are acting on the objects.

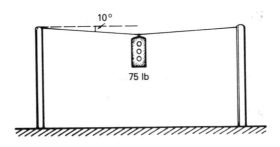

FIGURE 5.13 Sketch for Problem 5.

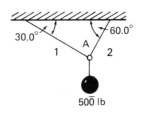

FIGURE 5.14 Sketch for Problem 6.

6. Solve for the force in each wire in Figure 5.14.

7. a. Solve for the force in each wire in Figure 5.15.
   b. Does the fact that the wires are different lengths make any difference in your answer?

8. Refer to Figure 5.16. Determine the horizontal force required to maintain the weight in the position shown.

9. Refer to Figure 5.7A. Solve for the force in each truss member and check the answer given in the text.

10. Refer to Figure 5.7A. Solve for the force in each truss member if the angle between members is $12\overline{0}°$.

11. Solve for the forces in the boom and wire rope shown in Figure 5.17.

12. Refer to Figure 5.2C. If the angle between the ropes is 70.0° and the load is 80.0 kg, solve for the force in each rope. (Remember, kilograms must be changed to force units.)

NEWTON'S SECOND LAW

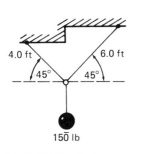

FIGURE 5.15 Sketch for Problem 7.

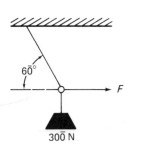

FIGURE 5.16 Sketch for Problem 8.

13. A velocity of 70.0 mi/hr = _____ ft/s = _____ m/s.

14. A robot arm is programmed to pick up a part and place it 5.0 ft away in 2.0 s. What is the average velocity in feet per second?

15. The piston stroke in an automobile engine is 3.65 in. At a given rpm, the piston travels from the top to the bottom of its stroke in 0.0150 s. What is the average velocity in (a) inches per second and (b) feet per second?

16. You are driving on a highway that has distance markers every one tenth of a mile. You pass 7 markers in 1 min. What is your speed in miles per hour and feet per second?

17. A conveyor belt for castings travels at a constant velocity of 6.00 ft/s. How far does a casting travel in 45.0 s?   *27o'*

18. A rock is dropped from some height (smooth acceleration). At time zero, it is released. At the 1 s tick, its velocity is measured at 9.80 m/s.   *$\frac{0+9.8}{2} =$*
    a. What is the average velocity (in meters per second) for this 1 s interval?   *4.9 m*
    b. How far (in meters) did the rock travel?   *9.8 m.*

19. A cam on a box folding machine increases the velocity of a pushrod smoothly from zero ft/s to 8.00 ft/s in 0.752 s. Find: (a) the average velocity of the pushrod and (b) the distance it travels.

   a. $\frac{0+8}{2} \times .752 = 3.01$

   b. $8 \times .752 = 6.016'$

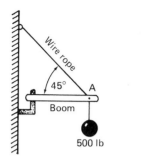

FIGURE 5.17 Sketch for Problem 11.

20. Refer to Sample Problem 5.8. If the crane weighs only $75\overline{0}0$ lb, how much force is required for the same acceleration?

21. A small commuter train with a mass of $10\,\overline{0}00$ slugs increases speed smoothly from zero mi/hr to 30.0 mi/hr in 8.00 s. Find (a) the acceleration in feet per second squared and (b) the force required to cause this acceleration (neglecting friction and air resistance).

22. a. A mass of 5.0 slugs has a weight of _____ lb.
    b. A mass of 5.0 kilograms has a weight of _____ N.

23. A rocket increases its velocity smoothly from a standing start to $18\,\overline{0}00$ mi/hr in 20.0 min (0.333 hr.). Find the acceleration in (a) miles per hour squared and (b) feet per second squared.

24. From a standing start, an automatic stock feeding machine exerts a 5.0 N force to push a metal part across a horizontal frictionless surface. If the velocity is 16 m/s at the end of 2.0 s, find: (a) the acceleration and (b) the mass of the object.

25. A $32\overline{0}0$ lb car traveling at 60.0 mi/hr brakes to a stop in a time of 6.30 seconds.
    a. What is the deceleration in feet per second squared?
    b. What force must the brakes exert on the car to stop it?

26. A drop forge hammer weighing $10\overline{0}$ lb falls (drops) through a certain distance, where it strikes the metal part to be formed at a velocity of 22.0 ft/s. Find: (a) the acceleration of the hammer, (b) the mass of the hammer, and (c) the average velocity.

27. Refer to Problem 26. The hammer is brought to a stop by the metal part in 0.00200 s.
    a. What is the deceleration in feet per second squared?
    b. What force does the hammer exert on the metal part? (*Hint*: $F = ma$)

28. Refer to Figure 5.9B. What is the average velocity of the car undergoing constant acceleration for the complete 4.0 s interval?

29. Refer to Figure 5.9B. What is the velocity of the car at the 4.0 s mark?

Computer Program

This program will determine the force in each of 2 wires supporting a load that is acting vertically down. You can use the program to check your answers to a number of chapter problems.

Also, you can vary the angles of the wires to see how the forces change. For instance, vary the given angles in Problem 5 concerning the stoplight. Please note that all angles must be measured in a counterclockwise direction from the right-hand side of the X-axis—as shown in Figure 5.18.

PROGRAM

```
10   REM   CH FIVE PROGRAM
20   PRINT "PROGRAM TO FIND THE FORCE IN EACH OF TWO WIRES"
21   PRINT "SUPPORTING A WEIGHT.  WIRES ARE LABELED A  AND B."
30   PRINT
40   PRINT "TYPE IN THE ANGLE OF WIRE A, THE ANGLE OF WIRE B,"
41   PRINT "AND THEN THE WEIGHT OF THE OBJECT IN POUNDS."
50   INPUT AA,AB,W
60   PRINT "ANGLE A = ";AA;" DEG."
70   PRINT "ANGLE B = ";AB;" DEG."
80   PRINT "WEIGHT = ";W;" LB."
90   PRINT
100  LET Z = 3.1416 / 180
110  LET A1 =  COS (AA * Z)
120  LET A2 =  SIN (AA * Z)
130  LET B1 =  COS (AB * Z)
140  LET B2 =  SIN (AB * Z)
150  LET DEN = (A1 * B2) - (A2 * B1)
160  LET A = ( - W * B1) / DEN
170  LET B = (W * A1) / DEN
180  PRINT "FORCE ON WIRE A = ";A;" LB."
190  PRINT "FORCE ON WIRE B = ";B;" LB."
200  END
```

NOTES

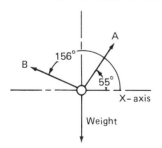

Lines 60–80 confirm the data placed in the computer.

Line 100 converts degrees into data the computer can use.

Lines 110–170 use trigonometry and algebra to solve the problems you obtain answers to graphically.

FIGURE 5.18   Example of Angle Measurement for the Computer Program.

The energy demands of modern civilization require sophisticated and large-scale power systems. The equipment in this photo is used to convert high-voltage direct current electrical energy to alternating current electrical energy. (Courtesy of Bonneville Power Administration)

Chapter 6

Objectives

When you finish this chapter, you will be able to:

- [ ] Explain that physical work results when a force moves an object through some distance.
- [ ] Recognize the work equation $W = FD$, where work (W) equals force (F) times distance (D).
- [ ] Explain how the work equation gives rise to the units for work in the metric and U.S. systems of measurement.
- [ ] Demonstrate that work and energy have the same units.
- [ ] Discuss various energy forms: mechanical, potential, kinetic, heat, light, sound, and so on. Describe and compare them, and demonstrate how they can do work.
- [ ] Recognize the gravitational potential energy equation, $PE = wH$, where potential energy (PE) equals weight (w) times height (H) through which weight is lifted.
- [ ] Recognize the kinetic energy formula $KE = 1/2mV^2$, where kinetic energy (KE) equals one-half times the moving mass (m) times the velocity of motion squared (V^2).
- [ ] Explain that power is defined by how fast (or slow) work is done.
- [ ] Recognize the power equation $P = W/t$, where power (P) equals work done (W) divided by the time (t) needed to perform the work.
- [ ] Demonstrate how the power equation gives rise to the familiar horsepower and watt units.

Work, Energy, and Power

The subjects of this chapter are somewhat abstract. That is, we can't point to "work" as we can to a house and say, "There it is." But we can see the results of work. Also, this knowledge is relatively new. Ever since the time of Isaac Newton, scientists have had a difficult time trying to define and measure energy and its relationship to work. But, after many years of research and thought, they have succeeded. Technicians must put this information to use in their daily tasks.

This chapter covers the areas of work, mechanical energy, and power, and includes definitions of such units of measurement as foot·pound, joule, watt, and horsepower. The distinction will be made between mechanical energy and other kinds of energy. The discussion on mechanical energy will be broken down into its two forms: potential energy and kinetic energy.

Machines require energy to perform their tasks (or do work). Since energy costs money, industry must be able to measure this quantity. Industry is continually looking for alternative manufacturing methods and service methods to reduce the use of energy. If you are going to forge a steel crankshaft or stamp out a sheet-metal part, you need to know the energy required. You have to know the energy needed to drive piles into the ground for a construction project. If you are going to specify the proper size of motor to raise an elevator or run a lathe, you need to know the energy and the power requirements. So, you must have a good understanding of just what energy is and how it is related to work and power.

6.1   What Is Work?

Work is done when a force is applied to an object and the object moves in the direction of the applied force. For example, a weight lifter raises 250 pounds from the floor to a position over his head. He does work while lifting the weight, but does not do any work while holding the weight over his head. Work is also done when you push a table across a room, as in Figure 6.1. Suppose the table moves a few feet and then stops suddenly when a leg gets caught in a rug. While the table is moving, you are doing work. When the table stops, you are no longer doing work, even though you may be pushing harder to try and start the table moving again.

---

* work

The product of a force and the distance through which motion takes place in the direction of the force.

The formula for work is:

---

* WORK FORMULA

work = force × distance

$W = FD$

where W = the work in foot·pounds (ft·lb) or joules (J)
(note that 1 J = 1 N·m)
F = the force in lb or N
D = the distance in ft or m

Technicians often must convert work units from one system of measurement to the other:

To convert foot·pounds to joules, multiply by 1.36.
To convert joules to foot·pounds, multiply by 0.738.

To raise a 100 pound stone 1 foot requires 100 foot·pounds of work (100 pounds × 1 foot = 100 foot·pounds). To raise it 2 feet, you must perform 200 foot·pounds of work (100 pounds × 2 feet = 200 foot·pounds). Also, in Figure 6.1, if you exert a force of 15 pounds to push the desk 30 feet across the room, you have done 15 pounds × 30 feet = 450 foot·pounds of work. Sample Problem 6.1 demonstrates the use of the work formula.

SAMPLE PROBLEM 6.1

It takes work to compress (or stretch) a spring. This work can be recovered when the spring is allowed to return to its normal position. The chief engineer needs a spring for a large oven-door return mechanism. He gives you 2 springs: Spring A will compress to a distance (D) of 5.80 in. under an average load (w) of 207 lb, and spring B will compress to a distance of 4.20 in. under an average load of 253 lb. Which spring requires more work (W) to compress to its limit?

SOLUTION

Wanted

W_A = ? and W_B = ?

Given

For spring A: F_A = 207 lb and D_A = 5.80 in. For spring B: F_B = 253 lb and D_B = 4.20 in.

FIGURE 6.1   An Example of Work. Work is performed on the table only when it moves.

Formula

Calculation

$W = FD$

$W_A = 207 \text{ lb} \times 5.80 \text{ in.} = 1200 \text{ ft·lb}$

To convert to joules, multiply by 1.36:

$1200 \text{ ft·lb} \times 1.36 = 1630 \text{ J}$

$W_B = 243 \text{ lb} \times 4.20 \text{ in.} = 1060 \text{ ft·lb}$

Converting to joules:

$1060 \text{ ft·lb} \times 1.36 = 1440 \text{ J}$

Spring A requires more work. The springs will, of course, store this energy, and the engineer may want spring A, depending on the size of the oven door.

6.2 What Is Energy?

Some old-fashioned, wall-mounted clocks operate by falling weights on chains. In the morning, you pull the weights up to the clock. As the weights slowly fall during the day, they do the work necessary to turn the clock mechanism. In another common instance, when a flame is applied to the bottom of a tea kettle containing water, the heat energy ultimately changes the water (a liquid) to steam (a gas). The motion of the molecules of the gas is greater than that of the molecules of the liquid, and the gas pressure performs work when it lifts the lid of the kettle.

PHOTO 6.1   Much of the electrical energy used in the northeastern United States and parts of Canada comes from the tremendous energy of Niagara Falls. The potential energy of the water changes to kinetic energy as it falls through the drop in height and moving water flows through turbines to generate electricity. (Courtesy of the N.Y. State Commerce Commission)

Both the falling weights and the flame were able to produce work. We say that they possess **energy.** The energy of the falling weights, which resulted from their weight and motion, is called **mechanical energy.** The *heat energy* of the flame increased the molecular motion of the water, which in turn lifted the kettle lid. Thus, heat energy can be used to produce mechanical energy. Heat energy will be discussed in more detail in the chapters devoted to heat.

---

* energy   The ability to produce work.

* mechanical   A kind of energy produced whenever a force is applied to an object
energy   and the object moves in the direction of the applied force.

There are other kinds of energy in addition to the two we just mentioned. Some of these are *light energy, sound energy, electrical energy, magnetic energy, chemical energy,* and *atomic* (or *nuclear*) *energy*. We shall discuss some of these various types of energy throughout this book.

In general, mechanical energy exists in two forms: potential and kinetic. In order to examine the difference between the two forms, let's go back to our example of the wall clock with the hanging weights. The weights are raised to the clock. But we do not want the clock to run just yet, so the mechanism is locked. While the weights are hanging motionless at their top position, they are performing no work. They merely possess the ability to do work—that is, they have energy. Thus, they possess position energy or **potential energy.**

✳ potential energy   The form of energy available to do work that an object has because of its position or condition.

The mechanism is then unlocked, and the weights are allowed to fall slowly. Now the weights not only possess the ability to do work, they are in fact doing work. Since the weights possess the ability to do work because of their motion, we say they have moving energy or **kinetic energy.** As the weights fall, their potential energy changes to kinetic energy. Note that if one weight broke loose and fell freely, it would possess kinetic energy while falling, but would be doing no useful work.

✳ kinetic energy   The form of energy available to do work that an object has because of its motion.

Water behind a dam possesses potential energy. This energy appears as kinetic energy as the water flows through a turbine to generate electricity. An object may possess potential or kinetic energy for reasons other than falling from some height. Thus, a coiled spring may possess energy because of the tension on its molecules. When the spring is released and flies apart, its potential energy changes to kinetic energy.

Energy may be changed from one form to another. Thus, when the clock weights fell, they caused the hour and minute hands to move (mechanical energy) and they also caused chimes to ring on the hour (sound energy).

For our purposes, we can think of work and mechanical energy as being similar. When you pushed the desk across the floor, you did 450 foot·pounds of work. We can also say that you used 450 foot·pounds of energy.

Mechanical energy is measured in the same units as work—namely, foot·pounds or joules. The clock weights at their top position possess the same number of foot·pounds of potential energy as the number of foot·pounds of work that were required to raise them from their bottom position. The amount of work is equal to the product of the total weight and the vertical height. Let's illustrate this last point another way. We are loading boxes onto the flatbed of a truck. If we disregard friction, it makes no difference whether we pick the box up in our arms and place it on the truck or if we place the box on rollers and roll it up an incline. The work done is the same because we must use the weight of the box and the vertical height for our calculations.

The interesting thing about energy is that although it can be changed from one form to another, it cannot be destroyed or created. For instance, if we were to add up all the energies resulting from the falling weights—the mechanical energy of the hands and the sound energy of the chimes—we would find them equal to the amount of potential energy possessed by the weights at their top position.

Modern scientific discoveries lead to the conclusion that energy may be created from matter and that matter, in turn, may be created from energy. This statement, however, does not mean that energy can be created or destroyed. Rather, it seems to indicate that matter and energy are two different phases of the same thing.

Let's look at one example of how industry makes use of potential energy. Figure 6.2 shows a piston-lift gravity-drop hammer, a machine used in drop forging metal parts. The ram is a massive piece of metal that is raised about 4 or 5 feet and then is dropped on the metal to be formed by the die. From handbook data and experiments, the engineers and technicians know how much energy is required to form a particular part. For our discussion, let's say the ram weighs 2000 pounds (including the upper half of the die) and the height of drop is 5 feet. The potential energy of the ram in the raised position is equal to the product of its weight and the height of drop. The formula used to find potential energy is the same as the one for work—only the symbols are changed:

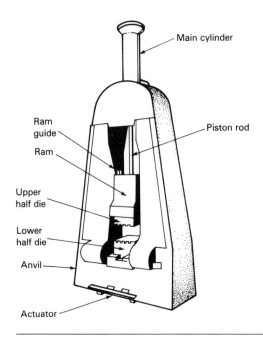

Main cylinder

Ram guide

Ram

Piston rod

Upper half die

Lower half die

Anvil

Actuator

FIGURE 6.2 A Piston-Lift Gravity-Drop Hammer. (From *Manufacturing Processes* by Begeman & Amstead. © 1969, John Wiley & Sons)

∗ POTENTIAL ENERGY FORMULA

$E_{pot.}$

potential energy = weight × height

$$PE = w \times H$$

where w = weight in lb or N
H = height in ft or m
PE = potential energy in ft·lb or J (1 J = 1 N·m)

At maximum height, the ram's potential energy is:

$$PE = w \times H = 2000 \text{ lb} \times 5 \text{ ft} = 10\ 000 \text{ ft·lb}$$

At the instant just before falling, the ram has no motion, and therefore no kinetic energy. As the ram falls, it picks up speed, and as it gets closer to the anvil, its potential energy decreases and its kinetic energy increases. At the halfway mark, half of the original potential energy has been converted to kinetic energy. At the instant just before hitting the anvil, all the initial, or original, potential energy is converted to kinetic energy—that is, 10 000 foot·pounds. We can check these two statements with the potential energy formula. At one-half the height (2.5 feet):

$$PE = 2000 \text{ lb} \times 2.5 \text{ ft} = 5000 \text{ ft·lb}$$

At zero height (the instant before hitting):

$$PE = 2000 \text{ lb} \times 0 \text{ ft} = 0 \text{ ft·lb}$$

As the ram hits the metal to be formed, the kinetic energy is used to form the part into the desired shape.

There are occasions when we must determine the kinetic energy (KE) of an object when the potential energy is not a factor. For example, your car engine has to do work (expend energy) to get your car moving at a speed of, say, 30 miles per hour. The work the engine performs gives the car kinetic energy. The car has no potential energy with respect to the ground because it is on the ground. If the car is brought to a stop, the brakes must dissipate the kinetic energy of the car. By the way, once the car is up to speed, the car does not accumulate more energy. The engine expends energy to overcome friction and wind resistance.

The laws of physics give us a formula for kinetic energy that includes the mass of the object and its velocity. The formula is:

---

* KINETIC ENERGY FORMULA

E_{KIN}

kinetic energy = 1/2 × mass × velocity squared

$$KE = \frac{1}{2}mV^2$$

where   KE = the kinetic energy in ft·lb or J
          m = the mass in slugs or kg
          V = the velocity in ft/s or m/s

Notice that the units formed in this equation are slugs × feet per second squared $(\text{ft/s})^2$, and these units must be the same as foot·pounds.

<table>
<tr><td>

SAMPLE PROBLEM 6.2

SOLUTION

Wanted

Given

Formula

Calculation

</td><td>

A pile-driver ram slams into a piling at a velocity (V) of $4\overline{0}$ ft/s. The weight (w) of the ram is 1500 lb. What is the kinetic energy (KE) of the ram at the instant before it hits the piling?

KE = ?

w = 1500 lb and V = $4\overline{0}$ ft/s

$$KE = \frac{1}{2}mV^2$$

The kinetic energy formula requires that we convert weight into mass. To change weight into slugs, multiply by 0.0311:

1500 lb $\times$ 0.0311 = 47 slugs

Now we can use the formula:

$$KE = \frac{1}{2}mV^2 = \frac{1}{2} \times 47 \text{ slugs} \times (40 \text{ ft/s})^2$$

$$KE = \frac{1}{2} \times 47 \times 1600 = 38\ 000 \text{ ft·lb}$$

To convert to joules, multiply ft·lb by 1.36:

38 000 ft·lb $\times$ 1.36 = 51 700 J = 52 kJ

This information will allow us to determine whether the energy to drive the pile is sufficient for the type of soil encountered.

</td></tr>
</table>

6.3 What Is Power?

Just as velocity is the time rate of travel and is specified as the distance traveled divided by the time ($V = \Delta D/t$), so **power** is specified as the amount of work performed divided by time.

---

* power　　　　　The time rate of doing work.

The formula for determining power is:

✳ POWER FORMULA

power = work ÷ time

$$P = \frac{W}{t}$$

The usual unit of time is the second. If work is in units of foot·pounds, the units of power are foot·pounds per second. If work is in joules, then the units of power are joules per second. In the SI metric system, a joule per second is called a watt (W). Therefore, 1 watt of power is equal to 1 joule per second. As you see, the symbol for watt is W. Be careful not to confuse this unit of

PHOTO 6.2 Thoroughbred horses have been bred for speed; they cannot pull heavy loads the way work horses can. Engineers have estimated that an "average" horse has the power of 550 ft·lb/s. (Courtesy of the N.Y. State Commerce Commission)

power with the symbol for work. The symbol for work (W) is an algebraic symbol used to express work in a formula, as in the power formula above. The symbol for watts (W) is a unit symbol, designating the units of power.

An example should help define power. Suppose it takes you 2 seconds to raise a 100 pound weight 2 feet above the floor. Another person might do the same thing in 4 seconds. In each case, the work is the same:

$$1\bar{0}0 \text{ lb} \times 2.0 \text{ ft} = 2\bar{0}0 \text{ ft·lb}$$

However, the time rate of doing the work—that is, the work divided by the time required to perform it—is twice as great for you. The power you developed is:

$$P = \frac{W}{t} = \frac{2\bar{0}0 \text{ ft·lb}}{2.0 \text{ s}} = 1\bar{0}0 \text{ ft·lb/s}$$

The power developed by the second person is:

$$P = \frac{2\bar{0}0 \text{ ft·lb}}{4.0 \text{ s}} = 5\bar{0} \text{ ft·lb/s}$$

Since the foot·pound per second is a rather small unit for practical use (such as use with automobile engines, locomotives, and industrial drive units), the horsepower (HP) has been adopted as a more practical unit of power. It has been estimated that the power of an "average horse" is equal to 550 foot·pounds per second, which is the accepted figure used for conversion to horsepower. The conversion to horsepower is accomplished by dividing foot·pounds per second by 550:

$$\text{HP} = \frac{P \text{ (ft·lb/s)}}{550 \text{ (ft·lb/s)/HP}}$$

Note that the units ft·lb/s cancel, since they are in the numerator and the denominator; and HP in the denominator is under a division sign, so it will end up in the numerator. In our previous example, your horsepower for lifting the 100 pound weight is:

$$\text{HP} = \frac{1\bar{0}0 \text{ ft·lb/s}}{550 \text{ (ft·lb/s)/HP}} = 0.18 \text{ HP}$$

The conversion from horsepower to watts, and vice versa, is rather common. From the Conversion Factors Table at the front, we note that 1 horsepower equals 746 watts. To convert from horsepower to watts, multiply by 746:

Number of horsepower $\times$ 746 W/HP = number of watts

To convert from watts to horsepower, multiply by 0.00134:

Number of watts $\times$ 0.00134 HP/W = number of horsepower

Sample Problems 6.3 and 6.4 illustrate the power formulas.

SAMPLE PROBLEM 6.3	A hoist at a building construction site must be capable of lifting a $5\overline{0}0$ lb load (force, F) to a height (distance, D) of $5\overline{0}$ ft in 8 s (t). A. How much energy (work, W) is used? B. How much horsepower (HP) is required?
SOLUTION A	
Wanted	$W = ?$
Given	$F = 5\overline{0}0$ lb, $D = 5\overline{0}$ ft, and $t = 8$ s
Formula	$W = F \times D$
Calculation	$W = 500$ lb $\times$ 50 ft $= 25\ 000$ ft·lb
SOLUTION B	
Wanted	$HP = ?$
Given	$W = 25\ 000$ ft·lb, $t = 8$ s, and HP $= P/550$
Formula	To find the horsepower, we must first use the power formula: $$P = \frac{W}{t} = \frac{25\ 000 \text{ ft·lb}}{8 \text{ s}} = 3125 \text{ ft·lb/s}$$
Calculation	Now we can calculate the horsepower: $$HP = \frac{P}{550} = \frac{3125}{550} = 5.7 \text{ HP (or 4.2 kW)}$$

The technician can now specify the required size of the motor that will be used to raise and lower the hoist.

SAMPLE PROBLEM 6.4

An average force (F) of $90\overline{0}$ N is exerted on the piston of an engine during the power stroke. The length (distance, D) of the stroke is $15\overline{0}$ mm, and 15 strokes occur each second (t). If there are 4 cylinders, find the power developed in W and HP.

SOLUTION

Wanted

$P = ?$

Given

$F = 90\overline{0}$ N, $D = 15\overline{0}$ mm, and $t = 15$ strokes/s

Formula

We need to use the work formula:

$$W = F \times D$$

Calculation

First we need to convert the distance to meters:

$$150 \text{ mm} = 0.15 \text{ m}$$

Then we need to find the work done (energy used) during each stroke:

$$W = F \times D = 900 \text{ N} \times 0.15 \text{ m} = 135 \text{ N·m}$$

Since a newton·meter is a joule, work for each stroke is:

$$W = 135 \text{ N·m/stroke} = 135 \text{ J/stroke}$$

To find the power for each piston, we can now use the power formula:

$$P = \frac{W}{t} = 135 \text{ J/stroke} \times 15 \text{ strokes/s} = 2025 \text{ J/s}$$

We can now find the total power for 4 cylinders:

$$P_{\text{total}} = 2025 \text{ J/s} \times 4 = 8100 \text{ J/s}$$

Since 1 J/s = 1 W:

$P = 8100$ W

To convert this answer to horsepower, we multiply (using the information from the Conversion Factors Table at the front):

HP = 8100 W × 0.00134 = 10.9 HP

This information can be compared with actual experimental data to help determine the losses due to friction, water pumps, fans, and so on.

Summary

This chapter discusses the three related terms of work, energy, and power. Work is performed only when a force moves some object through a distance. Work is measured in units of foot·pounds or joules.

An object or substance can do work if it possesses something called energy. There are many types of energy: mechanical, electrical, chemical, light, and so on. If an object is at rest but has energy available to do work, because of its condition or position, it has potential energy. If the object is in motion, and this motion can be used to do work, the object has kinetic energy. Both kinetic and potential energy are used by industry. Energy, like work, is measured in units of foot·pounds or joules.

The machine that can perform a given amount of work in the shortest time has the most power. If two machines are doing the same amount of work, say lifting elevators to the tenth floor of a building, the one that gets the elevator to the tenth floor first is more powerful (has more power). Power is measured in units of horsepower and watts.

Key Terms

* work
* work formula
* energy
* mechanical energy
* potential energy

* kinetic energy
* potential energy formula
* kinetic energy formula
* power
* power formula

PHOTO 6.3   A windmill is designed to use the kinetic energy of the earth's winds and convert it to useful work. Windmills are becoming more common as a source of power that does not use fossil fuels as energy. (NASA photo)

Questions

1. What is meant by work? What are the units of work?
2. What is meant by energy? What are the units of energy?
3. Name three different kinds of energy.
4. What is meant by potential energy? What is meant by kinetic energy?
5. Give two examples of useful applications where potential energy is converted to kinetic energy.
6. What is meant by power? What are the units of power?
7. Define horsepower.

Problems

WORK AND ENERGY

1. A force of $1\overline{0}$ lb is required to push a $50\overline{0}$ lb crate, on rollers, for 50 ft across a level floor.
   a. How much work is done?
   b. How much work is done lifting the crate 15 ft up to the next floor?

2. A hiker carries a 50.0 lb backpack up a $60\overline{0}\overline{0}$ ft mountain. How much work did the hiker do in carrying the backpack?

3. When an elevator is on the ground floor, its $60\overline{0}$ lb counterweight is 75.0 ft above ground. What is the counterweight's potential energy?

4. It is decided that a pile driver must have $5\overline{0}\,000$ ft·lb of potential energy when its ram is 25 ft above the piling. How much should the ram weigh?

5. Which requires more work: hoisting a $50\overline{0}$ lb load $2\overline{0}$ ft, or a $20\overline{0}$ lb load $6\overline{0}$ ft?

6. $68\overline{0}$ ft·lb = _____ J.

7. $70\overline{0}\overline{0}$ J = _____ ft·lb.

8. A counterweight on a drawbridge weighs $70\overline{0}$ meganewtons (MN). When the bridge is opened, the counterweight drops through a vertical distance of 5.00 m. What is the potential energy when the counterweight is at its maximum height?

9. The drop forge in Figure 6.2 has a ram and upper die with a combined mass of $40\overline{0}$ kg, and the drop height is 1.70 m. How much energy is available to form the metal part? (*Hint:* Don't forget to convert kilograms to newtons.)

10. Refer to the water tank in Figure 6.3. The tank has a capacity of $96\overline{0}\,000$ lb of water and has an average height above ground of $15\overline{0}$ ft. How much work is required to fill the tank from a reservoir at ground level?

11. A particular electric utility has a "pumped storage facility." That is, water is pumped up to a reservoir on a mountain during periods of low electricity demand and then the water is used to drive generators during periods of peak electricity demand. Such a storage facility has a capacity of 50.0 gigagrams (Gg) of water at an elevation of $30\overline{0}$ m. What total energy, in joules, is available (what is the potential energy)? (*Hint:* Don't forget to change grams to newtons.)

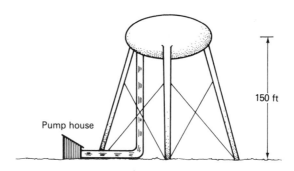

FIGURE 6.3   Sketch for Problem 10.

12. An automobile, having a mass of $10\overline{0}$ slugs, is traveling at a velocity of $6\overline{0}$ ft/s. What is the kinetic energy in foot·pounds?

13. Solve Sample Problem 6.2 if the velocity is 28 ft/s and the ram weighs $6\overline{0}0$ lb.

14. What is the kinetic energy of a $5\overline{0}$ g bullet traveling at $7\overline{0}0$ m/s? (Note: Change grams to kilograms for the formula.)

POWER

15. Solve Sample Problem 6.3 if the load is changed from 500 lb to $15\overline{0}0$ lb, the height is changed to 80.0 ft, and the time is changed to 10.0 s.

16. Refer to Problem 10.
    a. If it takes a pump $1\overline{0}$ hr to fill the tank, what horsepower is required?
    b. The city engineer says that water usage is increasing, and asks you to determine the horsepower required if the pump must fill the tank in 2-1/2 hr. (*Hint:* Convert hours to seconds.)

17. When fully compressed, the intake and exhaust valve springs on a particular engine store 15.0 ft·lb of energy. Each spring operates 10 times each second when the engine is running at rated speed. If there are 16 springs on the engine, how much horsepower is required to operate the springs? Give the answer in watts also.

18. Refer to Problem 11. If the water is used over a 3 hr period, what power in watts can be developed?

19. An elevator, with full load, weighs 45$\overline{0}$0 lb. It has no counterweight, so the motor must lift the total load. What horsepower is required if the maximum speed of rise is 40.0 ft/min? (*Hint:* Find the distance traveled in 1 s, then find the power; you cannot use the kinetic energy formula.)

Computer Program

This program will calculate the kinetic energy of an object from mass and velocity data. If you are using U.S. Customary units, you do not need mass in slugs but may use weight in pounds. In the SI system, supply the mass in kilograms. You can use the program to check some of the chapter problems and to generate data to plot curves of kinetic energy versus mass or velocity. One curve would be kinetic energy versus velocity and the other kinetic energy versus mass. Plot kinetic energy on the vertical Y-axis and mass or velocity on the X-axis.

As an example, hold the mass constant but keep doubling the velocity—say 10 ft/s, 20 ft/s, 40 ft/s, and so on. Record the kinetic energy for each velocity and plot this on a graph. Observe how the energy changes with velocity. Hold the velocity constant when plotting kinetic energy versus mass.

PROGRAM

```
10    REM   CH SIX PROGRAM
20    PRINT "PROGRAM TO FIND KINETIC ENERGY."
30    PRINT
40    PRINT "IF UNITS ARE U.S. CUSTOMARY-TYPE 1, IF SI UNITS-TYPE 2."
50    INPUT A
60    PRINT "YOU HAVE SELECTED -- ";A
70    IF A = 1 THEN   GOTO 100
80    IF A = 2 THEN   GOTO 200
90    END
100   PRINT "TYPE IN WEIGHT IN POUNDS AND THEN THE VELOCITY IN FT/S."
110   INPUT W,V
120   PRINT "WEIGHT = ";W;" LB."
130   PRINT "VELOCITY = ";V;" FT/S."
140   LET M = W * 0.0311
150   LET KE = 0.5 * M * (V ^ 2)
160   PRINT
170   PRINT "KINETIC ENERGY = ";KE;" FT.LB."
180   GOTO 90
200   PRINT "TYPE IN MASS IN KILOGRAMS AND THEN VELOCITY IN M/S."
210   INPUT M,V
```

```
220   PRINT "MASS = ";M;" KG."
230   PRINT "VELOCITY = ";V;" M/S."
240   LET KE = 0.5 * M * (V ^ 2)
250   PRINT
260   PRINT "KINETIC ENERGY = ";KE;" J."
270   GOTO 90
```

NOTES

Lines 70 and 80 select the subroutine based on the units used.

Line 140 converts weight to mass.

Lines 150 and 240 make use of the kinetic energy formula.

Note that in lines 150 and 240 the exponent symbol used is (^).

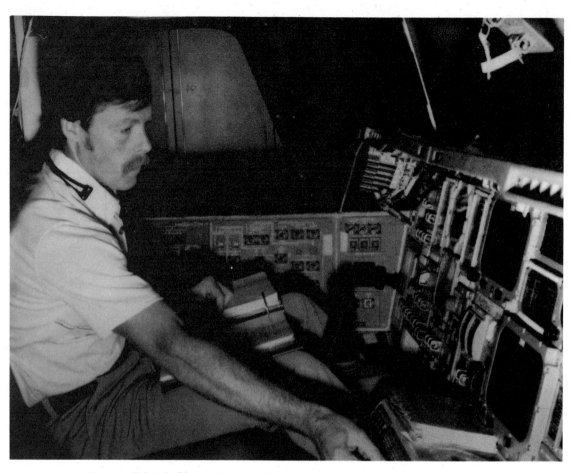

Astronaut Robert L. Gibson is shown in a Space Shuttle simulator. The Space Shuttle's orbit is governed by the laws of rotational motion. (NASA photo)

Chapter 7

Objectives

- ☐ Explain that angular velocity is the angular distance through which an object rotates in some time interval, divided by that time interval.
- ☐ Convert angular velocity to linear speed by using the formula $V = 2\pi RN$, where V is linear speed, R is the radius of rotation, and N is angular velocity.
- ☐ Explain how Newton's laws of motion for linear motion also apply to rotational (angular) motion.
- ☐ Explain that rotational motion must always begin with a force acting to deflect a moving object out of its straight-line path and that the name given such a deflecting force is centripetal force.
- ☐ Recognize the centripetal force equation $F_c = mV^2/R$, where centripetal force (F_c) equals mass being deflected (m) times linear speed of its motion squared (V^2) divided by the radius of the object's circular path (R).
- ☐ Understand that centrifugal force is a force reacting to centripetal force, according to Newton's Third Law of Motion.
- ☐ Recognize the torque equation $T = FR$, where torque (T) is equal to force (F) times the moment arm distance (R).

Rotational Motion

In today's technical world, the vast majority of machines employ rotational motion in some aspect of their operation. If for nothing else, most machines require electric motors or diesel or gasoline engines to supply the required energy through rotating shafts. Technicians must understand rotational motion and how to measure it. They must know why changing a pulley diameter changes the belt speed, and they must know how to measure the amount of twist needed to tighten a bolt.

This chapter covers introductory topics related to rotational, or circular, motion, including angular velocity, linear speed, centripetal force, centrifugal force, and torque. Let's start at the beginning with angular velocity.

7.1 Angular Velocity

Angular velocity is the time rate of change of the angle through which an object rotates. It is very similar to linear speed, or velocity, which is the time rate of change of distance through

PHOTO 7.1 Stereo tape machines illustrate the case of a drive pulley connected to a second pulley that is free to rotate at any given speed. The drive reel does not turn at a constant rate, but slows down as the radius of the tape on the reel increases. (KPNW photo)

which an object travels. Linear velocity is the same velocity that we discussed in Chapter 5 concerning Newton's laws. Here, we are using the term "linear" (meaning length or distance) to distinguish it from angular velocity.

---

* angular velocity   The angular distance through which an object rotates in some time interval, divided by that time interval.

We are already aware that angles are measured in degrees, so it shouldn't be surprising to learn that angular velocity could be specified in "degrees per second." For an example, we can refer to Figure 7.1. A fly is standing on the edge of a rotating long-playing record. If the fly (and the spot it is standing on) rotates through 200 degrees, from point A_1 to point A_2, in 1 second, we could say the angular velocity is 200 degrees per second. However, one of the most common measures of angular velocity that technicians encounter is *revolutions per minute* (rpm). So, instead of saying the fly and the phonograph record are rotating at 200 degrees per second, we say they are rotating at 33-1/3 rotations, or revolutions, or cycles every minute. The following represent typical angular velocities for some well-known devices:

Portable electric drills   1200 rpm
Portable circular saws   1500 rpm
Automobile engines   4000 rpm
Airplane jet engines   14 000–28 000 rpm, depending
  on the size

Let's look at our fly on the phonograph record in Figure 7.1 again. As the record rotates through one complete revolution,

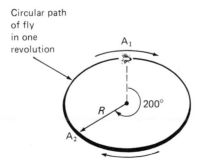

FIGURE 7.1   Illustration of Angular Velocity and Linear Speed.

the fly travels through a distance represented by the circular path shown. Our basic problem is to relate angular velocity (N) to linear speed, or velocity. It can be shown that for constant angular velocity, linear speed is equal to the magnitude of linear velocity at any instant. In our discussions, we will use the term "linear speed" (V). By measuring the length of the circular path, we can obtain the **linear speed** at which the fly is moving. The length of the path is determined by the formula for the circumference of a circle:

$$\text{circumference} = 2 \times \text{pi} \times \text{radius} \quad \text{or} \quad C = 2\pi R$$

The circumference has the same units as the radius, usually inches, feet, or meters. We will carry pi to just two decimal places, and therefore use the value 3.14.

---

* linear speed

The distance an object travels in 1 minute. If the object is rotating, then its linear speed is equal to the circumference of its circle of rotation multiplied by the angular velocity.

The distance the fly travels in 1 minute (linear speed) is simply the circumference multiplied by the angular velocity. In the formula for linear speed, we can substitute $2\pi R$ for C:

* LINEAR SPEED FORMULA

linear speed = circumference × angular velocity

$$V = \underbrace{2\pi R}_{C} N$$

where V = the linear speed in in./min, ft/min, or m/min, depending on the units for the radius
        R = the radius in in., ft, or m
        N = the angular velocity in rpm

Revolutions cancel out in this equation, because there is one circumference per revolution. For example, if the radius is in feet, we have $2\pi R$ feet per revolution. Sample Problem 7.1 demonstrates that even though the angular velocity remains the same, the linear speed can change, depending on the radius.

SAMPLE PROBLEM 7.1

As in Figure 7.2, a fly is standing on a record that is rotating at 33-1/3 rpm (angular velocity, N). Determine the linear speed (V) of the fly:

    A.   when it is standing 6.00 in. (radius, R_1) from the center of the record

    B.   after it walks to a position 3.00 in. (radius, R_2) from the record's center

SOLUTION A

Wanted

$V = ?$

Given

N = 33-1/3 rpm and R_1 = 6.00 in.

Formula

$V = 2\pi RN$

Calculation

$V = 2 \times 3.14 \times 6.00 \text{ in.} \times 33.3 \text{ rpm} = 1260 \text{ in./min}$

SOLUTION B

Wanted

$V = ?$

Given

N = 33-1/3 rpm and R_2 = 3.00 in.

Formula

$V = 2\pi RN$

Calculation

$V = 2 \times 3.14 \times 3.00 \text{ in.} \times 33.3 \text{ rpm} = 628 \text{ in./min}$

When working with pulleys and V-belts, a spot on the belt will stay lined up with a spot on the pulley (assuming no slippage)

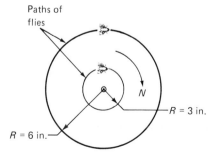

Paths of flies

N

R = 3 in.

R = 6 in.

FIGURE 7.2   Illustration for Sample Problem 7.1.

as the belt travels around the pulleys. Thus, the velocity of the V-belt is equal to the linear velocity of a point on the rim of the pulley. Sample Problem 7.2 illustrates a typical V-belt problem encountered by engineers and technicians.

SAMPLE PROBLEM 7.2	For a particular cooling fan operation, a handbook recommends the pulley speed (angular velocity, N) to be $50\bar{0}$ rpm and a linear belt speed (linear speed, V) to be $17\bar{0}0$ ft/min. What is the required driven pulley diameter (d)? Refer to Figure 7.3.

SOLUTION

Wanted

$d = ?$

Given

$N = 50\bar{0}$ rpm and $V = 17\bar{0}0$ ft/min

Formula

Our approach will be to find the pulley radius in feet, then convert the answer to the diameter, and then convert feet to inches because pulley diameters are generally listed in inches. To begin, we must rearrange the linear speed formula:

$$V = 2\pi RN \qquad \text{to} \qquad R = \frac{V}{2\pi N}$$

Calculation

$$R = \frac{1700 \text{ ft/min}}{2 \times 3.14 \times 500 \text{ rpm}} = 0.540 \text{ ft}$$

Since the diameter of a circle is twice the radius:

$$d = 2 \times 0.540 = 1.08 \text{ ft}$$

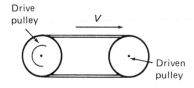

FIGURE 7.3   Pulleys and V-Belt for Sample Problem 7.2.

To convert to inches, multiply by 12 in./ft:

$$d = 1.08 \text{ ft} \times 12 \text{ in./ft} = 13 \text{ in.}$$

The linear speed formula also applies to a wheel rolling along the ground: The wheel travels a distance along the ground equal to one circumference when the wheel makes one revolution. Sample Problem 7.3 relates the speed of a locomotive to the angular velocity of the wheels:

SAMPLE PROBLEM 7.3

A locomotive is traveling at a speed of 88.0 ft/s (speed, V) and the wheels are 2.50 ft in diameter (d). What is the angular velocity (N) of the wheels in revolutions per minute?

SOLUTION

Wanted

$N = ?$

Given

$V = 88.0$ ft/s and $d = 2.50$ ft

Formula

First, we must change the diameter to the radius dimension by dividing by 2:

$$R = \frac{1}{2} \times 2.50 = 1.25 \text{ ft}$$

We can now use the linear speed formula, but rearranged:

$$V = 2\pi R N \qquad \text{to} \qquad N = \frac{V}{2\pi R}$$

Calculation

$$N = \frac{88.0 \text{ ft/s}}{2 \times 3.14 \times 1.25 \text{ ft}} = 11.2 \text{ rps}$$

To change from revolutions per second to revolutions per minute, we multiply by 60.0 s/min:

$$N = 11.2 \text{ rev/s} \times 60.0 \text{ s/min} = 673 \text{ rpm}$$

This information can help technicians determine gear ratios and motor speeds.

7.2 Centripetal Force

We will now apply Newton's First and Second Laws of Motion to an object traveling in a circular path. It is a shock to most of us, as it probably was to Newton's fellow scientists, to learn that a revolving object moving at constant speed is not in equilibrium.

PHOTO 7.2   This huge centrifuge at the Manned Spacecraft Center in Houston, Texas, swings a three-man gondola at the end of a 50 foot arm. The astronauts within the gondola experience centripetal force similar in magnitude to the forces they will feel during liftoff and re-entry. (NASA photo)

FIGURE 7.4 *Example of Centripetal Force.*

Newton's second law implies that if an object moving at constant speed is not in equilibrium, it is being accelerated by some unbalanced force. How do we know the object is not in equilibrium? Where is the unbalanced force?

To help us answer these questions, Figure 7.4 shows a child in a playground. The child has fastened a rock to the end of a rope and is whirling the rock around his head. The rock moves in a circle. If the rock were to break free of the rope, it would not continue to follow the circular path, but would fly off in a straight line tangent to the circle. Newton's first law says that an object in equilibrium either is at rest or is moving at a constant speed in a straight line. Since the rock is moving in a circle, not a straight line, it is not in equilibrium, even though its speed is constant. So, Newton's second law, $F = ma$, must apply. What does this force do? It does not change the amount of velocity of the rock, but it does change the *direction* of the velocity. That is, the rope must be exerting an unbalanced force on the rock to keep it moving in a circle. The force of the rope on the rock is directed toward the center of the circle. This type of force is called **centripetal force.**

* centripetal force The force that causes an object to move in a curved, or circular, path. The force is directed to the center of the circle.

To determine centripetal force (F_c), we must write Newton's law as:

$$\text{centripetal force} = \text{mass} \times \text{acceleration} \quad \text{or} \quad F_c = ma$$

But where is the acceleration if the rock is rotating at a *constant* rpm? Even though the linear speed (and angular velocity) is constant, the rock is continually *changing direction*. We can consider that for a very brief instant, the rock is traveling in essentially a straight line. Therefore, we can think of it as having an instantaneous velocity in some direction at each instant of time. The direction of the rock in each succeeding instant keeps changing, which thus means that acceleration is present, even though it is not apparent to us. Why? Remember that velocity is a vector, and has both amount and direction. In Chapter 5, we saw acceleration change the amount of velocity; here, we see changes in the direction of velocity. Since acceleration is defined as any change in velocity, there must be an acceleration present. The acceleration is in the same direction as the centripetal force—

toward the center of rotation—so it is called *centripetal acceleration.*

Newton's second law, as applied to rotating objects, seems to defy our observations. Think about it. Picture any rotating object: the rock the child is twirling, a small chunk of metal on the rim of a pulley, or a gear tooth. All may be rotating at a constant angular velocity, and each of their distances from the center of rotation is fixed. Yet, here Newton's laws say "These objects are accelerating toward the center!" Of course, Newton is right; but the explanation is rather involved, and so will be left for other texts.

Instead, we will proceed with the application of Newton's second law ($F = ma$) to circular motion. The formula will be easier to use if the acceleration toward the center of rotation (a in Newton's Second Law of Motion) is replaced with its equivalent value of V^2/R. The formula for centripetal force, then, is:

$$F_c = m \times \frac{V^2}{R}$$

We can rearrange this formula slightly to present it in its most common form:

* **CENTRIPETAL FORCE FORMULA**

$$F_c = \frac{mV^2}{R}$$

where
F_c = the centripetal force and must be in lb or N
m = the mass and must be in slugs or Kg
V = the linear speed and must be in ft/s or m/s
R = the radius of the object's path and must be in ft or m

Well, here we are using a term like "centripetal force," which hardly anyone has heard of, and no mention has been made of **centrifugal force,** which almost everyone has heard of. Is there a connection between the two forces?

* **centrifugal force**   The reaction to centripetal force.

In our example of the child in the playground with the rock, the rope exerts a centripetal force on the rock, and the rock, in accordance with Newton's Third Law of Motion, exerts an equal and opposite reactive force—that is, a centrifugal force—on the rope. Note that the centrifugal force is not acting on the rock. So, we see that the centrifugal force is not a balancing force and does not put the rock in equilibrium. As far as the rock is concerned, only the unbalanced centripetal force is acting on it. How can we understand this?

Let's go back to Chapter 5 on Newton's laws and talk about the time you pulled a small cart across a table with a force of 1 pound. The cart accelerated because of the unbalanced force. But, the cart was exerting an equal and opposite force on the spring scale. Circular motion is similar. The centrifugal force is due to the tendency of the rock to move in a straight line. Thus, this tendency creates a reaction to the force, causing the rock to move in a circle.

The centripetal force formula may be used for centrifugal force since they have the same magnitudes. Sample Problem 7.4 illustrates a technical situation requiring the use of the centripetal force formula.

SAMPLE PROBLEM 7.4

A centrifugal casting machine whirls a mass (m) of molten metal weighing 2.00 kg around a 0.750 m radius (R). From handbook data, we know that to produce a casting with a satisfactory density, we need a centripetal force (F_c) of 10.0 N.
  A. Find the linear velocity (V) required to obtain the 10.0 N force.
  B. Find the angular velocity (N) needed.

SOLUTION A

Wanted

$V = ?$

Given

$F_c = 10.0$ N, $m = 2.00$ kg, and $R = 0.750$ m

Formula

To find V, we must rearrange:

$$F_c = \frac{mV^2}{R} \quad \text{to} \quad V^2 = \frac{F_c R}{m}$$

Calculation

$$V^2 = \frac{10.0 \text{ N} \times 0.750 \text{ m}}{2.00 \text{ kg}} = 3.75 \overbrace{\text{(kg·m/s}^2\text{·m)/kg}}^{\text{N}} = 3.75 \text{ m}^2/\text{s}^2$$

$$V = \sqrt{V^2} = \sqrt{3.75} = 1.94 \text{ m/s}$$

SOLUTION B

Wanted

$N = ?$

Given

$R = 0.750 \text{ m}$ and $V = 1.94 \text{ m/s}$

Formula

The formula for angular velocity must be rearranged:

$$V = 2\pi RN \quad \text{to} \quad N = \frac{V}{2\pi R}$$

Calculation

$$N = \frac{1.94 \text{ m/s}}{2 \times 3.14 \times 0.750 \text{ m}} = 0.412 \text{ rps}$$

This answer must be changed to rpm by multiplying by 60 s/min:

$$0.412 \text{ rev/s} \times 60 \text{ s/min} = 24.7 \text{ rpm}$$

7.3 Torque

We continue our discussion of circular, or rotational, motion by turning to the type of force system that produces rotation. This force system is called **torque.** When you twist the cap off a peanut butter jar or tighten the nut on a bolt, you are applying a torque. Torque is the product of the force applied to the turning object and the perpendicular distance from the force to the center of rotation. This distance from the force to the center of rotation is called the torque arm or torque radius.

---

* torque   A twisting or turning action that tends to produce rotation.

The formula for torque is:

PHOTO 7.3 This device, called a *Foucault pendulum,* is located at the United Nations General Assembly Building. It demonstrates the rotation of the earth. The 200 pound metal sphere swings back and forth in a constant arc, regardless of any motion of the anchor to which it is attached by a wire. The metal ring beneath the sphere rotates along with the earth, and thus the sphere swings across the ring at different angles, rotating slowly in a clockwise direction. (United Nations photo)

∗ TORQUE FORMULA

torque = force × distance

$$T = F \times R$$

where

T = torque in units of lb·ft, lb·in., or N·m
F = the applied force in lb or N
R = the torque arm distance in ft, in., or m

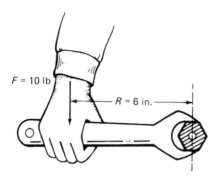

FIGURE 7.5   Example of Torque (Force × Radius).

In Figure 7.5, a force of 10 pounds is being applied to the wrench at a distance of 6 inches from the center of the bolt. We can use the formula to determine how much twisting action, or torque, is being applied:

$$T = F \times R = 10 \text{ lb} \times 6 \text{ in.} = 60 \text{ lb·in.}$$

Note that the torque arm must be perpendicular to the force.

The torque formula and terms (force, distance, and units) are similar to those for work. However, they are not interchangeable because work and torque are not the same thing. When dealing with work, the force moves *through* the distance; whereas, for torque, the force is *perpendicular* to the distance. To help identify the difference, engineers have arbitrarily agreed to write work units as "foot·pound" and torque units as "pound·foot." In the SI metric system, the joule is the unit of work, and the newton·meter is the unit of torque.

Now let's take another look at Figure 7.3. If the left-hand pulley is the drive pulley, it is pulling on the bottom portion of the belt with some force. This force is transmitted to the right-hand pulley and appears as a torque (the belt force times the pulley radius) to turn (twist) the pulley. Sample Problem 7.5 shows how to determine the torque of a belt and pulley arrangement.

SAMPLE PROBLEM 7.5

Figure 7.6 indicates the right-hand pulley of Figure 7.3 and a section of the V-belt that is turning the pulley. Notice that there are forces on both ends of the V-belt, since it must be tight to transfer motion. If the pulley is 8.00 in. in diameter, what torque (T) is being applied?

SOLUTION

Wanted

$T = ?$

Given

$F = 200$ lb and $F = 400$ lb. We must first find the net force causing rotation, which is the difference between the forces pulling on the ends of the V-belt. The lower section of the V-belt supplies this net force, which is $F = 400$ lb $- 200$ lb $= 200$ lb. The torque arm (R) is the pulley radius, or $R = 8.00$ in. dia/2 $= 4.00$ in.

Formula

$T = F \times R$

Calculation

$T = 200$ lb $\times 4.00$ in. $= 800$ lb·in.

This problem is rather usual for technicians because V-belts are used on such a variety of machinery for transmitting power. Among the reasons for V-belt popularity are smoothness of operation, dependability, ease of maintenance, and inexpensive replacement.

In the next chapter, we will use the term "moment" instead of "torque." They both mean the same thing, but each term developed through long usage in different industries. Torque is generally used to indicate the twisting action applied to or by a rotating shaft, or applied to nuts and bolts. If you have ever tightened the nuts on the head bolts on your car's engine, you know that a specified torque must be applied (with a torque-wrench) so all the nuts are tightened equally and properly. Moment is used when working on problems dealing with lever systems and beams.

FIGURE 7.6   Sketch for Sample Problem 7.5.

Summary

This chapter covers some basic terms and formulas that not only are necessary for an understanding of rotational motion, but are important to industrial applications. Rotational velocity is called angular velocity. The most common units of angular velocity are revolutions per minute (rpm).

An object making one revolution covers a certain distance represented by the circumference of its circular path. This distance is a linear, not an angular, distance, and therefore we can obtain a linear speed for a rotating object. The units of linear speed can be inches per minute, feet per minute, or meters per minute.

The force that causes an object to travel in a curved or circular path is called centripetal force. This force changes the direction of motion of a rotating object so that it travels in a circle instead of in a straight line. Centripetal force is directed to the center of the object's circular path. The reactive force to centripetal force is called centrifugal force. Although centripetal and centrifugal forces are equal in amount and opposite in direction, they are not balanced forces because they are not applied to the same object, or body.

Torque is a term applied to a force system that produces a twisting, or turning, action. Another term that means the same thing is moment. In industry, the term "torque" is usually applied to the twisting action of rotating equipment, such as shafts and nuts and bolts. The term "moment" is used when discussing lever systems and beams.

Key Terms

* angular velocity
* linear speed
* linear speed formula
* centripetal force
* centripetal force formula
* centrifugal force
* torque
* torque formula

Questions

1. What is angular velocity?
2. What is linear speed?

3. Are all particles on a rotating disk traveling at the same (a) angular velocity? (b) linear speed? (*Note:* Do not consider the center, as it does not rotate.)

4. Newton's second law says that an unbalanced force produces a change in velocity (acceleration) on a given object. How does this law apply to an object rotating at constant angular velocity?

5. What is centripetal force?

6. What is centrifugal force? Is it an action or a reaction?

7. In inexpensive tape recorders, the take-up reel is driven at a constant rpm. As the tape winds up on the take-up reel, is the tape's velocity through the recording head constant or changing? Explain.

8. Define "torque." What are its units?

9. Name some common everyday devices or machines that develop a torque to perform their tasks.

Problems

ANGULAR VELOCITY AND LINEAR SPEED

1. A carpenter uses a portable circular saw that rotates at 5500 rpm, and the blade radius is 4.00 in. (0.333 ft). What is the linear speed of the blade tips in (a) feet per minute and *1832'/min* (b) miles per hour? (*Hint:* Multiply ft/min by 0.01136 to get mi/hr.) *20.8*

2. A router turns at 30 000 rpm. If the router bit has a radius of 0.500 in. (0.0417 ft), what is the linear speed of the edge of the bit in (a) feet per minute and (b) miles per hour?

3. A 45 rpm phonograph record has a 6-7/8 in. diameter.
   a. What is the linear speed of a point on the rim? Answer in feet per minute.
   b. What is the linear speed of a point 1/3 of the distance to the rim in feet per minute?

4. A jet engine compressor rotates at 15 000 rpm and is 2.00 ft in diameter. What is the tip speed (linear speed) in feet per minute?

5. Solve Sample Problem 7.2 if the pulley speed is changed to 1000 rpm.

6. An elevator at a construction site is raised and lowered by a cable wrapped around a 4.00 ft diameter drum. How fast

$$V = 2\pi R N$$

$30.0 \text{'/min} \stackrel{?}{=} 3.14 \times 2.0 \quad N$

$$\frac{30.0}{2 \cdot 3.14 \times 2.0} = N$$

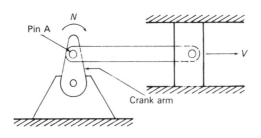

FIGURE 7.7 Crank and Piston for Problem 11.

does the drum have to turn in order to raise the elevator at the rate of 30.0 ft/min?

7. A standard roller chain (as on bicycles and motorcycles) is used to drive a blower. The engineer needs to know the velocity of the chain to help calculate the maximum tensile force that can be applied. The sprocket diameter is 1.30 ft, and it rotates at $30\bar{0}$ rpm.

8. Some types of automatic elevators may reach speeds of $30\bar{0}$ ft/min. If the elevator cable is wrapped around a drum that rotates at 50.0 rpm, what is the drum diameter in feet?

9. Because of excessive bending stresses and other factors, the handbook states that for a particular size of fabric-rubber belt, the minimum pulley diameter is 18.0 in. at a belt speed of $50\bar{0}$ ft/min. What is the pulley speed in rpm?

10. You are given a pulley arrangement similar to that in Figure 7.3. The belt speed is $320\bar{0}$ ft/min and the pulley diameter is 13 in. What is the rpm of the pulley?

11. Figure 7.7 shows an 80 mm crank arm (measured from the shaft center to the center of pin A), rotating clockwise at 450 rpm and operating a pump piston. What is the velocity of the piston at the instant shown, in meters per second? (*Hint:* At this instant, it is the same as the linear speed of pin A.)

12. A car is traveling at $6\bar{0}$ mi/hr (88 ft/s). What is the rpm of the wheels if they are $3\bar{0}$ in. in diameter?

13. Figure 7.8 shows a V-belt drive. The small pulley is the driving pulley. It is $20\bar{0}$ mm in diameter and is rotating at $120\bar{0}$ rpm. What is the angular velocity of the large pulley if its diameter is $60\bar{0}$ mm? (*Hint:* The belt speed [linear velocity] is the same on the small pulley as on the large pulley.)

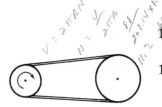

FIGURE 7.8 Pulleys and V-Belt for Problem 13.

CENTRIPETAL FORCE

14. A $32\bar{0}0$ lb car traveling at $3\bar{0}$ mi/hr (44 ft/s) makes a turn having a radius of $6\bar{0}$ ft.
    a. What centripetal force is acting on the car? (This force is supplied by the friction between the tires and the road.)
    b. Repeat 14a if the car makes a wider turn with a radius of 150 ft. (*Hint:* Don't forget to change the object weight to mass.)

15. A rule of thumb for safe flywheel operation is that the rim speed for cast iron wheels should not exceed $60\bar{0}0$ ft/min. What is the rpm for a wheel with a diameter of 3.00 ft?

16. A particular centrifuge generates a force 13 $\bar{0}00$ times that of gravity. This means the centripetal force on a 1 lb object is 13 $\bar{0}00$ lbs. If the diameter is 18.0 in., what is: (a) the linear speed in feet per second? (b) the angular velocity in revolutions per minute?

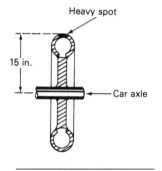

Heavy spot

15 in.

Car axle

FIGURE 7.9 Sectional View of Car Wheel and Tire for Problem 17.

17. A test is being made on an automobile tire, shown in Figure 7.9. The tire is mounted on a fixed spindle in a laboratory. It is out of balance because it has a heavy spot on it amounting to an extra weight of 1 oz (1/16 lb), located 15 in. from the center of the spindle. When the tire is rotated at $5\bar{0}0$ rpm, what is the centripetal force on this extra weight?

18. Refer to Sample Problem 7.2. The centrifugal force acting on a belt as it goes around a pulley must be accounted for or the belt may slip. What is the centrifugal force (equal to centripetal force) on a 1.00 in. length of V-belt weighing 0.220 lb? (*Hint:* To get velocity in ft/s, divide the velocity in ft/min by 60 s/min.)

TORQUE

19. Refer to Figure 7.5. What torque can be developed if the torque arm is 9.5 in.?

20. Refer to Figure 7.10. We are shown a 4-way wheel wrench used to remove automobile wheel nuts. If each of your hands can exert a force of 25.0 lb, how much torque can be applied? Answer in (a) pound·inches and (b) pound·feet.

21. Specifications call for a nut to be tightened to $2\bar{0}$ N·m of torque. What force is required if the wrench is 210 mm long—that is, the torque arm is 210 mm?

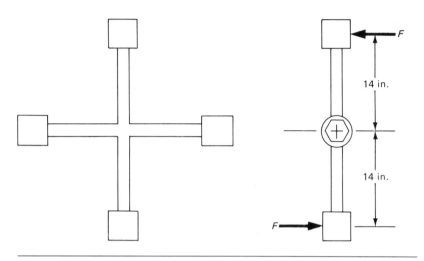

FIGURE 7.10   Four-Way Wheel Wrench for Problem 20.

22. Refer to Figure 7.6. If the forces are $15\overline{0}$ lb and 275 lb, what pulley diameter is required if a torque of $70\overline{0}$ lb·in. must be applied?

23. Refer to Figure 7.7. Let the crank arm be 3.00 in. If the piston is driving the crank and is exerting a force of $31\overline{0}$ lb on pin A, what torque is being developed at the instant shown?

24. Refer to Problem 6, concerning the elevator and drum at the construction site. If the elevator exerts a pull on the wire rope of $80\overline{0}$ lb, what torque must the drum overcome to lift the elevator?

Computer Program

You may use this program to find torque, or the applied force, or the torque arm from the formula $T = F \times R$. Use units of inches and pounds.

This program may be used to solve some of the chapter problems or to assist in experiments of your own.

PROGRAM

```
10    REM   CH SEVEN PROGRAM
20    PRINT "TORQUE FORMULA, T=F*R. USE INCHES AND POUNDS. TO FIND"
21    PRINT "T-TYPE 1, TO FIND F-TYPE 2, AND TO FIND R-TYPE 3."
30    INPUT A
40    PRINT "YOUR SELECTION IS --- ";A
45    PRINT
50    IF A = 1 THEN  GOTO 100
60    IF A = 2 THEN  GOTO 200
70    IF A = 3 THEN  GOTO 300
80    PRINT
90    END
100   PRINT "TYPE IN THE FORCE AND THEN THE TORQUE ARM."
110   INPUT F,R
120   PRINT "FORCE = ";F;" LB."
130   PRINT "TORQUE ARM = ";R;" IN."
140   LET T = F * R
150   PRINT
160   PRINT "TORQUE = ";T;" LB.IN."
170   GOTO 90
200   PRINT "TYPE IN THE TORQUE AND THEN THE TORQUE ARM."
210   INPUT T,R
220   PRINT "TORQUE = ";T;" LB.IN."
230   PRINT "TORQUE ARM = ";R;" IN."
240   LET F = T / R
250   PRINT
260   PRINT "FORCE = ";F;" LB."
270   GOTO 90
300   PRINT "TYPE IN THE TORQUE AND THEN THE FORCE.D"
310   INPUT T,F
320   PRINT "TORQUE = ";T;" LB.IN."
330   PRINT "FORCE = ";F;" LB."
340   LET R = T / F
350   PRINT
360   PRINT "TORQUE ARM = ";R;" IN."
370   GOTO 90
```

NOTES

Lines 50, 60, and 70 select the proper subroutine.

Lines 120 and 130, 220 and 230, and 320 and 330 confirm your data input.

This articulated front end loader uses levers, beams, and gears to perform its tasks. In the coal stockpiling operation shown here, 30 tons of coal are moved on each machine pass. (Courtesy of Clark Equipment Company)

Chapter 8

Objectives

When you finish this chapter, you will be able to:

- [ ] Identify machines as devices that transfer or multiply forces, and give examples of machines.
- [ ] Describe a lever. List and compare the three kinds of levers, and give examples of each.
- [ ] Recognize, first, that Newton's First Law of Motion describing equilibrium applies to a body at rest as well as in motion and, second, that the equations $\Sigma F = 0$ and $\Sigma M = 0$ describe the equilibrium state.
- [ ] Apply to various situations the equations $\Sigma F = 0$ and $\Sigma M = 0$—namely, that the sum of forces (F) acting on a body may be set equal to zero, and the sum of moments (torques) (M) about any point may be set equal to zero, for any object in equilibrium.
- [ ] Use beams in a variety of problem applications.
- [ ] Describe and explain the operation of gear wheels, drivers, followers, idlers, gear trains, and bevel and worm gears.

Machines

Most large, modern machines are considered to be composed of no more than six basic or "simple" machines: the lever, the wheel and axle (now, commonly called gear wheels), the inclined plane, the screw, the pulley, and the wedge. Though time and space limit our discussion to levers and gear wheels, we will see how to make use of the characteristics of these machines.

The major part of this chapter is on levers and beams (a cousin of the lever) and is devoted to determining the relationship between the load and the forces required to support the load. Technicians use this information to aid in the design and servicing of supporting structures and to aid in specifying the proper size of lever or beam. The section on gears emphasizes the speed relationships between mating gears, because the primary purpose of gears is to allow a driven device to operate at a speed different from that of the driving device.

8.1 What Is a Machine?

As we saw in Chapter 6, when we talk of work, we take into consideration two factors: force and motion. We stated that work is the product of the force applied and the distance through which it acts. Consequently, a small force acting through a large distance may produce the same amount of work as a large force acting through a short distance.

Often it is to our advantage to do work by applying a small force over a long distance. For example, suppose you wished to lift a 2000 pound automobile 1 foot in order to change a tire. The amount of work required for this task is 2000 foot·pounds (2000 pounds × 1 foot). Obviously, your muscles are not strong enough to apply the 2000 pounds of force required. But suppose you had a device that would permit you to do the work by applying a force of 50 pounds through a distance of 40 feet. The amount of work accomplished would still be 2000 foot·pounds (50 pounds × 40 feet)—enough to raise the 2000 pound automobile 1 foot—but you would have the advantage of needing to apply a force of only 50 pounds, an amount well within the capabilities of your muscles. The device is called a **machine**.

---

* machine | A device used for multiplying or transferring an applied force.

In the performance of its job, a machine may increase or decrease an applied force; transfer the applied force to another

PHOTO 8.1   This "piggy-packer" can lift a fully loaded truck and place it on a flatcar for transportation by train. The work done by this machine is enormous, but even more work must be done by the hydraulic levers and motors that put the machine in operation. (Union Pacific Railroad photo)

position or direction; change the motion of the force from slow to fast, or vice versa; or convert rotating motion to or from reciprocating (back-and-forth) motion. Because work is put into a machine, work comes out of the machine. The machine does not create work. The amount of work output is never greater than the work input. In fact, the work output always is less, since friction within the machine robs some of the useful mechanical energy, converting it into heat and other types of energy. However, the machine always makes work easier, since, as we have seen, it is easier for us to apply a force of 50 pounds through a distance of 40 feet than to apply a force of 2000 pounds through a distance of 1 foot.

8.2 The Lever

A lever is a machine that is commonly used where muscle power is required. For example, appliances are often moved by a dolly, which is a type of lever. Levers are also found in machines that perform a number of different operations. Packaging machinery is an example of such a machine. The automobile makes use of numerous levers: The rocker arms that operate the valves in the engine are levers, and levers are used in the connecting linkage between brake pedal and master cylinder, between gear shift and gear box, and between accelerator pedal and carburetor.

To understand how levers work, consider a 6 foot iron bar pivoted at its center like a seesaw, as shown in Figure 8.1. We call the pivot on which the bar rests a **fulcrum;** the combination of bar and fulcrum is called a **lever.** For the purposes of our discussion, let us neglect the weight of the bar and the friction between it and the fulcrum. (A fulcrum can be a "knife-edge," as shown in Figure 8.1, or an axle about which the lever arm may pivot, as the wheelbarrow in Figure 8.4 will show.)

---

* fulcrum   The point about which the lever rotates or turns. The fulcrum is the "pivot-point" for the lever.

* lever   A rigid object that is capable of turning about one point and that has two or more points where forces can be applied.

Levers are usually designed to amplify a force or a motion. To understand how levers are able to amplify a force, let's begin with a situation we are already familiar with. Since the 6 foot iron bar in Figure 8.1 is pivoted at its center, pushing down with a force of 60 pounds on one end will balance a weight of 60 pounds on the other end. We know this instinctively from our past experiences, but why is this so? Also, if the weight were moved closer to the fulcrum, what force would balance it?

In order to handle these problems, and more complicated situations that we will discuss later, Newton's first law must be expanded. Remember, Newton's first law said that if a body is

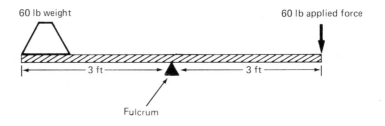

FIGURE 8.1   A Simple Lever with a Fulcrum at the Center.

in equilibrium, all the forces acting on it are in balance. While we did a lot of work on concurrent forces, we see that the forces on the lever are parallel and do not meet at a point of concurrence. The weight and the force each act straight down and are 6 feet apart; the fulcrum exerts a force straight up between the two. So, to take care of this situation for parallel forces, we must restate **Newton's First Law of Motion** in two parts:

1. The sum of all parallel forces must be zero.
2. The sum of all the moments with respect to a center of rotation must be zero.

According to the first part of Newton's law, all parallel forces must balance. In equation form, the capital Greek letter sigma (Σ) will indicate "the sum of." Thus, for parallel forces, we can write the formula:

$$\Sigma F = 0$$

As an example, we can use the lever in Figure 8.1. By drawing a force-vector diagram, as in Figure 8.2, we see all the forces acting on the lever. By convention, technicians and engineers label forces acting down as negative and those acting up as positive. The weight and applied force are acting down, so they have negative signs. The fulcrum, of course, is holding up the lever, the weight, and the force, thereby preventing the lever and weight from falling to the ground. The fulcrum force is given a positive sign. We can use the formula to confirm that the fulcrum is exerting a 120 pound upward force:

$$\Sigma F = 0 = (-w) + (-F) + (+F_{\text{fulcrum}})$$
$$= (-60 \text{ lb}) + (-60 \text{ lb}) + (+F_{\text{fulcrum}})$$

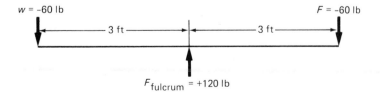

FIGURE 8.2 Force-Vector Diagram of the Lever in Figure 8.1.

Rearranging the formula, we have:

$$F_{fulcrum} = 0 + 60 \text{ lb} + 60 \text{ lb} = 120 \text{ lb}$$

The second part of Newton's first law says that the sum of the moments with respect to a center of rotation must also be zero. The previous chapter on torque stated that torque and **moment** are the same thing: A twisting action that tends to produce rotation and that is the product of a force and the perpendicular distance (moment arm) to the center of rotation. Thus:

$$M = F \times R$$

In the previous chapter, the center of rotation, or moment center, was pretty well defined, and we had only one force to deal with. Where is the moment center on a lever? Actually, if you take a statics course, you will find that you can pick your moment center anywhere you choose. For our discussion, though, we will pick the fulcrum as the moment center. Thus, to apply Newton's first law, at the fulcrum, we have:

$$\Sigma M = 0$$

If the 60 pound force were not applied to the lever in Figure 8.1, the weight would cause the lever to tilt down on the left end, thus rotating the lever counterclockwise (CCW) about the fulcrum. If the weight were removed, and the 60 pound force applied, the lever would tilt down on the right end, causing rotation in a clockwise (CW) direction. In order to handle clockwise and counterclockwise rotations (moments) in formulas, we will show CW moments as plus (+) and CCW moments as minus (−). In other words, we add CW moments, and subtract CCW moments. Summing moments about the fulcrum, we have:

$$\Sigma M = 0 = -(60 \text{ lb} \times 3 \text{ ft}) + (60 \text{ lb} \times 3 \text{ ft})$$
$$= (-180) + (+180) = 0$$

Since the moments did sum to zero, we have confirmed that the lever is in equilibrium. Any slight increase in the force will move the lever. The force of the fulcrum is not in the moment formula because its moment arm is zero, and therefore its moment is zero.

There are quite a few things to remember when solving problems involving levers. The steps to follow are listed below.

Steps for Determining Forces on Levers

Step 1 Locate the moment center. For levers, the moment center is at the fulcrum.

Step 2 Draw a force-vector diagram. Label all forces and their distances from the moment center.

Step 3 Sum the moments about the moment center and set the equation equal to zero ($\Sigma M = 0$). Moment arms must be measured from the force involved to the moment center. Clockwise moments are labeled plus ($+$), and counterclockwise moments, minus ($-$).

Step 4 Rearrange the moment equation, if necessary, to solve for the unknown force.

Step 5 Sum the parallel forces to find the reaction force at the fulcrum. Set the equation to zero and rearrange, if necessary, to solve for the unknown force ($\Sigma F = 0$). Forces acting down are labeled minus ($-$), and forces acting up are labeled plus ($+$).

SAMPLE PROBLEM 8.1

Suppose we were to shift the fulcrum in Figure 8.1 so that it is no longer at the center of the lever arm, but at a point 2 ft from the 60 lb weight and 4 ft from where the force will be applied, as in Figure 8.3A.

A. What force (F) is required to balance the weight (w)?
B. What is the fulcrum ($F_{fulcrum}$) reaction?

SOLUTION A

Wanted

$F = ?$

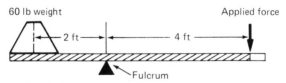

60 lb weight · Applied force
2 ft · 4 ft
Fulcrum

A. Sketch of lever

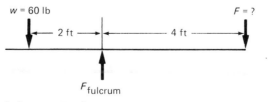

$w = 60$ lb · $F = ?$
2 ft · 4 ft
$F_{fulcrum}$

B. Force-vector diagram

FIGURE 8.3 Simple Lever for Sample Problem 8.1.

Given	We draw a force-vector diagram (Figure 8.3B) with all given data labeled.
Formula	We use the formula for summing moments first to find the unknown force F:

$$\Sigma M = 0$$

Calculation	$\Sigma M = 0 = -(60\ \text{lb} \times 2\ \text{ft}) + (F \times 4\ \text{ft})$

$$0 = (-120\ \text{lb·ft}) + (+4F)$$
$$4F = +120\ \text{lb·ft}$$
$$F = \frac{120\ \text{lb·ft}}{4\ \text{ft}} = 30\ \text{lb down}$$

SOLUTION B

Wanted	$F_{\text{fulcrum}} = ?$
Given	$w = 60\ \text{lb and } F = 30\ \text{lb}$
Formula	We can now use the formula for summing parallel forces to determine the force at the fulcrum:

$$\Sigma F = 0$$

Calculation	$\Sigma F = 0 = (-60\ \text{lb}) + (-30\ \text{lb}) + (+F_{\text{fulcrum}})$

$$= (-90\ \text{lb}) + (+F_{\text{fulcrum}})$$
$$F_{\text{fulcrum}} = 90\ \text{lb up}$$

This information can be used to determine whether the devices supplying the force and supporting the fulcrum are strong enough.

In each of the levers just discussed, the fulcrum lies between the force and the load. This type of lever is known as a **first-class lever**. The wheelbarrow in Figure 8.4 is an example of a **second-class lever**, and Figure 8.5 illustrates a **third-class lever**. Figure 8.5A shows that the forearm is a third-class lever.

* first-class lever	A lever that has the fulcrum located between the force and the load.

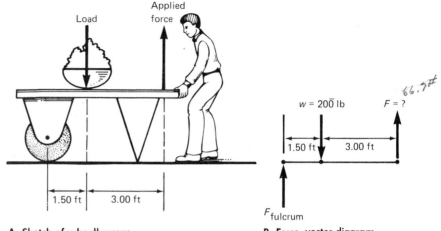

A. Sketch of wheelbarrow

B. Force-vector diagram

FIGURE 8.4   Example of a Second-Class Lever.

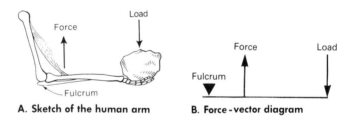

A. Sketch of the human arm

B. Force-vector diagram

FIGURE 8.5   Example of a Third-Class Lever.

∗ second-class lever

A lever that has the fulcrum and the applied force at opposite ends and the load in between.

∗ third-class lever

A lever that has the fulcrum at one end, and the applied force is closer to the fulcrum than the load is.

SAMPLE PROBLEM 8.2

Refer to Figure 8.4. The load has a weight of 200.0 lb down and is located 1.50 ft from the fulcrum. The person exerting the lifting force grasps the handles 3.00 ft from the load.
A. What force (F) is required just to lift the wheelbarrow legs off the ground?
B. What is the force on the wheel axle ($F_{fulcrum}$)?

SOLUTION A	
Wanted	$F = ?$
Given	We draw a force-vector diagram of the wheelbarrow (Figure 8.4B) with all given data labeled. Note that the moment arm of the lifting force is the distance to the fulcrum, 4.50 ft.
Formula	We use the formula for moment summation:
	$$\Sigma M = 0$$
Calculation	$\Sigma M = 0 = +(200.0 \text{ lb} \times 1.50 \text{ ft}) - (F \times 4.50 \text{ ft})$
	$0 = (+300.0 \text{ lb·ft}) + (-4.50 \ F)$
	$4.50 \ F = 300.0 \text{ lb·ft}$
	$F = \dfrac{300 \text{ lb·ft}}{4.50 \text{ ft}} = 66.7 \text{ lb up}$
SOLUTION B	
Wanted	$F_{\text{fulcrum}} = ?$
Given	$w = 200.0 \text{ lb}; F = 66.7 \text{ lb}$
Formula	$\Sigma F = 0$
Calculation	$\Sigma F = 0 = (-200.0 \text{ lb}) + (+66.7 \text{ lb}) + (+F_{\text{fulcrum}})$
	$0 = (-133.3 \text{ lb}) + (+F_{\text{fulcrum}})$
	$F_{\text{fulcrum}} = 133 \text{ lb up}$

8.3 The Beam

Beams are devices related to levers. Usually, beams are placed in a horizontal position, and forces (loads) are applied vertically. Beams support loads in various arrangements, some of which are shown in Figure 8.6. Examples of beams are the floor joists (and ceiling joists) in your home or in a commercial structure, beam-type bridges that allow a road to pass over a major highway, and the support frame of a car.

---

* beam An object, or body, that normally spans two supports and, in turn, supports one or more loads.

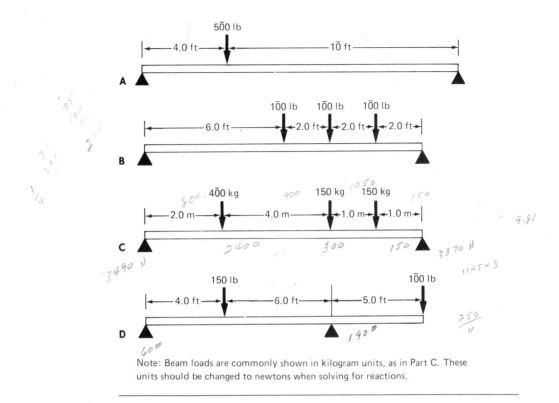

Note: Beam loads are commonly shown in kilogram units, as in Part C. These units should be changed to newtons when solving for reactions.

FIGURE 8.6   *Examples of Beams and Various Loadings.*

The beams we plan to discuss will have several loads and two supports. The first step in designing a beam to support a load is to find the **reaction force** at the supports. Look at the sketches in Figure 8.6. All the beams are resting on knife edges. These knife edges were called fulcrums when applied to levers, but for beams we call them supports.

* reaction force

The upward force supplied by the support that reacts to the load or loads on the beam.

An actual beam does not rest on knife edges. More practical supports are needed. If you look at the floor joists in your home, you will probably find that the joists are 10 to 12 feet long and rest on sills about 4 inches wide. However, when the 4 inches are compared with the 12 feet of beam, we can consider the support as a knife edge. Knife edges are used in drawings so that we know exactly where the support is located.

The object of our beam problems will be to determine the reaction forces at the beam supports. When you work in industry, or take the appropriate course later in your technical program, you will need to know these reaction forces to design a proper support and to specify the correct size of beam.

A beam problem is handled in a manner similar to that used for solving lever problems. The moment center for a lever was taken at the fulcrum, or support. For a beam, you have a choice of either support for a moment center. In fact, it is best to solve two moment equations, each with a different support as a moment center, and then check your answer by summing the parallel forces.

The list of steps for solving beam problems is essentially the same as that used for solving lever problems. The steps are repeated below with the appropriate wording for beams included.

Steps for Determining the Reaction Forces of Beams

Step 1 Draw a force-vector diagram. Show the support (reaction) forces. Label all known forces and distances.

Step 2 Locate the moment center at either the left- or right-hand support. Sum the moments of the forces and set the equation equal to zero: $\Sigma M = 0$.

Step 3 Rearrange the moment equation and solve for the unknown reaction.

Step 4 Repeat steps 2 and 3 with the other support as the moment center.

Step 5 Check by summing the parallel forces. They should equal zero: $\Sigma F = 0$.

Sample Problem 8.3 shows how to solve for the support reactions on a beam. For our solutions, we will neglect the weight of the beam.

SAMPLE PROBLEM 8.3 Refer to Figure 8.6B. Given the beam, loading, and dimensions shown, find the right reaction force (R_r) and the left reaction force (R_l).

SOLUTION

Wanted

$R_r = ?$ and $R_l = ?$

Given

We draw a force-vector diagram of the beam (Figure 8.7) with all given data labeled.

Formula

$\Sigma M = 0$ and $\Sigma F = 0$

Calculation

First, we use the left support as the moment center and measure all moment arms to the left support to find the right reaction force (R_r):

$$\Sigma M = 0 = +(100 \text{ lb} \times 6 \text{ ft}) + (100 \text{ lb} \times 8 \text{ ft})$$
$$+ (100 \text{ lb} \times 10 \text{ ft}) - (R_r \times 12 \text{ ft})$$

$$0 = +2400 \text{ lb·ft} - 12R_r$$

$$R_r = \frac{2400 \text{ lb·ft}}{12 \text{ ft}} = 200 \text{ lb}$$

Next, the right support is used as the moment center to find the left reaction (R_l):

$$\Sigma M = 0 = -(100 \text{ lb} \times 2 \text{ ft}) - (100 \text{ lb} \times 4 \text{ ft})$$
$$- (100 \text{ lb} \times 6 \text{ ft}) + (R_l \times 12 \text{ ft})$$

$$0 = -1200 \text{ lb·ft} + 12R_l$$

$$R_l = \frac{1200 \text{ lb·ft}}{12 \text{ ft}} = 100 \text{ lb}$$

We can check the answers by summing all parallel forces:

$$\Sigma F = 0 = +R_l \quad -100 \text{ lb} -100 \text{ lb} -100 \text{ lb} +R_r$$
$$0 = +100 -100 \text{ lb} -100 \text{ lb} -100 \text{ lb} +200 \text{ lb}$$

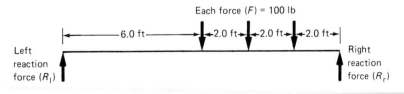

FIGURE 8.7 Force-Vector Diagram for Sample Problem 8.3.

8.4 Gear Wheels

Although **gear wheels** can be used to increase a force, their primary purpose in modern industry is to change shaft speed. An electric motor, for instance, may have its most effective speed at 3600 rpm. If it is used to drive a centrifugal pump that operates most efficiently at 1200 rpm, gears may be used to change the speed. The automotive engine may operate best at 3000–4000 rpm, but the wheels of the car run from zero up to about 1000 rpm. Gears are used to keep the engine running at a fairly constant speed as the wheel speed is varied. It will be easier to understand the discussion about gear speeds if we first consider rollers.

---

* gear wheels

Toothed wheels used to transmit rotational motion.

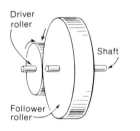

Driver roller

Shaft

Follower roller

FIGURE 8.8   A Set of Rollers.

Look at the two rollers illustrated in Figure 8.8. The rollers are in contact with each other, and if one rotates, it causes the others to do so too by reason of the friction between them. Let's call the smaller one the *driver* and the larger one the *follower*. (The larger one could have been chosen as the driver.) As the driver turns, the follower turns too.

If a point on the circumference of the driver travels a certain rotational distance, it will cause a point on the circumference of the follower to travel the same distance. But since the circumference of the driver is smaller than the circumference of the follower, for every rotation of the driver, the follower will make only a part of a rotation. Thus, the rotational speed of the follower is less than that of the driver.

Let's be specific and say the circumference of the driver is 10 inches and the circumference of the follower is 30 inches. This means that the driver must rotate three times to cover 30 inches, while the follower rotates just once. Since the smaller roller rotates faster than the larger one, the angular velocity (N) of the smaller roller will be greater. By the way, although the term "angular velocity" is technically correct, handbooks have been using the term "speed" for years. They may say "the speed of the roller is 45 rpm." So, we will also use the term "speed" to mean angular velocity.

Notice that the speed of the smaller roller is three times as fast as the larger roller, while the circumference of the smaller roller is one-third the circumference of the large roller. *The speeds*

PHOTO 8.2   Much of the work in strip mining operations is accomplished by huge machines that make extensive use of gears, pulleys, and levers. (Courtesy of Clark Equipment Company)

of the rollers are inversely proportional to their circumferences. That is, the larger roller has the smaller speed, and the smaller roller has the larger speed. This is true of all rollers and can be shown algebraically:

$$\frac{N_d}{N_f} = \frac{C_f}{C_d}$$

The term N_d/N_f is called the **speed ratio** of the rollers. The subscripts "d" and "f" denote driver and follower, respectively.

* **speed ratio**

The ratio of the angular velocity of one roller to the angular velocity of the other contacting roller.

This inverse proportion can be extended to the roller diameters (D):

$$\frac{N_d}{N_f} = \frac{C_f}{C_d} = \frac{\pi D_f}{\pi D_d}$$

Since the π's cancel, we are left with this more useful formula:

* **SPEED RATIO FORMULA**

$$\frac{N_d}{N_f} = \frac{D_f}{D_d} \qquad \text{or} \qquad N_d D_d = N_f D_f$$

where N_d = speed of driver roller
           N_f = speed of follower roller
           D_f = diameter of follower roller
           D_d = diameter of driver roller

SAMPLE PROBLEM 8.4

A phonograph turntable 8.75 in. in diameter (D_f) is driven by a spindle 0.1875 in. in diameter (see Figure 8.9). If the turntable rotates at a speed (N_f) of 33.3 rpm, what is the speed (N_d) of the driving spindle?

SOLUTION

Wanted

N_d = ?

Given

D_f = 8.75 in., D_d = 0.1875 in., and N_f = 33.3 rpm

Formula

$$\frac{N_d}{N_f} = \frac{D_f}{D_d}$$

Calculation

$$\frac{N_d}{33.3 \text{ rpm}} = \frac{8.75 \text{ in.}}{0.1875 \text{ in.}}$$

$$N_d = \frac{8.75 \text{ in.} \times 33.3 \text{ rpm}}{0.1875 \text{ in.}} = 1560 \text{ rpm}$$

With this information, a motor with the proper driving speed can be selected.

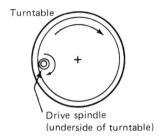

Turntable

Drive spindle
(underside of turntable)

FIGURE 8.9 A
Phonograph Turntable Drive.

The speed ratio of rollers is closely related to that of gears. The gear wheel (or just gear) is a special type of roller that is widely used in many mechanical devices, ranging from egg beaters to ship propulsion mechanisms. A gear wheel is really a roller with teeth. Figure 8.10 shows a pair of **spur gears**, a common type of gear wheel. The advantage of the gear wheel over the roller is that since the teeth of the gear wheels mesh together, they cannot slip, thus ensuring positive action.

✱ spur gears

Gears that have their teeth projecting radially from the axle, like spokes on a wheel.

In order to convert the rollers previously mentioned into gears that have the same speed ratio, the teeth on each gear must extend beyond the roller circle, as shown in Figure 8.10. Roughly half the tooth height is beyond the circle and half is inside it. This circle is called the **pitch circle.** The pitch-circle diameter is the one used to determine speed ratios when using the formula:

$$\frac{N_d}{N_f} = \frac{D_f}{D_d}$$

✱ pitch circle

An imaginary circle on a gear that is tangent to the pitch circle of a mating gear. If the gears were replaced with rollers of the same size as the pitch circles, the rollers would have the same speed ratio as the gears.

The number of teeth on each gear can also be used to determine the speed ratio since the teeth have to be the same size

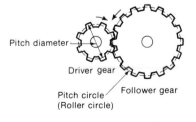

Pitch diameter

Driver gear

Pitch circle
(Roller circle)

Follower gear

FIGURE 8.10 A Set of Spur Gears.

on each gear or they won't mesh. The ratios are:

$$\frac{N_d}{N_f} = \frac{D_f}{D_d} = \frac{T_f}{T_d}$$

where T is the number of teeth on the gear.

We can demonstrate the use of gear teeth for finding speed ratios by imagining that the tooth spacing is such that there is one tooth for every inch of circumference. Therefore, if the pitch circles of two gears are the same sizes as the rollers mentioned previously (Figure 8.8), the gears will have 10 teeth and 30 teeth, respectively. This means the smaller gear rotates three times faster than the larger gear—the same answer we obtained for the rollers. Gear wheels must be cut carefully to produce a perfect mesh between their teeth.

If we examine the sets of rollers and gear wheels shown in Figures 8.8 and 8.10, we see that when the driver rotates in one direction, the follower rotates in the other direction. Sometimes it is desirable that both the driver and the follower rotate in the same direction. As Figure 8.11 shows, to accomplish this change in direction, a third gear, called an *idler,* is inserted between the driver and the follower. The motion then undergoes a double change, and as a result, the driver and the follower rotate in the same direction. We now have more than two gears in our drive system, a situation called a **gear train.**

✳ gear train   An assembly in which more than two gears are employed in a drive.

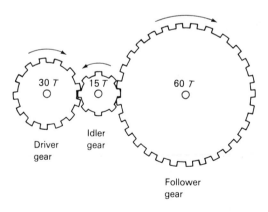

FIGURE 8.11   How an Idler Gear Changes the Direction of Rotation in a Gear Train.

The idler gear need not be used in determining the speed ratio of the system. Let's see why. First, we consider the speed ratio of the driver and the idler:

$$\frac{N_d}{N_i} = \frac{T_i}{T_d}$$

where the subscript "i" denotes the idler gear. Next, let's look at the speed ratio of the idler and the follower:

$$\frac{N_i}{N_f} = \frac{T_f}{T_i}$$

If we multiply these two speed ratios together, we obtain the speed ratio for the gear train:

$$\frac{N_d}{N_i} \times \frac{N_i}{N_f} = \frac{T_i}{T_d} \times \frac{T_f}{T_i} = \frac{T_f}{T_d}$$

or in the form of a formula, we have:

---

* IDLER GEAR TRAIN SPEED RATIO
$$\frac{N_d}{N_f} = \frac{T_f}{T_d} \qquad \text{or} \qquad N_d T_d = N_f T_f$$

The teeth on the idler gear cancel out. But how do we know the teeth ratios should be multiplied and not added? To find out, we can use the gear train in Figure 8.11 as an example. The ratio between the driver and idler gear is:

$$\frac{N_d}{N_i} = \frac{T_i}{T_d} = \frac{15}{30} = \frac{1}{2}$$

The driver rotates once and the idler twice. If the driver rotates twice, the idler rotates four times. The ratio between idler and follower is:

$$\frac{N_i}{N_f} = \frac{T_f}{T_i} = \frac{60}{15} = \frac{4}{1}$$

The idler rotates four times for every rotation of the follower. In analyzing this information, we find that if the driver revolves twice, the idler revolves four times and the follower once. This gives an overall ratio of 2/1. Now, let's use the formula with just the driver and follower:

$$\frac{N_d}{N_f} = \frac{T_f}{T_d} = \frac{60}{30} = \frac{2}{1}$$

Idler gear trains are used on many engine lathes so that the direction of rotation of the work can be changed without changing the direction of rotation of the driving motor. This application is shown in Figure 8.12. A more complex gear train is illustrated in Figure 8.13. Each pair of gears is a stage; therefore, the figure shows a **two-stage gear train.**

* two-stage gear train	A gear train that has two pairs of gears meshing, with one gear from each pair mounted on the same shaft and locked together.

In Figure 8.13, let gear A be the driver that causes gear wheel B to rotate. Since B has more teeth than A, the speed of rotation of B is reduced. Gear wheel C is locked to the same shaft as gear wheel B, and consequently rotates at the same speed. Gear wheel D is driven by C, and since D has more teeth than C, the speed of rotation is further reduced. If the entire gear train is reversed, and gear wheel D becomes the driver while A becomes the final driven wheel, the speed of rotation is greatly increased.

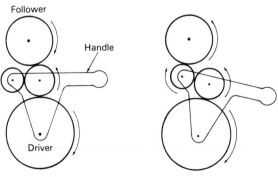

A. Forward rotation   B. Reverse rotation

FIGURE 8.12   How an Idler Gear Shifts the Direction of Rotation of the Work on an Engine Lathe.

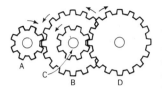

FIGURE 8.13 A Two-Stage Gear Train.

Of course, we could use just two gears to get the same speed change. In general, though, when there is a big difference in gear size, the teeth don't mesh as smoothly, and it takes extra machining to get smooth operation. Also, the gear assembly may not be as compact as with the smaller gears; what's more, the large gear may be nonstandard and therefore more costly.

In the type of gear train shown in Figure 8.13, gears B and C are not idler gears; therefore, they must be included in speed ratio calculations. In the development of the two-stage gear train formula that follows, the subscripts refer to the gears in Figure 8.13. First:

$$\frac{N_a}{N_b} = \frac{T_b}{T_a}$$

Rearranging this formula, we have:

$$N_b = \frac{N_a T_a}{T_b}$$

Also:

$$\frac{N_c}{N_d} = \frac{T_d}{T_c}$$

Rearranging this, we have:

$$N_c = \frac{N_d T_d}{T_c}$$

Since gears B and C are locked to the same shaft, their speeds are the same; we therefore have:

$$N_b = N_c$$

Now, we substitute the rearranged formulas:

$$\frac{N_a T_a}{T_b} = \frac{N_d T_d}{T_c}$$

We rearrange again for our speed ratio:

---

* TWO-STAGE GEAR TRAIN SPEED RATIO

$$\frac{N_a}{N_d} = \frac{T_b T_d}{T_a T_c}$$

Note that pitch-circle diameters can be used in place of gear teeth, as mentioned in the original speed ratio formula.

Sample Problem 8.5 shows the application of the two-stage gear train speed ratio formula.

SAMPLE PROBLEM 8.5

Figure 8.13 represents a speed reduction drive between a motor and the cutting-tool spindle on a milling machine. The gears have the following number of teeth: $T_a = 12$, $T_b = 30$, $T_c = 16$, and $T_d = 40$. If the driver gear A rotates at a speed (N_a) of $15\overline{0}0$ rpm, find the speed (N_d) of gear D.

SOLUTION

Wanted

$N_d = ?$

Given

$N_a = 1500$ rpm, $T_a = 12$, $T_b = 30$, $T_c = 16$, and $T_d = 40$

Formula

$$\frac{N_a}{N_d} = \frac{T_b T_d}{T_a T_c}$$

Calculation

Rearranging to solve for N_d:

$$N_d = \frac{N_a T_a T_c}{T_b T_d} = \frac{1500 \times 12 \times 16}{30 \times 40} = 240 \text{ rpm}$$

Sometimes, we may want to convert a rotary motion to another rotary motion at right angles to the first. For example, the drive shaft on an outboard motor is vertical while the propeller shaft is horizontal. A common method for doing so is to use the kind of gears shown in Figure 8.14. The teeth of the gears are beveled and thus will mesh when the wheels are turned at right angles to each other. This type of gearing is called a set of **bevel gears**. As in the case of ordinary spur gears, the speed ratio depends on the ratio between the number of teeth in the two gears.

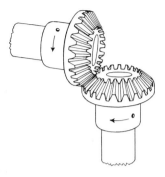

FIGURE 8.14   A Set of Bevel Gears.

---

* bevel gears

A pair of gears that have their teeth cut on a conical surface (or bevel), allowing their axles to be set at an angle to each other.

Another type of gear mechanism that can produce the same change of motion as the bevel gear is the **worm gear,** illustrated in Figure 8.15. If the worm is rotated, it will cause the gear to rotate also. For every complete revolution of the worm, the gear will be advanced a single tooth. We can readily see that, in addition to changing the direction of motion, the worm gear is capable of producing a tremendous speed ratio. Some trucks use a worm gear in the differential housing on the rear axle. Normally, the gear cannot drive the worm, which may be a desirable locking feature in some applications.

---

* worm gear

A screw whose threads engage the teeth of a second gear.

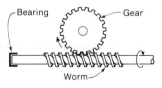

FIGURE 8.15   The Worm Gear.

Summary

A machine is a device for multiplying a force, transferring a force to another position or direction, changing motion from slow to fast, or changing rotating motion to reciprocating motion (and vice versa).

The technician must know the forces involved when working with levers and beams. Levers are usually designed to amplify a force (and sometimes to amplify a motion). Beams are designed to span a gap between supports and, in turn, support a load. For both levers and beams, Newton's first law must be presented in two parts:

1. The sum of the parallel forces must be zero, or $\Sigma F = 0$.
2. The sum of the moments about a moment center must be zero, or $\Sigma M = 0$.

The speed ratio for a pair of gears, or for a gear train with idler gears, is inversely proportional to the gear diameters and also inversely proportional to the number of teeth on each gear. The speed ratio formula is: $N_d/N_f = D_f/D_d = T_f/T_d$. The two-stage gear train speed ratio is: $N_a/N_d = T_b T_d/T_a T_c$.

Key Terms

* machine
* lever
* fulcrum
* Newton's First Law of Motion
* moment
* first-class lever
* second-class lever
* third-class lever
* beam
* reaction force
* gear wheels

* speed ratio
* speed ratio formula
* spur gears
* pitch circle
* gear train
* idler gear train speed ratio
* two-stage gear train
* two-stage gear train speed ratio
* bevel gears
* worm gear

Questions

1. Define the word "machine."
2. Define the word "lever."
3. Sketch the three types of levers.
4. Suggest an application, not already mentioned, for each type of lever, and identify the type of lever for those applications mentioned.
5. What is a beam?
6. How is Newton's first law presented for use in lever and beam problems?
7. Explain the use of gears.
8. What is the advantage of an idler gear?
9. How many gears does a three-stage gear train have?
10. Describe spur gears, bevel gears, and worm gears.

Problems

LEVERS

1. Refer to Figure 8.1. Double the length of the lever so that the fulcrum is 6.0 ft from the 60.0 lb weight and from the applied force. Determine the applied force and the reaction at the fulcrum.
2. Do Sample Problem 8.1 with the weight changed to $35\overline{0}$ lb.
3. Refer to Figure 8.16 showing a rocker arm on a truck engine. If the valve lifter exerts a force of 115 lb at the position shown, how much force is applied to the valve stem?

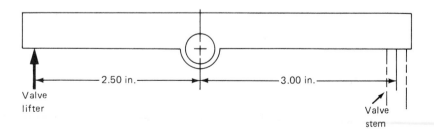

FIGURE 8.16   Rocker Arm for Problem 3.

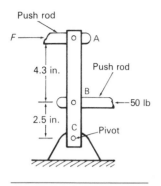

Push rod

F ——→

A

4.3 in.

Push rod

B

←—50 lb

2.5 in.

C

Pivot

FIGURE 8.17 Control
Lever Mechanism for
Problem 6.

4. Refer to Figure 8.3. Change the distance from the fulcrum to the applied force to 6.0 ft, and solve for the applied force and the reaction at the fulcrum.

5. Refer to Figure 8.4, but use the following data: load equals 175 kg; distance from load to fulcrum equals 0.700 m; and distance from the applied force to fulcrum equals 1.60 m. What is the applied force? Remember to change kilograms (mass units) to force units.

6. Refer to Figure 8.17. A speed-control device on a steam turbine operates a push rod attached to a lever at point A. The lever, in turn, operates a second push rod at point B. This second push rod is connected to a steam valve and must work against a 50 lb spring load.
   a. What force must the push rod at point A exert?
   b. What are the reaction and its direction at the fulcrum point (point C)?

7. Refer to Figure 8.5. If the load is 8.0 kg, the elbow-to-hand distance equals 0.40 m; and the elbow-to-muscle distance equals 25 mm. Solve for the applied force (applied by the muscle).

BEAMS

8. Refer to Figure 8.18A. The reaction at B equals 153 lb. Solve for the reaction at A by (a) summing vertical forces and by (b) summing moments about B.

9. Refer to Figure 8.18B. The reaction at B equals 125 lb. Solve for the reaction at A by (a) summing vertical forces and by (b) summing moments about B.

10. Refer to Figure 8.18C. The reaction at A equals 267 lb. Solve for the reaction at B by (a) summing vertical forces and by (b) summing moments about A.

11. Solve for the reactions of the beam in Figure 8.6A.

12. Refer to Figure 8.6A. Move the load to within 2 ft of the left support and solve.

13. Solve for the reactions of the beam in Figure 8.6C.

14. Solve for the reactions of the beam in Figure 8.6D.

15. Figure 8.19 shows two pulleys on a shaft, and the forces applied by the belt tensions. Indicate the magnitude and direction of the reaction at each bearing.

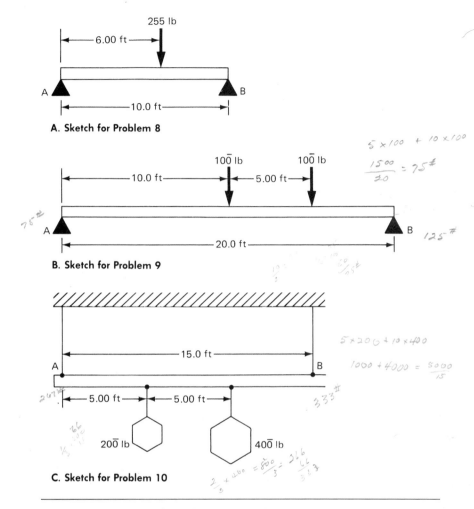

A. Sketch for Problem 8

B. Sketch for Problem 9

C. Sketch for Problem 10

FIGURE 8.18   Beams for Problems 8, 9, and 10.

16. Figure 8.20 shows a small hoist mounted on a truck used in a lumber yard. When the hoist is in the position shown, what is the maximum load that can be lifted without tilting the truck? (*Hint:* Consider the reaction on the left-hand tires to be zero and the moment center to be at the right-hand set of tires.)

Gears

17. Solve Sample Problem 8.4, but change the speed of the turntable to 45 rpm.

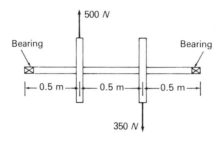

FIGURE 8.19 Pulleys Mounted on a Shaft for Problem 15.

18. A small shop lathe has a drive gear with a diameter of 76.0 mm rotating at a speed of 25$\overline{0}$ rpm. A follower gear has a diameter of 113 mm. What is its speed of rotation?

19. A milling machine has a 2.50 in. diameter gear mounted on the motor shaft, which rotates at 1725 rpm. The meshing gear that drives the milling bit spindle is 6.50 in. in diameter. What is this meshing gear's speed in rpm?

20. A pump with a recommended operating speed of 400 rpm is geared to a motor that operates at 1725 rpm. The gear on the motor is 5.00 in. in diameter. What is the size of the gear on the pump?

21. A small centrifugal pump is geared to the shaft of a turbine. The gear mounted on the turbine shaft has a diameter of

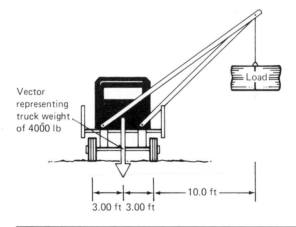

FIGURE 8.20 Truck and Hoist for Problem 16.

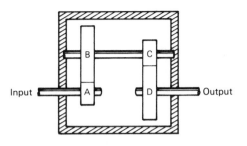

FIGURE 8.21 Gear Train for Problem 26.

5.0 in. and a speed of 3600 rpm. The pump gear has a diameter of 8.5 in. What is the speed of the large gear?

22. Two rollers in a steel mill must turn in the same direction, but not at the same speed. Roller A rotates at $40\bar{0}$ rpm and has a gear with 36 teeth. The idler gear has 23 teeth, and the other roller (roller B) has a gear with 42 teeth. What is the speed of roller B?

23. Relatively small speed differences are required between rollers of some types of rubber or plastic mixing machines. Suppose we are given two pairs of gears: one pair has 16 and 22 teeth, the other pair has 22 and 28 teeth. Which pair will provide the smallest speed change?

24. Do Sample Problem 8.5 with gear B changed to 16 teeth and gear C changed to 30 teeth.

25. The speed reduction between a motor and a drive shaft on a large boring mill is from 1750 to 200 rpm. If the gear on the motor shaft is 3.5 in. in diameter, what is the diameter of the gear on the drive shaft?

26. Refer to Figure 8.21. A particular gear train assembly must have the input and output shafts in line (the assembly can be changed or updated without disturbing the rest of the machine). One way is to have gears A and C the same diameter, and have gears B and D the same diameter. If the diameter of gear A is 3.5 in., the diameter of gear B is 9.2 in., and the input speed to A is $50\bar{0}$ rpm, what is the output speed?

27. What is the gear train speed ratio of an electric clock measured from the shaft of the electric motor, with a speed of 60 rev/s, to the hour hand?

Computer Program

This program will calculate the reactions at the supports of a beam for a beam with a single load placed anywhere between the supports. You may use this program to solve some of the chapter problems and some problems of your own design. For instance, keep the load and beam span constant, but move the load to different locations, and see how this change affects the support reactions.

PROGRAM

```
10   REM   CH EIGHT PROGRAM
20   PRINT "PROGRAM FOR A BEAM SUPPORTED AT ITS ENDS WITH A SINGLE"
21   PRINT "LOAD PLACED ANYWHERE BETWEEN THE SUPPORTS."
30   PRINT
40   PRINT "UNITS ARE TO BE IN FEET AND POUNDS."
50   PRINT
60   PRINT "TYPE IN THE LENGTH OF THE SPAN, THEN THE LOAD, AND THEN"
61   PRINT "THE DISTANCE OF THE LOAD FROM THE LEFT SUPPORT."
70   INPUT S,W,D
80   PRINT "THE SPAN = ";S;" FT."
90   PRINT "THE LOAD = ";W;" LB."
100  PRINT "THE DISTANCE FROM LOAD TO LEFT SUPPORT = ";D;" FT."
110  LET L = (W * (S - D)) / S
120  LET R = (W * D) / S
130  PRINT
140  PRINT
150  PRINT "LEFT SUPPORT = ";L;" LB."
160  PRINT "RIGHT SUPPORT = ";R;" LB."
170  END
```

NOTES

Lines 30, 50, 130, and 140 leave spaces between lines on the screen.

Lines 80–100 confirm your input.

Lines 110 and 120 determine the left and right reactions.

Part III

Heat

In electric arc welding, energy of the electric arc heats the metal to its melting point. The metal parts then fuse together as the metal cools. In some cases, the cooling rate must be controlled, because if the cooling metal contracts too rapidly, a crack may develop.

Chapter 9

Objectives

When you finish this chapter, you will be able to:

- [ ] Understand the difference between heat and temperature and be able to define each.
- [ ] Describe and compare the Celsius, Fahrenheit, and Kelvin temperature scales and convert values among them.
- [ ] Describe and compare the three heat transfer methods: conduction, convection, and radiation.
- [ ] Recognize the specific heat equation, $Q = mc\,(\Delta T)$, showing that the quantity of heat absorbed is equal to the amount of material being heated (m) times its specific heat capacity (c) times the change in temperature (ΔT).
- [ ] Explain that heat may change an object's state (solid, liquid, or gas). Recognize the change of state formulas $Q = mL_f$ and $Q = mL_v$, which include the terms "heat of fusion" (L_f) and "heat of vaporization" (L_v).
- [ ] Recognize the heat expansion equation $\Delta L = \alpha L(\Delta T)$, showing how the change in an object's length (ΔL) is equal to its thermal coefficient of linear expansion (α) times its initial length (L) times the change in temperature (ΔT).
- [ ] Explain that when dealing with gases, we must use the ideal gas law, $P_1 V_1 / T_1 = P_2 V_2 / T_2$, showing how initial values of pressure (P), volume (V), and temperature (T) relate to some final state values of the same variables.

Introduction to Heat

In order to understand how heat is utilized to do work and how it is controlled, we must understand what heat is, how it is measured, and how it affects different substances. This chapter discusses heat and temperature and the difference between the two; the various temperature scales (Fahrenheit, Celsius, and Kelvin); the units of heat such as calories, Btu, and joules; a constant called "specific heat capacity" used in heat-transfer formulas; and the effects of heat on liquids, solids, and gases.

In Part II, we dealt with mechanics—that is, the effects produced by bodies that possess potential or kinetic energy. In Part III, we continue to study the effects produced by kinetic energy—the energy of motion. But now we will study the kinetic energy of the molecules that make up matter.

9.1   What Is Heat?

In Chapter 3, we saw that the molecules of all substances are constantly moving and that each individual molecule has kinetic energy. Because of their attraction for one another, molecules also may have a small amount of both kinetic and potential energy. The sum of these energies is called **internal energy**.

---

* internal energy

The sum of the kinetic and potential energies of an object's atoms and molecules.

All substances possess internal energy, but in varying amounts. The kinetic energy possessed by a moving body depends on its mass and velocity as discussed in Chapter 6. Every substance possesses internal energy to a degree that depends on the number of molecules, the kind of molecules, and their velocities. As Figure 9.1 indicates, if the motion of the molecules increases, the substance has more internal energy, and it becomes "hotter." If the motion of the molecules decreases, the substance has less internal energy, and it becomes "cooler."

One way to increase the internal energy of an object is to heat it over a fire. When a flame is applied to the bottom of a water boiler, for example, the internal energy of the flame flows into the water. This flow of energy (heating of the water) increases the water's internal energy. At the same time, the flame tends to become cooler and must be replenished with more burning fuel. For another example, consider the hot radiator in your

PHOTO 9.1   This computer-enhanced photograph, taken by the Skylab space vehicle, shows a solar flare on the sun's surface. The temperature at the interior of the sun has been estimated to be 20 million degrees Celsius, and this intense heat causes the gases that make up the sun to be in constant motion, often expanding in a flare. (NASA photo)

home. As the **heat** (internal energy) flows from the radiator into the room, the room air heats up and the radiator cools. To replenish the heat lost by the radiator, more hot water is pumped to the radiator as the cooled water flows back to the furnace. The molecules of the air in the room have their velocities increased, and the radiator molecules have their velocities decreased. Note that the radiator cools down by losing heat, not by gaining cold. Heat is a form of energy; "cold" does not exist except as the absence of heat.

∗ heat	The amount of internal energy that flows from one substance to another.

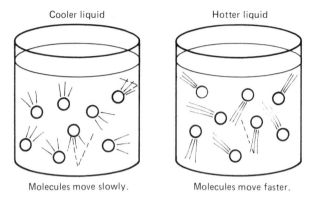

Cooler liquid Hotter liquid

Molecules move slowly. Molecules move faster.

FIGURE 9.1 Random Motion of Molecules in a Liquid. For clarity, only a few molecules are shown.

When we speak of the velocity of the molecules of a substance, we do not mean that all the molecules have the same velocity. Since they are constantly bumping into each other at random (in gases and liquids), some travel faster than others. It is the average velocity of the molecules that we are discussing. This average velocity affects the kinetic energy of the molecules. Remember:

$$\text{KE} = \frac{1}{2}mV^2$$

The higher the velocity, the higher the kinetic energy and the hotter the substance becomes. The measure of the "hotness" or "coldness" is called **temperature**. Since the mass of a molecule is constant, any change in velocity of the molecules results in a change in temperature.

---

* temperature

A measure of the average kinetic energy of the molecules of a substance.

The difference between heat and temperature is not easy to understand, so let's perform an experiment. In the laboratory, we fill a small flask and a large flask with water. Our object is to apply the same amount of heat to each flask and then measure the water temperatures. The flasks are placed over identical burners and heated for the same length of time. We are careful not to boil the water, just heat it a bit. After heating both flasks for, say, one minute, we measure their temperatures. The water in the small flask is found to have the higher temperature. What happened? Why weren't the temperatures the same?

The small flask held fewer molecules of water (less mass), so more of the heat from the burner went to increasing the velocities of those molecules—that is, to increasing their kinetic energies. The same amount of heat added to the large flask had to be distributed among more molecules. Therefore, each molecule received a smaller amount of energy, and its increase in velocity was less, as was indicated by a lower reading on the thermometer.

9.2   The Transfer of Heat from One Substance to Another

Now, how is heat transferred? Suppose that you hold an iron nail in a flame. Soon the end you are holding becomes so hot that you are forced to drop it. The end held in the flame received heat energy produced from the chemical energy of the burning material. The molecules of iron in that end were speeded up. These molecules, in turn, bumped against their neighbors with greater force, causing them to speed up also. All the molecules of the nail soon were vibrating very rapidly. This rapid vibration, in turn, caused the molecules of your fingers to speed up, too. You experienced the sensation of "hot" and dropped the nail.

We call this method of heat transfer **conduction.** Not all materials are equally good heat conductors. Metals, generally, are good conductors. Nonmetals, such as wood, paper, glass, and so forth, are poor conductors. As a matter of fact, you can hold a stick of wood in a flame until it catches fire and yet not feel the heat at your end. Liquids and gases, generally, are very poor conductors of heat.

---

* conduction   The transfer of heat energy from one end of a body to the other, or from one object to another, through direct contact.

There are other methods of heat transfer. If you hold your hand over a hot stove, you feel the heat, even though your hand is not in contact with the stove. How is the heat transferred from the stove to your hand? First of all, the air just above the stove

PHOTO 9.2 The high temperatures in the earth's interior create convection currents in the liquid rock. When the liquid rock finds a crack in the layers of solid rock near the earth's surface, it sometimes forces its way to the surface and erupts as a volcano. The Irazu volcano in Costa Rica has erupted several times, growing larger each time. (United Nations photo)

becomes hot by contact. As it becomes hot, its molecules travel faster, and as a result, the molecules travel farther from their neighbors. Because the molecules of hot air are spaced farther apart, a cubic foot of hot air weighs less than a cubic foot of cold air. Now, being less dense, the hot air rises and comes in contact with the hand held above the stove. Thus, the heat from the stove is transferred to the hand by a rising current of hot air. We call this method of heat transfer **convection.**

* convection

The transfer of heat energy through the bodily movement of a substance.

Heat transfer in a gas takes place chiefly by this method of convection. Convection takes place in liquids, too, as the warm, and hence, lighter liquid rises in currents to the top. Actually, this discussion has been about *natural convection*. If a fan (or pump for liquids) were used to force the gas or liquid in other directions, we would have *forced convection*.

Finally, we come to a third means for transferring energy. The sun transfers its light and heat energy through 90 million miles of empty space—a complete vacuum. Since conduction and convection cannot account for this transfer, there is another method called **radiation,** in which heat energy is transferred by means of electromagnetic waves.

---

* radiation

The transfer of heat energy by means of electromagnetic waves radiating in all directions.

All bodies radiate, or give off, heat. The hotter the body, the more heat is radiated. Thus, we have another way in which the heat energy of a substance changes—by radiation from, or to, surrounding bodies. The heat energy radiates in all directions, which you can prove by holding your hand near the side of a hot stove. You will feel the heat, even though the stove is not in contact with your hand and the rising currents of hot air cannot reach it.

Regardless of how the heat is transferred, if two bodies, one hot and the other cold, are placed near each other, heat will be transferred from the hot body to the cold one. This process will continue until the molecules of both bodies are moving at the same rate. As Figure 9.2 illustrates, heat is *conducted* through the metal stove body and also through the metal bottom of the kettle to the water. Heat is distributed around the room by rising currents of hot air, which is heating by *convection*. Heat is also distributed to objects in the room by the *radiation* of electromagnetic waves.

9.3   The Measurement of Temperature

THE FAHRENHEIT AND CELSIUS TEMPERATURE SCALES

We have seen that all substances possess internal energy, since all substances are composed of molecules in motion. If we touch

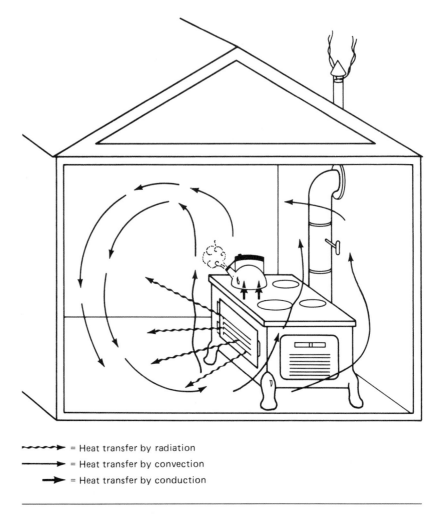

~~~~► = Heat transfer by radiation

——► = Heat transfer by convection

—► = Heat transfer by conduction

FIGURE 9.2 Heat Transfer by Conduction, Convection, and Radiation.

a kettle that has been standing over a flame, we say it is "hot." If we touch a piece of ice, we say it feels "cold."

In the case of the hot kettle, the internal energy of the flame has speeded up the molecules of the kettle, which, in turn, speeds up the molecules of the hand, causing the sensation of "hot." When we touch the ice, whose molecules are moving more slowly than those of the hand, some of the energy of the molecules of the hand is transferred to speed up the molecules of the ice. This loss of heat energy gives us the sensation of "cold." (Remember "coldness" does not exist and is, therefore, never transferred.)

PHOTO 9.3   A glacier consists of ice in motion. The temperature of the ice is always below 0 degrees Celsius (32 degrees Fahrenheit), the freezing point of water. (National Park Service photograph by M. Woodbridge Williams)

Here, then, is a crude method of indicating the intensity of molecular motion of a body. Unfortunately, the hand is not a reliable indicator and is easily confused. For example, place one hand in a bowl of hot water and the other in a bowl of cold water. After a little while, place both hands in a bowl of lukewarm water. To the hand that was in the hot water, the lukewarm water feels cold. To the hand that was in the cold water, the lukewarm water feels hot.

A more reliable device than the hand is the thermometer, invented by the Italian physicist Galileo Galilei about 1592. His thermometer was improved in 1706 by Daniel Fahrenheit, a German scientist, who invented the mercury thermometer.

Fahrenheit blew a small bulb at the end of a fine glass tube and poured in a small amount of mercury. He then pumped all the air out of the tube and sealed it. When the mercury was heated, it expanded and rose in the tube. The more the mercury was heated, the more it expanded and the higher it rose. When the mercury was cooled, it contracted and dropped toward the bulb. The more the mercury was cooled, the more it contracted and the lower it dropped.

Fahrenheit noted that when his thermometer was placed in ice that had been heated to the point where it began to change its state to water (the *melting point*), the mercury always dropped to a certain point in the tube. He marked this point 32. When he placed the thermometer in water that had been heated to the point where it began to change its state to steam (the *boiling point*), the mercury rose to a certain height in the tube. He marked this point 212.

The tube between these two points was divided into 180 equal divisions, called *Fahrenheit degrees* (°F). He now had a thermometer that could measure the amount of heat of a substance between the melting point of ice and the boiling point of water. By carrying these divisions beyond these two points, a greater range of intensity can be measured.

In 1742, Anders Celsius, a Swedish scientist, designed a mercury thermometer on which the melting point of ice was designated as 0 degrees and the boiling point of water was 100 degrees. The intervening space was divided into 100 equal divisions, called *Celsius degrees* (°C). This scale is used in most scientific work today and is the common metric temperature scale. Figure 9.3 shows a mercury thermometer with both Fahrenheit and Celsius scales.

On occasion, it becomes necessary to change from one scale to another. We know that there are 100 degrees on the Celsius scale between the freezing and boiling points of water, and 180 degrees on the Fahrenheit scale between the same two points.

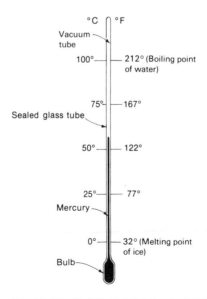

FIGURE 9.3   The Mercury Thermometer: Celsius and Fahrenheit Scales.

Therefore, 100 Celsius degrees equal 180 Fahrenheit degrees, and the ratio of the scales to each other is:

$$\frac{T_C}{T_F} = \frac{100}{180}$$

Thus

$$T_C = \frac{100}{180} \times T_F = \frac{5}{9} T_F$$

where T_C is the Celsius temperature and T_F is the Fahrenheit temperature.

This ratio means that the Celsius scale has only 5/9 as many degrees as the Fahrenheit scale between any two given temperatures. However, we cannot stop here, because if we take (5/9) of 212 degrees Fahrenheit, which is the boiling point of water, we get 118 degrees, instead of the correct value of 100 degrees Celsius. What did we do wrong? We did not account for the fact that the freezing point on the Fahrenheit scale is 32 degrees, not 0 degrees. Therefore, if we subtract 32 from 212 and then multiply by 5/9, we get the correct answer. The number 32 has to be subtracted from every Fahrenheit value we wish to convert to the Celsius scale. On the other hand, when we wish to convert from the Celsius scale to the Fahrenheit scale, we must first take (9/5) of the Celsius temperature and then add 32 degrees Fahrenheit. The final formulas become:

* TEMPERATURE CONVERSION FORMULAS

$$T_C = \frac{5}{9} (T_F - 32°F) \quad \text{and} \quad T_F = \frac{9}{5} T_C + 32°F$$

SAMPLE PROBLEM 9.1

You are asked to help prepare some specifications for overseas customers as well as for those in the United States. All temperatures must be expressed in both Celsius and Fahrenheit units.

A. Convert the annealing temperature of carbon steel from 1650°F to the correct Celsius temperature.

B. Find the comparable Fahrenheit temperature for a carbide tool bit that will maintain a cutting edge at temperatures over 1200°C.

SOLUTION A

Wanted

$T_C = ?$

Given

$T_F = 1650°F$

Formula

We want to convert from T_F to T_C, so we use the formula:

$$T_C = \frac{5}{9}(T_F - 32°F)$$

Calculation

$$T_C = \frac{5}{9}(1650° - 32°F) = \frac{5}{9}(1618) = 899°C$$

SOLUTION B

Wanted

$T_F = ?$

Given

$T_C = 12\overline{0}0°C$

Formula

We want to convert from T_C to T_F, so we use the formula:

$$T_F = \frac{9}{5}T_C + 32°F$$

Calculation

$$T_F = \frac{9}{5}12\overline{0}0°C + 32°F = 2160 + 32 = 2190°F$$

With these calculated temperatures, the specifications for annealing carbon steel and for maintaining carbide tool bits can be listed in both the Celsius and the Fahrenheit temperature scales.

THE KELVIN (ABSOLUTE) TEMPERATURE SCALE

In both the Fahrenheit and the Celsius scales, temperatures below zero can be reached. In parts of Canada and Alaska, temperatures as low as -80 degrees Fahrenheit have been reported; and liquefied natural gas is stored in very large tanks at 260 degrees Fahrenheit *below* zero (-162 degrees Celsius). Scientists in the past wondered whether there was a limit to how low temperatures could go.

PHOTO 9.4 Hydrogen becomes a liquid at a temperature of −253 degrees Celsius or 20 kelvin. In this form, it is used as a rocket fuel. (NASA photo)

About the middle of the nineteenth century, the English scientist Lord Kelvin (1824–1907) investigated this matter. He placed gas in a sealed container, cooled the gas to 0 degrees Celsius (the temperature of melting ice), and observed the pressure produced by the gas molecules bombarding the sides of the container. He then cooled the gas 1 degree below 0 degrees. Having lost some of their internal energy, the molecules of the gas slowed up and the pressure dropped 1/273 of the original pressure. Cooling the gas further, he observed that the gas pressure was reduced by 1/273 of the initial pressure at 0 degrees for every degree he went below 0 degrees. From this observation, he reasoned that if he could cool the gas to a temperature 273 degrees below 0 degrees Celsius, **absolute zero** would be attained, at which point all molecular motion (and the pressure resulting from it) would cease. Scientists have since learned that molecules of a substance still have some kinetic energy (motion energy) at absolute zero.

---

* absolute zero   The lowest limit of temperature.

Using these facts, Kelvin designed the temperature scale that bears his name. Absolute zero is 0 kelvin (273 degrees below 0 degrees on the Celsius scale). The melting point of ice (0°C) is 273 kelvin (273 K). The boiling point of water (100°C) is 373 kelvin (373 K). Note that we do not say "373 degrees kelvin" but "373 kelvin." The conversion formula from one scale to the other is:

$$T_K = T_C + 273°C$$

SAMPLE PROBLEM 9.2	Change $-2\overline{0}°C$ to the kelvin temperature.
SOLUTION	
Wanted	$T_K = ?$
Given	$T_C = -20°C$
Formula	$T_K = T_C + 273°C$
Calculation	$T_K = -20°C + 273°C = 253 \text{ K}$

Why so many temperature scales? When we finally go metric, there will be just two scales, Celsius and Kelvin. The absolute temperature scale is needed for some formulas that we will discuss later. This situation is similar to the one we handled earlier concerning pressure. As you will recall, absolute pressure was needed in the pressure–volume formula.

Although the temperature of absolute zero has never been attained, scientists have produced temperatures that are only fractions of a degree away from it. Absolute zero is the lowest limit of temperature, but we know of no upper limit. It has been estimated that the internal temperature of the sun is about 20 000 000 kelvin, and the internal temperatures of some of the other stars are thought to be as high as 500 000 000 kelvin.

9.4 The Measurement of Heat

So far, we have considered the measurement of temperature—that is, the average intensity of motion of the molecules. But temperature is not a measure of the total internal energy possessed by a body, which, as we now know, depends on the mass and velocity of the molecules and the number and kind of molecules present in the body.

Since internal energy is the total potential and kinetic energy of the atoms and molecules of a substance, the mechanical energy units foot·pound and joule could be used. In fact, the joule (J) has recently been adopted as the unit for the quantity of heat in the SI system. However, as a technician, you will encounter heat units still being used that were developed before it was realized that heat could be converted into mechanical energy, and vice versa. These units are the **calorie** (cal) and the **British thermal unit** (Btu). When speaking, we rarely say "British thermal unit"—we usually say "bee-tee-yew."

PHOTO 9.5 It takes a great deal of heat energy to raise the temperature of steel to its melting point—about 400 Btu per pound of steel. This basic oxygen furnace produces tons of steel each day. (Courtesy of Bethlehem Steel Corp.)

* **calorie**	The amount of heat needed to raise the mass of 1 gram of water 1 degree Celsius from 14.5 degrees Celsius to 15.5 degrees Celsius.
* **British thermal unit**	The amount of heat needed to raise the mass of 1 pound of water 1 degree Fahrenheit from 63 degrees Fahrenheit to 64 degrees Fahrenheit.

Note that we have defined the calorie as the amount of heat needed to raise the temperature of 1 gram of *water* 1 degree Celsius. Not all substances require the same amount of heat to raise their temperatures 1 degree. For example, to raise the temperature of 1 gram of mineral oil 1 degree Celsius requires only about one-half of a calorie. An equal temperature rise for an equal mass of aluminum requires only about one-quarter of a calorie; for glass, about one-fifth of a calorie; for iron, about one-eighth of a calorie. Thus, for a given mass and a given temperature rise, water requires twice as much heat as mineral oil; four times as much as aluminum; five times as much as glass; and eight times as much as iron. Similarly, if the water is cooled, it will give off twice as much heat as the mineral oil; four times as much as aluminum; five times as much as glass; and eight times as much as iron.

If we consider the number of calories required to raise the temperature of 1 gram of a substance 1 degree Celsius, we obtain the **specific heat capacity** of the substance. Thus, the specific heat capacity of these five substances is:

Water   1
Mineral oil   0.5
Aluminum   0.22
Glass   0.20
Iron   0.11

* **specific heat capacity**	The amount of heat that must be added (or subtracted) from 1 gram of a substance to raise (or lower) the temperature 1 degree Celsius. It is also the heat added or subtracted from the mass of 1 pound of a substance to change the temperature 1 degree Fahrenheit.

The units of specific heat capacity are calorie per gram per degree Celsius (cal/g·°C). Since it has already been stated that 1 calorie raises 1 gram of water 1 degree Celsius (1 cal/g·°C), the specific heat capacity of water is 1. Again, it takes just two-tenths of a calorie to raise 1 gram of glass 1 degree Celsius (0.2 cal/g·°C), so the specific heat capacity of glass is 0.2. The specific heat capacities of other substances are listed in Table 9.1. Note that the units for specific heat capacity can be either calorie per gram per degree Celsius (cal/g·°C) or Btu per pound per degree Fahrenheit (Btu/lb·°F). Also, you will note that other heat properties are listed in Table 9.1. These properties will be described later.

Why do we need to know the specific heat capacity of a substance? Well, industry must know the amount of fuel required for certain heating operations, and in order to calculate the fuel needed, the amount of heat must be known. For instance, we

TABLE 9.1 Heat Properties of Various Substances

Substance	Specific Heat Capacity (cal/g·°C or Btu/lb·°F)	Melting Point (°C)	Heat of Fusion (L_f)		Boiling Point (°C)	Heat of Vaporization (L_v)	
			cal/g	Btu/lb		cal/g	Btu/lb
Solids							
Aluminum	0.22	66$\overline{0}$	94	170	2490		
Concrete	0.16	—	—	—	—		
Copper	0.092	1080	42	76	2590		
Ice	0.50	0.00	80	144			
Steel	0.11	—	—	—	—		
Wood	0.42	—	—	—	—		
Glass	0.20	—	—	—	—		
Liquids							
Methyl Alcohol	0.60	−97	16	29	65		
Water	1.00	0.00	80	144	100.00	540	972
Ethylene Glycol (Prestone, Zerex)	0.575						
#12 Freon (Liquid)	0.23	—	—	—	−29.4		
Gases*							
Air	0.24						
Steam	0.48						
#12 Freon (gas)	0.10						

*The specific heat capacity of gases can vary, depending on pressure and temperature. The specific heat capacity for steam is for atmospheric pressure and temperatures near 100°C, and the specific heat capacity for air is for atmospheric pressure (14.7 psia) and 70°F.

might want to know how much heat is needed to raise a certain quantity of steel to its annealing temperature. Specific heat capacity is helpful in determining the heat required.

Let us summarize the relationships among heat, specific heat capacity, and temperature. Heat is the amount of internal energy flowing into or out of a substance. Internal energy depends on the kind of molecules in the substance, the number of molecules, and their velocities. The symbol for heat is Q, and the units of heat are the calorie (cal) or the British thermal unit (Btu). The kind of molecule and its relation to temperature change is represented by the specific heat capacity. Its symbol is c, and its units are calories per gram per degree Celsius (cal/g·°C) or Btu per pound per degree Fahrenheit (Btu/lb·°F). The number of molecules is measured by mass, with symbol m, or by weight, whose symbol is w. The units for mass must be in grams (g), and the units for weight must be in pounds (lb). The change in the average velocities of the molecules, due to heat being added or subtracted, is indicated by a change in temperature. The symbol for a *change* in temperature is ΔT, and the units of temperature are degrees Celsius (°C) or degrees Fahrenheit (°F).

There is a direct relationship among these quantities. When using calories, degrees Celsius, and mass, the formula is:

*** HEAT–
TEMPERATURE
FORMULA 1**

heat = mass × specific heat capacity
     × temperature change

$$Q = m \times c \times \Delta T$$

where Q = heat in units of calories (cal)
        m = mass and the units must be in grams (g)
        c = specific heat capacity and the units are in cal/ g·°C
        ΔT = the temperature change in °C

When using Btu, weight, and degrees Fahrenheit, the formula is:

*** HEAT–
TEMPERATURE
FORMULA 2**

heat = weight × specific heat capacity
     × temperature change

$$Q = w \times c \times \Delta T$$

where Q = the heat in units of Btu
w = the weight in units of pounds (lb)
c = the specific heat capacity in units of Btu/lb·°F
ΔT = the change in temperature in °F

Before proceeding with the next sample problem, see Table 9.2, which presents a list of conversions to and from Btu, calories, and joules.

TABLE 9.2 Conversions to and from Btu, Calories, and Joules

To Convert from	To	Multiply by
Btu	joule	1055
Btu	calorie	252
calorie	joule	4.19
calorie	Btu	0.00397

SAMPLE PROBLEM 9.3

How much heat is needed to raise the temperature from $7\overline{0}$°F to $98\overline{0}$°F for a particular heat-treating operation for a piece of steel weighing $15\overline{0}$ lb? Give the answer in Btu and joules.

SOLUTION

Wanted Q = ?

Given   We are given $w = 15\overline{0}$ lb, and we can calculate $\Delta T = T_2 - T_1 = 98\overline{0} - 7\overline{0} = 910$°F. From Table 9.1, we find that the specific heat capacity (c) for steel = 0.11 Btu/lb·F.

Formula   We write out the formula and solve for Q:

$$Q = w \times c \times \Delta T$$

Calculation

$$Q = 150 \text{ lb} \times 0.11 \frac{\text{Btu}}{\text{lb·°F}} \times 910\text{°F} = 15\ 000 \text{ Btu}$$

To convert from Btu to joules, we multiply by 1055:

15 000 Btu × 1055 = 15 800 000 J, or 16 MJ (megajoule)

After determining the heat needed to raise the temperature of a substance, the time required to supply the heat may be found if we know the amount of heat supplied each minute. The amount of heat supplied per minute is called the heat rate. The formula is:

---

* HEAT–TIME FORMULA

time = heat required ÷ heat rate

$$t = \frac{Q}{H_r}$$

where H_r = the heat rate in units of cal/min or Btu/min. Note that time could also be in s or hr if appropriate to the problem.

t = the time in units determined by the heat rate.

SAMPLE PROBLEM 9.4

It takes 85 700 cal to melt a certain amount of solder. If the heating element supplies 15 $\bar{0}$00 cal/min, what time is required to melt the solder?

SOLUTION

Wanted

t = ?

Given

Q = 85 700 cal and H_r = 15 $\bar{0}$00 cal/min

Formula

$$t = \frac{Q}{H_r}$$

Calculation

$$t = \frac{85\ 700\ \text{cal}}{15\ \bar{0}00\ \text{cal/min}} = 5.71\ \text{min}$$

We can also calculate the relationship between chemical and heat energies. It has been found that burning a pound of coal will produce between 10 000 and 15 000 Btu, and if 1 pound of gasoline is burned in air, about 20 000 Btu will be produced. Fuel oil can supply about 19 000 Btu for each pound burned. Thus, different substances have a different amount of energy available for each pound of substance burned. This amount is called the heat of combustion, measured in units of British thermal unit per pound (Btu/lb). The relationship between the amount of fuel needed to produce heat and the amount of heat required is:

* **AMOUNT OF FUEL FORMULA**

amount of fuel = heat required ÷ heat of combustion

$$f = \frac{Q}{H_c}$$

where f = the amount of fuel in lb
Q = the heat needed in Btu
H_c = the heat of combustion in Btu/lb

SAMPLE PROBLEM 9.5

Sample Problem 9.3 indicated that it takes 15 000 Btu to heat a given amount of steel. How much fuel oil is needed?

SOLUTION

Wanted

$f = ?$

Given

We are given Q = 15 000 Btu, and we know H_c = 19 000 Btu/lb.

Formula

$$f = \frac{Q}{H_c}$$

Calculation

$$f = \frac{15\ 000\ \text{Btu}}{19\ 000\ \text{Btu/lb}} = 0.79\ \text{lb}$$

9.5 The Effects of Heat

If we continue to add heat to a substance—paper, for example—we know that a point is reached where it bursts into flames at about 450 degrees Fahrenheit (230 degrees Celsius), depending on the kind of paper. This point, known as the *kindling point,* is the point at which the paper combines chemically with the oxygen in the air, forming new substances.

There are a number of other changes produced by heat. Some of these are changes in the electrical properties of substances, or changes in the color of incandescent bodies. Some of these effects will be discussed later in this book in the appropriate sections. Here, however, we will confine ourselves to the physical changes produced by heat: changes of state and changes due to expansion and contraction.

CHANGES OF STATE

Perhaps the most striking effects of heat are the changes of state that it can produce in a substance. By this we mean that the application of heat can change a substance from a solid to a liquid and, finally, to a gas—the three common states of matter.

Consider water, for example. If its heat energy is low—that is, if the molecules move comparatively slowly—the attraction between neighboring molecules causes them to form the rigid crystalline structure of ice, a solid. As heat is added, increasing the heat energy of the ice, the molecules move faster, until their velocities are great enough to allow them to escape from the rigid structure and form water, a liquid. In the liquid, the molecules move in swarms because their mutual attractions—although not great enough to hold them in a rigid pattern—are still strong enough to hold them together. As more heat is added, the molecules move faster and faster, until their velocities are great enough for them to overcome their mutual attractions and fly off in different directions. The liquid then changes to steam, a gas.

So when the solid (ice) changes to the liquid (water), heat must be added. On the other hand, when water changes to ice, heat must be taken away. Similarly, heat must be added to change water to steam, and it must be taken away to change steam into water.

Let us examine the heat required to produce changes in a common substance such as water. Figure 9.4 is a graphical display of the information in the following paragraphs.

PHOTO 9.6   Icebergs floating in water illustrate that the solid and liquid forms of the same substance can exist at the same temperature. Heat must be supplied or withdrawn to convert one form to the other. (Official U.S. Coast Guard photo)

Suppose we start with 1 gram of water in its solid state, ice, that has been cooled to a temperature of 10 degrees below zero on the Celsius scale. Since ice has a specific heat of about 0.5— that is, 0.5 calorie is required to raise 1 gram of ice 1 degree Celsius—we need 5 calories of heat to raise the gram of ice the 10 degrees from − 10 degrees to 0 degrees.

At 0 degrees Celsius, we reach a crucial temperature for ice. Any added heat will change the ice to water, a liquid. We call this temperature the *melting point* of ice. The melting point is not the same for all substances. Ice melts at 0 degrees Celsius; lead melts at 327 degrees; aluminum melts at 660 degrees; iron melts at 1530 degrees.

However, before the ice changes completely into water, the molecules must be released from their rigid structure so that they may take on the freer motion found in liquids. Breaking the rigid structure requires energy. Heat is therefore used *not* to raise the temperature, but to free the molecules. We call the amount

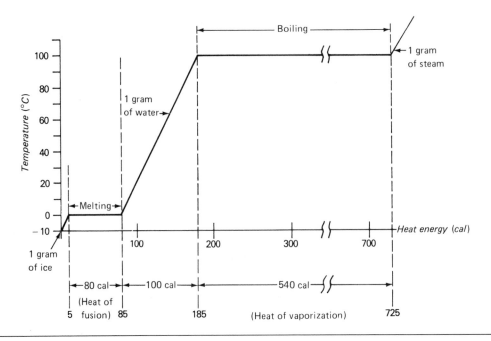

FIGURE 9.4   Graph of Temperature versus Heat Energy for 1 Gram of Water.

of heat required to free the molecules of 1 gram of a substance the **heat of fusion.** The units of heat of fusion are calorie per gram (cal/g) or British thermal unit per pound (Btu/lb); the symbol for it is L_f.

---

* heat of fusion

The amount of heat required to convert 1 gram (or 1 pound) of a substance from a solid to a liquid (at its melting point). It is also the heat that must be removed to convert a substance from a liquid to a solid.

Not all substances have the same heat of fusion. The heat of fusion of 1 gram of ice has been shown to be 80 calories, which means that 80 calories of heat are required to change 1 gram of ice at 0 degrees Celsius to 1 gram of water at 0 degrees Celsius (refer to Figure 9.4). After that, 1 calorie of heat is required for every degree rise in temperature. To raise the temperature of 1 gram of water from 0 degrees to 100 degrees Celsius, 100 calories are required.

At 100 degrees Celsius, another crucial point of water is reached: the temperature at which liquid water changes to steam, a gas. We call this the *boiling point* of water. Not all liquids have the same boiling point. Thus, ether boils at 35 degrees Celsius; glycerine boils at 291 degrees; mercury boils at 357 degrees. Boiling points of some other substances are given in Table 9.1.

As is true with the change from a solid to a liquid, so do we find during the change from a liquid to a gas that heat energy is required to break the attraction of the molecules. Again, the water does not change temperature during this change of state. The heat required to free the molecules of 1 gram of a liquid is called the **heat of vaporization.** The units of heat of vaporization are calorie per gram (cal/g), or British thermal unit per pound (Btu/lb); the symbol for it is L_v.

---

* heat of vaporization

The amount of heat required to convert 1 gram (or 1 pound) of a substance from a liquid to a gas (at its boiling point). It is also the heat that must be removed to convert a substance from a gas to a liquid.

Again, not all liquids have the same heat of vaporization. For water, the heat of vaporization has been demonstrated to be 540 calories per gram. Thus, 540 calories of heat must be added to change 1 gram of liquid water at 100 degrees Celsius to 1 gram of steam at 100 degrees Celsius (as shown in Figure 9.4). After that, since steam has a specific heat of about 0.5, half a calorie of heat is required again for every 1 degree rise in temperature. Assuming that we wish to raise the temperature of 1 gram of steam to 110 degrees Celsius, 5 more calories of heat must be added. (We are assuming, of course, that the gas is in a closed container.)

Now suppose we start taking heat away from 1 gram of steam at 110 degrees Celsius. To reach the temperature of 100 degrees, the steam must lose 5 calories of heat. If we continue to remove heat at this point, the steam starts to change back to liquid water. We say that the steam condenses, and we call this temperature the *condensation point*. The condensation point of steam and the boiling point of water are the same temperature— that is, 100 degrees Celsius.

However, just as 1 gram of water at 100 degrees Celsius must take in 540 calories to change to steam at 100 degrees, so 1 gram of steam at 100 degrees must lose 540 calories to change back to 1 gram of water at 100 degrees. The motion of the molecules of the liquid is less free than that of the molecules of the gas, and thus heat energy is thrown off. After that, 1 calorie of heat must be lost for every fall of 1 degree Celsius of the water until 0 degrees is reached. To reach 0 degrees Celsius, 100 calories must be taken away.

As we continue to remove heat at 0 degrees Celsius, the water starts to change to ice. We say that the water freezes, and we call this temperature the *freezing point*. The freezing point of water and the melting point of ice are the same temperature— that is, 0 degrees Celsius.

Again, heat energy is lost when the molecules of the liquid assume less free motion in the solid. Since the gram of ice required 80 calories of heat to free the molecules, the gram of water must lose 80 calories to assume the rigid form of ice. The gram of water at 0 degrees Celsius must lose 80 calories to become a gram of ice at 0 degrees. After that, the gram of ice must lose one-half calorie for every drop of 1 degree Celsius.

We emphasize again that ice does not change temperature while melting, and water remains at a constant temperature while boiling. However, this is not true of all materials. Generally, mixtures of two or more substances, such as glass, paraffin, and most compositions of steel, melt through a range of temperatures.

The amount of heat change in a substance being converted from one state to another (such as from a solid to a liquid or from a liquid to a gas) is obtained by multiplying the mass or weight of the substance by the appropriate heat of fusion or heat of vaporization constant shown in Table 9.1. That is:

∗ CHANGE-OF-STATE FORMULAS

heat = mass (or weight) × heat of fusion or heat of vaporization

$$Q = m \times L_f \quad \text{or} \quad Q = m \times L_v$$
$$Q = w \times L_f \quad \text{or} \quad Q = w \times L_v$$

where L_f = the heat of fusion in cal/g or Btu/lb
       L_v = the heat of vaporization in cal/g or Btu/lb

As Table 9.1 shows, the symbols for the heat of fusion and heat of vaporization are L_f and L_v. The "L" is used because some scientists refer to these constants as the "Latent heat of fusion" and "Latent heat of vaporization." Latent means hidden, or not visible, and a temperature change cannot be observed during a change of state.

If the heat supplied raises a material's temperature as well as changes its state, then the heat–temperature formula mentioned earlier is simply added to the change-of-state formula. For example:

$$Q = m \times c \times \Delta T + m \times L_f$$
$$\text{or} \quad Q = w \times c \times \Delta T + w \times L_f$$

Sample Problem 9.6 illustrates this point.

SAMPLE PROBLEM 9.6	The foundry you work for is submitting a bid to cast some copper parts. To arrive at a price, the engineer in charge must know the cost of producing the part. One of the expenses is the fuel that will be used. The engineer asks you to determine how much heat in calories is required to raise the temperature of 2.00 kg of copper from 25.0°C to 1080°C, which is the melting temperature of copper, and completely melt the copper.
SOLUTION	
Wanted	$Q = ?$
Given	We are given $m = 2.00$ kg, which we must convert to grams, or 2.00 kg = 2000 g. We can calculate $\Delta T = T_2 - T_1 = 1080 - 25.0 = 1055°C$, and we see from Table 9.1 that $c = 0.092$ and $L_f = 42$ cal/g.
Formula	$Q = m \times c \times \Delta T + m \times L_f$
Calculation	$Q = 2000 \text{ g} \times 0.092 \text{ cal/g·°C} \times 1055°C + 2000 \text{ g} \times 42 \text{ cal/g}$ $= 194\,120 + 84\,000 = 278\,120 \text{ cal} = 278 \text{ kcal}$

Before leaving this section, we would like to make a few more comments concerning the actions of a substance changing state. Again, let us assume that we heat some solid to its melting point. As stated earlier, with the addition of heat, the molecules are speeded up. As a result, when the liquid state is reached, the molecules are freed from the rigid structure of the solid and thus travel farther apart. We may therefore expect a substance to expand when it changes from a solid to a liquid. This statement is true of most substances. However, a few substances—such as ice or type metal (an alloy of tin, antimony, and lead)—contract when they pass from the solid to the liquid state. These exceptions occur because of a peculiar molecular arrangement in the solid that occupies more space than the same molecules in the liquid state. Water, for example, occupies about nine-tenths the volume of the ice from which it was produced.

When a substance passes from the liquid to the gaseous state, the expansion is enormous. There are no exceptions to this rule. Water, for example, expands about 1700 times when changing to steam. By the same token, all substances contract when passing from the gaseous to the liquid state, and most liquids contract when they solidify. Water, and other substances that contract on melting, expand on freezing.

EFFECTS OF HEAT ON THE LENGTH OF A SOLID

Now let us consider the effects produced by heat other than those involving a change of state. As heat is added to most solids, the motion of the molecules is speeded up, and they travel farther apart; as a result, the volume is expanded. When heat is taken away, the molecules move more slowly and come closer together; the volume, therefore, contracts. Not all solids expand and contract at the same rate. Aluminum and brass, for example, expand and contract much more than iron or glass.

Although the *volumes* expand or contract with the addition or subtraction of heat, when we consider solids, it is the expansion or contraction along their *lengths* that is most noticeable, especially if the solid is shaped as an elongated bar or wire. When steel rails are laid down, space must be left between the joints, or else the linear expansion due to the heat of summer may cause these rails to buckle. A 30 foot section of rail expands about one-half inch when heated from 0 to 100 degrees Celsius. For the same reason, cracks are left between the slabs of concrete of a highway. Provision is also made to permit a steel bridge to expand in the summertime without buckling. One way of doing this is to allow one end of the bridge to move freely. This method is illustrated in Figure 9.5. A short beam bridge is rigidly pinned on the left end and is resting on a rocker at the right end. The rocker allows the right end of the beam to move, so a gap, or

PHOTO 9.7 All gases expand upon heating. The Old Faithful geyser erupts every 64 minutes, when the temperature and pressure of the water become high enough. (Courtesy of the U.S. National Park Service)

clearance, is necessary between the bridge and the connecting structure. This expansion may total several feet in length for long bridges.

The unequal expansion and contraction of different substances poses a number of problems. For example, in the ordinary incandescent lamp, it is necessary to have several metallic con-

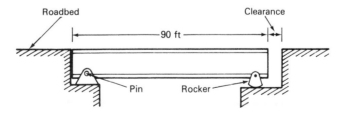

FIGURE 9.5 Clearance of a Beam Bridge to Allow for Expansion of the Steel in Hot Weather.

ductors of electricity sealed into the glass bulb. Since the flow of current heats these conductors, if the metal and glass have different rates of expansion and contraction, either the glass may crack or spaces may be left for air to seep in and destroy the filament. This problem was solved in early lamps by making the conductors of platinum, which has the same rate of expansion and contraction as the glass in which it is sealed. Today, instead of expensive platinum, we use wires of a nickel-and-steel alloy that also has an expansion rate similar to that of glass.

A device that makes good use of the unequal expansion and contraction of different metals is the thermostat. It is used to turn equipment on or off, depending on the temperature. Two substances that have unequal rates of expansion and contraction are fastened together to form a strip. These may be two pieces of different metals—for example, brass and steel as is shown in Figure 9.6. In this case, the steel is riveted to the brass. One end of the strip is fixed, and the other end is free to move up or down. As the strip is heated, the brass expands more than the steel. Since the two metals are fastened together, they tend to become curved rather than straight, as shown in the figure. The free end of the strip is forced up and the two upper contact points touch. As the strip is cooled, the brass contracts more than the steel. The free end of the strip bends down, separating the two upper contact points. As the strip is cooled further, it bends down still more until the two lower contact points are touching.

Now let us see how the thermostat is used to operate a device such as the oil burner that heats our home. As the temperature of the house drops, the strip bends down until the two bottom contact points touch. By means of a suitable electrical circuit (which you will study in the section on electricity), when these contact points touch, a relay operates that, in turn, closes a switch that starts the oil burner. The temperature at which the contact points touch may be regulated by the bottom adjustment screw that sets the contact points closer together or farther apart.

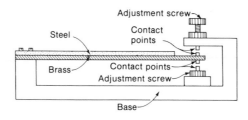

FIGURE 9.6   Construction of a Thermostat.

As the oil burner heats the house (and the thermostat), the strip starts to bend up. The two bottom contact points are separated, but the oil burner continues to raise the temperature until the two upper contact points touch. This operates a second relay, which turns the burner off. The upper adjustment screw regulates the temperature at which the burner is turned off. The oil burner remains off until the temperature drops sufficiently for the two bottom contact points to touch once again. Then the cycle is repeated. Generally, the motion of the thermostat strip opens or closes electrical circuits, as just described. However, it may also be used to operate mechanical valves or levers.

As technicians, you will be expected to make calculations relating to the expansion of solids. So, let's consider the change in length of a solid. Recall from Chapter 3 the discussion of deformation and strain on a steel rod when loaded. The deformation for each inch of length is called strain. Temperature has a similar effect on the dimensions of a metal rod, but no stress is developed if the rod is unrestrained. An increase in temperature increases the length of the rod. This change in length is identified by the symbol ΔL, just as in Chapter 3. However, the change in length for each inch of length cannot be called strain because that is reserved for loading applications and is not related to temperature. So, we call this change in length per inch of length, for each degree of temperature change, the **thermal coefficient of linear expansion.**

∗ thermal coefficient of linear expansion

The change in length of an object, per unit length (in./in. or mm/mm), for each degree of temperature change (°F or °C).

The symbol for the thermal coefficient of linear expansion is the lowercase Greek letter alpha (α). Handbooks, and Table 9.3 in this chapter, contain information on the thermal coefficients for different materials. In the U.S. customary system, we will use inches for length units and degrees Fahrenheit for temperature. In the SI metric system, millimeters and degrees Celsius are standard. The units for the thermal coefficient of linear expansion are inch per inch per degree Fahrenheit (in./in.·°F) or millimeter per millimeter per degree Celsius (mm/mm·°C). As you can see, the dimension units cancel and are not shown in Table 9.3

TABLE 9.3 Thermal Coefficients of Linear Expansion*

Solid	α	
	°F	°C
Aluminum alloy (average)	0.000 013	0.000 023
Brass	0.000 011	0.000 020
Bronze	0.000 010	0.000 018
Concrete	0.000 008	0.000 014 4
Copper	0.000 009 1	0.000 016
Glass, plate	0.000 005	0.000 009
Steel (average)	0.000 006 5	0.000 012

Note that since the Celsius degree is ⅖ of the Fahrenheit degree, the Celsius values are ⅖ of the Fahrenheit values.

The formula for the change in length is:

* LINEAR EXPANSION FORMULA

change in length = coefficient × original length × temperature change

$$\Delta L = \alpha \times L \times \Delta T$$

Sample Problem 9.7 illustrates the use of the thermal coefficient of linear expansion.

SAMPLE PROBLEM 9.7

A ski tow has a $30\overline{0}0$ ft steel cable. What is the difference in the cable length between a day with a $-10.0°F$ temperature and a day with a $+40.0°F$ temperature?

SOLUTION

Wanted

$\Delta L = ?$

Given

We are given $L = 30\overline{0}0$ ft, which we must convert to inches, or $L = 3000$ ft × 12.0 in./ft = 36 000 in. We can calculate $\Delta T = 40.0°F - (-10.0°F) = 50.0°F$, and we see from Table 9.3 that $\alpha = 0.000\ 006\ 5$.

Formula	$\Delta L = \alpha \times L \times \Delta T$
Calculation	$\Delta L = 0.000\ 006\ 5$ in./in.$\cdot°$F $\times\ 36\ 000$ in. $\times\ 50.0°$F $= 11.7$ in.

More so than solids, liquids expand on heating and contract on cooling. In the case of liquids, however, it is the change in volume that we must consider. Changes in length are meaningless, since liquids have no definite shape. As in the case of solids, not all liquids expand and contract at the same rate.

The volumes of gases exhibit the greatest rate of expansion and contraction as heat is added or taken away. A peculiarity of gases is that all expand or contract at the same rate. Recall that the volume of a gas at 0 degrees Celsius expands 1/273 of its volume for every degree it is heated and contracts 1/273 of its volume for every degree it is cooled.

EFFECTS OF HEAT AND PRESSURE

An increase in pressure raises the melting points of solids that expand on melting and lowers the melting points of solids, like ice, that contract on melting. We can see why if we recall that pressure tends to reduce the volume of a substance. If a substance tends to expand when it melts, the pressure will resist the expansion, and more heat must be applied before the solid melts— that is, the melting point is raised. On the other hand, since ice contracts on melting, the pressure will tend to favor this reduction in volume and the ice will melt at a lower temperature— that is, the melting point is lowered. You can see an example of this if you compress a snowball between your hands. The pressure may reduce the melting point low enough to turn the snow to water, even though the temperature of the air may be 0 degrees Celsius.

An increase in pressure raises the boiling points of liquids. The greater the pressure against the surface of a liquid, the greater the kinetic energy of the molecules has to be to enable them to escape from the liquid to form a gas. If the pressure is increased, more heat must be added to the liquid before the molecules can escape or boil off. On the other hand, if the pressure is reduced, less heat is needed to boil off the molecules—that is, the boiling point is reduced.

We have stated that the boiling point of water is 100 degrees Celsius, but only at sea level, where the mercury in the barometer indicates about 30 inches of air pressure. If we go to a higher altitude, the air pressure becomes less and the boiling point drops. It has been estimated that the boiling point of water drops about 1 degree Celsius for every 1000 feet of altitude.

A combination of heat and pressure also affects the volume of a substance. However, for most of our technical applications, solids and liquids are considered incompressible, and there is no pressure–temperature–volume relationship. The effects of heat and pressure on a gas, though, are related to its volume. Let us now discuss this relationship.

At the beginning of the nineteenth century, the French scientists Joseph Gay-Lussac (1778–1850) and Jacques Charles (1747–1823) investigated the behavior of gases as they are heated. They found that if a gas is kept under a constant pressure, it expands in equal proportions for equal increases in temperature. Thus, if the gas is unconfined, it expands on heating.

Let us consider a gas at two different temperatures and with a constant pressure. The temperatures must be measured on the Kelvin scale. The volume and temperature at the first condition have the symbols V_1 and T_1, and at condition two, V_2 and T_2. In the form of algebraic formulas:

* CHARLES' LAW FORMULAS

$$\frac{V_1}{V_2} = \frac{T_1}{T_2} \quad \text{or} \quad \frac{V_1}{T_1} = \frac{V_2}{T_2}$$

However, gases generally are confined in some container, where there is only one volume, not two. What happens to a confined gas when it is heated and cannot expand? The heated molecules are speeded up and strike the walls of the container with greater force. Consequently, the gas pressure goes up. If the volume is kept constant, heating a gas increases the pressure. Also, when a gas is compressed, the mechanical energy of compression is changed into heat energy, and the gas becomes hotter. When the molecules bombard the walls of the container, they are reflected back with varying speeds. If the gas is compressed, the pressure increases and the molecules strike the walls of the container with greater force. Thus, they are reflected with greater velocities, and the heat of the gas increases. On the other hand, if a gas is expanded, the molecules strike the walls with less force. They are reflected with smaller velocities, and the heat decreases.

We can now see that there are three factors to be considered when we deal with gases:

> If the *volume* is kept constant, then the greater the temperature, the greater the pressure.
>
> If the *pressure* is kept constant, then the greater the temperature, the greater the volume.
>
> If the *temperature* is kept constant, then the greater the pressure, the smaller the volume.

We can also see now why we had to qualify our statement of Boyle's Law back in Chapter 4. Recall that the volume of a gas varies inversely as the pressure applied to it if the temperature remains constant:

---

* BOYLE'S LAW FORMULAS

$$\frac{V_1}{V_2} = \frac{p_2}{p_1} \quad \text{or} \quad p_1 V_1 = p_2 V_2$$

By combining Boyle's and Charles' laws, we have a relationship for an ideal gas:

---

* IDEAL GAS LAW FORMULA

$$\frac{p_1 V_1}{T_1} = \frac{p_2 V_2}{T_2}$$

What is an ideal gas? Well, actually, an ideal gas is one that behaves the way the formula says it does. There is no true "ideal" gas, but for most everyday technical applications, most gases behave like an ideal gas. However, at extremely low temperatures, gases start to liquefy (an ideal gas does not), so the ideal gas law cannot be used at these temperatures.

Pressure and temperature must be measured at absolute values when using the ideal gas law. The temperatures must be in the Kelvin scale, and the pressures must be absolute, not gage. We must use absolute scales because there are no negative numbers in the absolute scales. For example, suppose we were using

Charles' Law to find an unknown volume and T_1 was $-30°F$. The negative sign would give us a negative volume, which is impossible. Thus, the formulas may only be used with absolute scales. Volumes may be in various units, including liters, cubic meters, and cubic inches. Also, pressures may be in various units; but to avoid confusion, we will use only the following units: volumes in cubic meters (m^3), absolute pressure in pascals (Pa), and temperatures on the Kelvin scale (K). Sample Problem 9.8 illustrates the use of the ideal gas law.

SAMPLE PROBLEM 9.8

A given mass of carbon dioxide fills 5.50 m^3 at a temperature of 70.0°C and a pressure of 61.0 MPa absolute. Find the volume if the temperature drops to $-10.0°C$ and the pressure drops to 7500 kPa absolute.

SOLUTION

Wanted

$V_2 = ?$

Given

$V_1 = 5.50$ m^3, $p_1 = 61.0$ MPa, $p_2 = 7500$ kPa, $T_1 = 70.0°C$, and $T_2 = -10.0°C$. We must change the pressures to pascals, or $p_1 = 61\ 000\ 000$ Pa and $p_2 = 7\ 500\ 000$ Pa. We must also calculate the absolute temperatures, or $T_1 = 70.0°C + 273°C = 343$ K and $T_2 = -10.0°C + 273°C = 263$ K.

Formula

We rearrange the formula:

$$\frac{p_1 V_1}{T_1} = \frac{p_2 V_2}{T_2} \quad \text{to} \quad V_2 = \frac{p_1 V_1 T_2}{T_1 p_2}$$

Calculation

$$V_2 = \frac{61\ 000\ 000\ \text{Pa} \times 5.50\ \text{m}^3 \times 263\ \text{K}}{343\ \text{K} \times 7\ 500\ 000\ \text{Pa}} = 34.3\ \text{m}^3$$

This information is useful when designing storage tanks, refrigeration systems, or pneumatic control systems.

Summary

Heat and temperature are two different, but related, measurements. Heat is the name we apply to the amount of internal energy that flows from one substance to another. The internal energy of a substance is the sum of the kinetic and potential energies of the molecules. In order to determine the amount of heat added to or removed from a substance, we must know the kind of substance and its mass (or weight), as well as the temperature change.

Heat can be transferred from one object, or substance, to another by means of conduction, convection, and radiation. All three methods are used by industry to perform various heating tasks. Heat is measured in units of calories (cal), British thermal units (Btu), or joules (J).

Temperature is not a measure of heat, but a measure of the average amount of motion of the molecules. The temperature scales are the Fahrenheit scale (°F), the Celsius scale (°C) and the absolute scale, called the Kelvin scale (K).

In order to relate temperature to heat, we must employ a measurement called the specific heat capacity. The specific heat capacity is the amount of heat required to raise the temperature of a specified amount of a given substance 1 degree. The units for specific heat capacity are calories per gram per degree Celsius (cal/g·°C) or Btu per pound per degree Fahrenheit (Btu/lb·°F).

If we know the rate at which heat is supplied or removed from a substance, we can calculate the time required for a particular industrial heating or cooling operation. We can also calculate the amount of fuel required.

Heat may cause a variety of changes in a substance. This chapter discusses change of state (solid, liquid, and gas) and expansion. When converting a solid to a liquid, or a liquid to a gas, heat is required to make this change of state. For the substances we discussed, no temperature change accompanies the change of state. Of course, when changing a gas to a liquid, or a liquid to a solid, heat must be removed from the substance. The heat required to change a solid to a liquid, at its melting temperature, is called the heat of fusion (L_f). The heat needed to change a liquid to a gas at its boiling temperature is called the heat of vaporization (L_v).

To aid in determining the linear expansion of a solid, handbooks list the thermal coefficient of linear expansion (α). This is the change in length per unit of length, per degree change in temperature. In the SI system, the units most commonly used are millimeters per millimeters per degree Celsius (mm/mm·°C). In the U.S. customary system, the units most commonly used are inch per inch per degree Fahrenheit (in./in.·°F).

Liquids and gases also tend to expand with the application of heat. The energy in a hot gas can be very useful in driving machinery (see Chapter 11 on heat engines). The effects of temperature, pressure, and volume must be combined when working with gases. These three variables, measured in absolute units, are related by the ideal gas law.

Key Terms

* internal energy
* heat
* temperature
* conduction
* convection
* radiation
* temperature conversion formulas
* absolute zero
* calorie (cal)
* British thermal unit (Btu)
* specific heat capacity
* heat–temperature formula 1
* heat–temperature formula 2
* heat–time formula
* amount of fuel formula
* heat of fusion
* heat of vaporization
* change-of-state formulas
* thermal coefficient of linear expansion
* linear expansion formula
* Charles' Law formulas
* Boyle's Law formulas
* ideal gas law formula

Questions

1. What is meant by the term "heat"? What is the difference in the internal energy and the motion of the molecules of a substance when it is said to be "hot" and when it is "cold"?

2. What term describes the measure of hotness and coldness of a substance?

3. Why might different objects be raised to different temperatures when the same amount of heat is applied to each?

4. Give three methods by which heat may be transferred from one substance to another.

5. Describe the Celsius thermometer and explain how it is used to measure the temperature of a substance.

6. How is the Kelvin temperature scale related to the Celsius scale?

7. Define absolute zero temperature.

8. Define the calorie; the British thermal unit; and the specific heat capacity of a substance.

9. Discuss the effect of heat upon the state of matter of a substance.

10. What is meant by melting point? By boiling point? By heat of fusion? By heat of vaporization?

11. Aside from changes in state, explain the effects of heat on a solid; on a liquid; on a gas.

12. What device takes advantage of the unequal expansion of solids?

13. Define the term "thermal coefficient of linear expansion."

14. What is the effect of pressure on the melting point of a solid? On the boiling point of a liquid?

15. The ideal gas law is the combination of what two laws?

Problems

TEMPERATURE MEASUREMENT

1. Convert the following Fahrenheit temperatures to Celsius: $32°$, $72°$, $10\bar{0}°$, $-2\bar{0}°$, and $-46\bar{0}°$.

2. Convert the following Celsius temperatures to Fahrenheit: $-10\bar{0}°$, $5\bar{0}°$, $78\ 1/2°$, $50\bar{0}°$.

3. At what temperature do both Fahrenheit and Celsius scales give the same reading?

4. Convert the Celsius temperatures in Problem 2 to the Kelvin scale.

5. Aluminum melts at $66\bar{0}°C$. Give the value in degrees Fahrenheit and kelvin.

MEASUREMENT OF HEAT

For those problems asking for heat, give the answer in calories, British thermal units, and joules.

6. How much heat is needed to raise 10.0 lb of steel $40\bar{0}°F$?

7. How much heat is needed to raise $70\bar{0}$ g of methyl alcohol 50.0 °C?

8. A 2.00 kg copper sleeve must be heated from 20.0°C to 115°C so that it will expand enough to slip over a shaft. How much heat is needed?

9. Solve Sample Problem 9.3 if the steel is raised from 70.0°F to 15$\overline{0}$0°F.

10. A quart of water weighs 2.1 lb. How much heat must the water absorb to raise its temperature from 70.0°F to 212°F?

11. A solid with a mass of 36 g requires 452 cal to raise its temperature 57°C. What is the substance?

12. Solve Sample Problem 9.4 if the amount of heat needed to melt the solder is 98 700 cal.

13. Solve Sample Problem 9.5 if it takes 125 000 Btu to heat a given amount of steel.

14. Your industrial plant has a water heater to make hot water for a cleaning process, and you are buying coal that will produce 11 000 Btu/lb. How many gallons of water can be raised in temperature from 7$\overline{0}$°F to 20$\overline{0}$°F with 1 lb of coal? (A gallon of water weights 8.34 lb.) We are assuming that all the coal's heat energy goes into heating the water—that is, there are no heat losses.

15. A particular industrial oven for drying paint can produce heat at the rate of 175 Btu each minute. We will assume that all the heat is absorbed by the part placed in the oven. The part weights 25 lb and is made of steel. If the initial temperature of the part is 7$\overline{0}$°F, how long can the part be left in the oven if its temperature is not to exceed 150°F?

16. In a heat-treating operation, a hot copper part is "quenched" in a tank of water. The temperature drops from 35$\overline{0}$°C to 2$\overline{0}$°C, and the part loses 7$\overline{0}$ 000 cal. What is its mass?

17. A truck engine cooling system holds 5.00 gal. If the system is completely filled with ethylene glycol, how much heat is required to raise its temperature from 70.0°F to 250°F? One gallon of ethylene glycol weighs 9.2 lb.

18. When burned, fuel oil produces 19 000 Btu/lb (called the heat of combustion). A certain industrial process requires 4$\overline{0}$0 gal of water per hour at 195°F. The initial temperature of the water is 55°F. How many pounds of fuel oil are used in an 8 hr day? One gallon of water weighs 8.34 lb.

19. A steel casting weighing 7$\overline{0}$0 lb is to be brought from 7$\overline{0}$°F to 1250°F for annealing. How much coal, with a heat of combustion of 12 000 Btu/lb, is required?

20. Using a heat source that can produce 5 cal/s, 350 g of water are to be heated from 20°C to 80°C. How long will it take to heat the water?

CHANGE OF STATE

21. Plot the graph of temperature versus heat energy for 1.00 g of aluminum from 20.0°C through the melting point.

22. Plot the graph of temperature versus heat energy for 1.0 g of methyl alcohol from a solid at −97°C to the boiling point.

23. A small block of ice has a mass of 535 g. How much heat is required to change the ice to water?

24. A particular car radiator holds 30.0 lb of water. If the water has already been brought up to a temperature of 212°F, how much heat is required to change all the water to steam?

25. A 1.00 lb kettle of steel holds 3.00 lb of water. Both are at 70.0°F.
    a. How much heat is required to bring the water and the kettle to 212°F?
    b. How much extra heat is required to boil off all the water?

26. A foundry has an electric furnace that can completely melt 1500 lb of copper. If the initial temperature of the copper is 65°F, how much heat is required?

27. An ice-making machine can make $3\bar{0}$ lb of ice an hour. If the incoming water is at $6\bar{0}$°F, and the ice is at 32°F, how much heat is removed from the water each hour?

28. A commercial building heats its offices with steam. On cold days, the boiler must supply $1\bar{0}$ lb of steam per hour at 275°F. The incoming water is $4\bar{0}$°F.
    a. How much heat is required to change the $1\bar{0}$ lb of water to $1\bar{0}$ lb of steam?
    b. If fuel oil with a heat of combustion of 19 000 Btu/lb is used, how many pounds of fuel oil are required per 24 hr day?

LINEAR EXPANSION OF SOLIDS

29. Solve Sample Problem 9.7 if the ski tow is 10 200 ft long.

30. When laying out hot-water or steam-pipe systems, the expansion must be considered for proper clearance in walls and floors. Support brackets must be designed to allow this expansion. A copper hot-water pipe has a 40.0 ft length (between bends).
    a. What is its change in length, in inches, from $7\bar{0}$°F to 180°F?
    b. Answer the same question for a steel pipe.

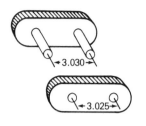

FIGURE 9.7 Sketch of Steel Link for Problem 33.

31. An aluminum rod 9.0 in. long (free to expand in only one direction) is to make contact with a mating part, and sound an alarm, when the temperature changes from $7\overline{0}°$F to 550°F. How big should the gap be?

32. A 100 ft steel surveyors' tape measures exactly 100 ft at 70.0°F. What length does the tape measure when the tape is 115°F?

33. Our problem is to assemble an industrial drive chain (similar to a bicycle chain). One side of the steel link shown in Figure 9.7 has the pins fixed in place 3.030 in. apart. The other side of the link has two holes spaced 3.025 in. apart. How much above $7\overline{0}°$F must the side with the holes be heated to fit over the pins?

34. How long must a bronze rod be if its length is to change 36 mm with a temperature change of $2\overline{0}0°$C?

35. A particular gasoline engine has a steel engine block and aluminum pistons. At the maximum operating temperature of 150°C, both piston and cylinder measure exactly 100.0 mm in diameter. What are their diameters at 20.0°C? Give the answer to three decimal places. (*Note:* You can treat a diameter the same way as a length.)

36. The side view of a solar collector is shown in Figure 9.8 (piping is omitted). The frame is brass and the collector has a glass cover. The glass is 96.000 in. long at exactly $-20°$F, and is butted against the brass seat as shown. How much clearance will there be at the operating temperature of exactly 260°F? Give the answer in inches to three decimal places. (*Note:* The length of brass to be used in the formula is also 96.000 in.)

37. A particular roller in a coal pulverizing operation is 50.000 in. in diameter. To prevent frequent replacement of the ex-

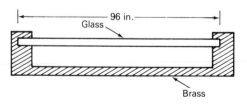

FIGURE 9.8 Sketch of Solar Collector for Problem 36.

pensive roller due to wear, a steel tire is shrunk fit in place over the roller. The inner diameter of the tire (when off the roller) is 49.800 in. at 70.0°F. To what temperature must the tire be raised so that it can be placed over the roller? (*Note:* Treat the diameter as you would a length.)

38. The steel beam in Figure 9.5 represents a highway overpass 90.0 ft long. The beam is fixed to a pin at the left end and on a rocker at the right end (to allow for expansion and contraction). The expected temperature range is from −20.0°F to 10$\overline{0}$°F. What should the clearance on the right end be at −20.0°F so that the clearance is zero at 100°F? (Assume the steel beam is the only object changing length.)

EFFECTS OF PRESSURE

39. Solve Sample Problem 9.8 if the initial temperature is 20.0°C.

40. What is the new volume when 6.5 m³ of air at 22°C is raised to 83°C at constant pressure. (*Hint:* Let p_1 and p_2 each equal 1 Pa [absolute].)

41. An automobile tire has an air pressure of 272 kPa (absolute) at 2$\overline{0}$°C. What is the pressure, assuming constant volume, when the temperature is raised to 5$\overline{0}$°C (caused by driving on a hot day)?

42. The container shown in Figure 9.9 is opened to let in air at 1 atmosphere of pressure (101.3 kPa). The volume of the container is 1.00 m³, and the atmospheric temperature is 20.0°C. A piston is placed into the top of the container, and its weight exerts a pressure of 175 kPa (absolute).
    a. What is the new volume, assuming the temperature remains constant?
    b. What is the new volume if the temperature increases to 10$\overline{0}$°C?

43. In Problem 42a, to what degree should the temperature of the air be raised in order to position the piston back at the top of the container?

44. An air compressor operates for a certain length of time, and we know it takes in 2.00 m³ of air at atmospheric pressure (101.3 kPa) at 20.0°C. The compressor discharges into a 0.200 m³ tank at a pressure of 1.40 MPa (absolute). What is the temperature of the discharged air?

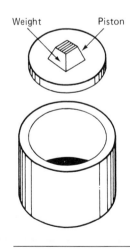

FIGURE 9.9 Sketch of Container for Problem 42.

Computer Program

This program will convert from the Fahrenheit to the Celsius scale, or vice versa. The program will also provide the answer in the Kelvin scale. You may use this program to check your answers to some of the chapter problems; what's more, you can make a thermometer similar to Figure 9.3. If you go 100 degrees higher and 100 degrees lower, you should discover the answer to Problem 3.

PROGRAM

```
10   REM   CH NINE PROGRAM
20   PRINT "PROGRAM FOR CONVERTING FROM THE FAHRENHEIT TO CELSIUS"
21   PRINT "SCALE AND VICE VERSA."
30   PRINT
40   PRINT "IF YOU WANT TO CONVERT TO CELSIUS-TYPE 1."
41   PRINT "IF YOU WANT TO CONVERT TO FAHRENHEIT-TYPE 2."
50   INPUT A
60   PRINT "YOU HAVE CHOSEN NUMBER--";A
70   IF A = 1 THEN   GOTO 100
80   IF A = 2 THEN   GOTO 200
90   END
100   PRINT
110   PRINT "TYPE IN FAHRENHEIT DEGREES."
120   INPUT F
130   PRINT "YOU HAVE TYPED   ";F;" DEG. F."
140   LET C = (F - 32) * 5 / 9
150   LET K = C + 273
160   PRINT
170   PRINT "TEMP. = ";C;" DEG. C - OR ";K;" K."
180   GOTO 90
200   PRINT
210   PRINT "TYPE IN CELSIUS DEGREES."
220   INPUT C
230   PRINT "YOU HAVE TYPED   ";C;" DEG. C."
240   LET F = (C * 9 / 5) + 32
250   LET K = C + 273
260   PRINT
270   PRINT "TEMP. = ";F;" DEG. F - OR ";K;" K."
280   GOTO 90
```

NOTES

Line 60 confirms your conversion selection.

Lines 130 and 230 confirm your temperature selection.

Lines 140, 150, 240, and 250 contain the formulas for conversion.

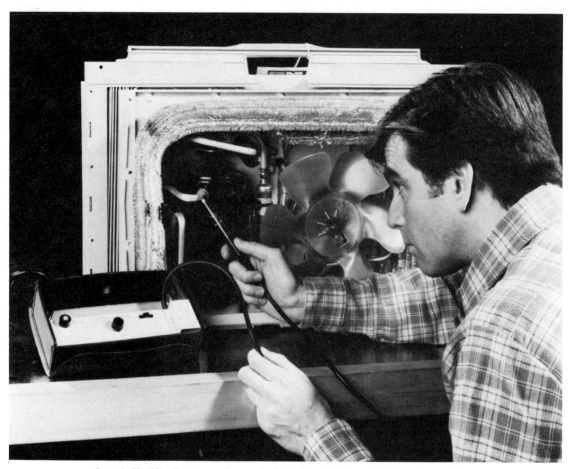

Control of building heating, cooling, and relative humidity is essential in modern society. Here, a technician services an air conditioning unit. (Courtesy of General Electric Co.)

Chapter 10

Objectives

When you finish this chapter, you will be able to:

☐ Diagram and compare hot air, hot water, radiant, and solar heating systems, and explain their sources of energy.

☐ Recognize the flow-rate formula $R_f = VA$: The rate of flow (R_f) equals velocity (V) times the cross-sectional area of a pipe (A).

☐ Diagram and discuss air conditioning systems and compare them with heating systems.

☐ Define the coefficient of heat transfer of a material as the amount of heat that can be conducted through the material under standardized conditions of temperature, area, and thickness. The symbol is "U."

☐ Recognize the heat transmission formula $H_{tr} = Q/hr = UA(\Delta T)$: The quantity of heat ($Q$) transmitted per hour equals the heat transfer coefficient (U) times the cross-sectional area of the flow (A) times the temperature difference on opposite sides of the material affording conduction (ΔT).

☐ Recognize the resistance to heat flow (R value) and understand that it is the reciprocal to the coefficient of heat transfer: $U = 1/R$.

☐ Calculate the overall coefficient of heat transfer if two or more materials are involved, using the formula: $U = 1/(R_1 + R_2 + R_3 \cdot \cdot \cdot)$.

☐ Discuss and compare purposes and types of insulation.

Building Heating and Cooling

The building industry is one of our largest industries, and with the national concern about energy, technicians are becoming more involved with the control of "climate" in buildings. To perform their tasks, technicians must know the various heating systems available; how to calculate fluid flow; what is meant by air conditioning; and the effectiveness of various types and thicknesses of insulation.

In this chapter, we will discuss hot air and hot water heating systems, furnaces, radiant and solar heating, the flow rate of fluids, and the fundamentals of insulation. We will also see how to calculate the flow rate of a fluid and the heat flow through a wall or ceiling.

10.1   Heating Systems

Most homes, commercial buildings, and, frequently, industrial buildings have *central heating,* in which a furnace is located in one part of the building and heat is piped to various other parts of the building. Central heating is in contrast to heating with fireplaces or stoves in every room. There are two major methods of piping heat throughout a building: the *hot air heating system* and the *hot water heating system*. (Steam is used for heating in many industrial and commercial buildings, but we will leave this matter for one of your future courses.)

HOT AIR HEATING SYSTEM

As Figure 10.1 shows, in a **hot air heating system,** the furnace is located in the lowest portion of the house, generally the basement. The fire in the furnace heats air, which is then forced up by a blower (not shown) in pipes, or *ducts,* located in the walls of the house. These ducts terminate in openings, or *registers,* located near the floors of the various rooms. The warm air circulates in the room, forcing the cooler, and heavier, air out through similar registers located near the floor. The cool air flows down other ducts back to the furnace where it is heated before rising once again.

The smoke and other gaseous products of combustion produced by the fire in the furnace enter the chimney. This brick or stone column with a hollow portion running through its center extends from the basement to well above the roof of the house. It ends in an opening through which the gases escape into the air.

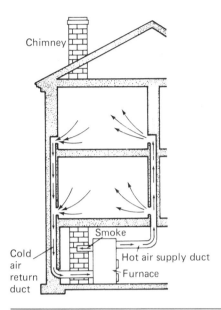

FIGURE 10.1   The Hot Air Heating System.

The hot air heating system has two main advantages over other central heating systems: It is usually less expensive to install than other types, and air conditioning equipment can be attached to the system easily. However, some people feel that the system creates drafts and distributes undesirable odors.

HOT WATER HEATING SYSTEM

In a **hot water heating system,** heated water is used to convey the heat from the furnace to the various rooms of the house. Obviously, it would not do to permit hot water to pour from registers. Instead, as Figure 10.2 shows, the hot water flows through *radiators* located at suitable places in the building. Note that the hot water heating system consists of a closed network of pipes. The hot water flows up one arm of this network because hot water is less dense than cold water and tends to rise. However, in more modern systems, the hot water usually is pumped through the system of pipes. The water flows from the supply pipe to the radiators, which are generally made of hollow metal pipes with heat-radiating fins. They are called *baseboard radiators* because they are placed on the floor next to a wall. In this position, the room air can flow over the pipes and fins. The hot water gives

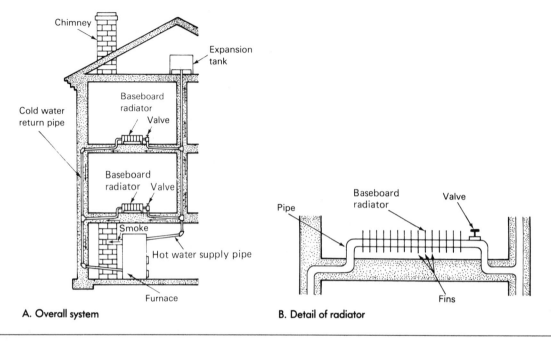

FIGURE 10.2 The Hot Water Heating System.

up its heat to the radiators and becomes cool in the process. The radiators, in turn, give up their heat to the room by means of convection currents in the air coming into contact with the hot metal surfaces and by radiation to surrounding objects. The fins serve to increase the surface area of the radiator pipe in order to facilitate heating (Figure 10.3). The cooled water flows down the other arm of the network and enters the furnace, where it is heated once again.

A valve at each radiator controls the amount of hot water entering that radiator. If the valve is completely closed, no hot water will enter that radiator, and as a result, it will remain cold. The entire system must be kept completely filled with water if circulation is to take place. Any air that gets in the system will form air pockets that prevent the water from circulating.

Note the *expansion tank* in Figure 10.2A. When water is heated, it expands. If there were no provision made to take care of this expansion, the pipes might burst. To avoid this possibility, the heating system is connected to the expansion tank, which is partially filled with air. The expanding water forces its way into the tank, compressing the air in the process. Thus, the danger of bursting pipes is avoided.

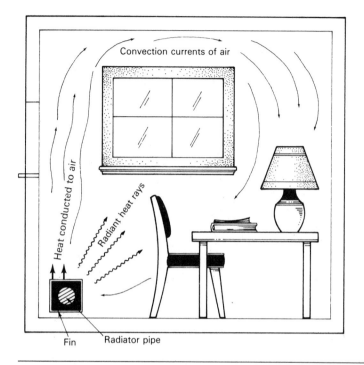

FIGURE 10.3   Heating via a Baseboard Radiator.

The hot water system may cost more than a hot air system, but many people feel it offers a more even heat (less drafts) and, therefore, a more comfortable heat.

RADIANT HEATING SYSTEM

In the hot air and hot water systems, most of the heating in a room is done by convection, either by warm air that is circulated directly in the room or by air that is warmed by contact with the hot radiators. There is another system, using radiant heat primarily, that is becoming increasingly popular.

In the **radiant heating system,** pipes bearing hot water or steam are embedded in the ceilings, floors, or walls of the rooms of the building. The hot pipes heat the ceilings, floors, or walls, and these, in turn, radiate their heat into the room.

There are a number of advantages claimed for this system. First, there are no unsightly radiators or registers. Then, because of the very large heating surfaces—that is, the ceilings, floors, or walls—heating is more uniform. Also, because there is very little convection, there are no drafts. Most important, perhaps, is that heating takes place only in bodies absorbing these radiations. Thus, a person can be warm even though the air, which absorbs little of these radiations, may be comfortably cool.

The furnaces that provide the hot water for the radiant heating system are similar to those previously described. In certain areas, where electricity is particularly cheap, electric heating wires are embedded in the floors, ceilings, or walls to provide the radiant heat, thus eliminating the need for a furnace.

FURNACES

There are a number of different furnaces that can be used with the various types of heating systems. The furnace employed in the hot air system, as shown in Figure 10.4, generally consists of a stove surrounded by a jacket in which the air circulates. The

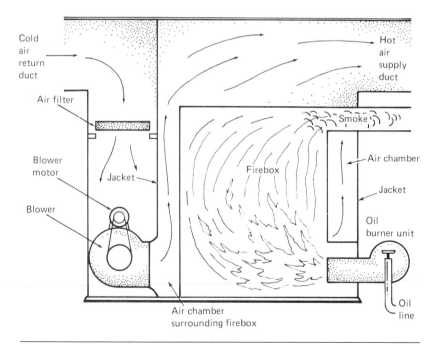

FIGURE 10.4 Oil-Burning Furnace Used with a Hot Air Heating System.

stove itself (called the *firebox*) consists of a metal box in which the fuel burns. The type illustrated is an oil-burning firebox. The burner unit contains a pump (run by an electric motor) that sprays a mixture of oil and air into the firebox, where the oil is ignited and burns fiercely. Cold air enters the bottom of the jacket through the return duct. This cold air is heated by contact with the hot walls of the firebox and then rises and passes up through the duct at the top of the jacket to registers in the various rooms.

In most hot air systems, a large blower, located near the furnace, blows the hot air through the ducts. The blower provides for more positive action than if the hot air were allowed to rise merely because of its lightness. Often, this circulating air is passed through filters to remove any dust that may be present and over pans of water to increase the amount of water vapor in the air. (This matter will be discussed a little later in the chapter.)

Another type of burner uses natural gas or propane to produce heat. Generally, both oil and gas burners are automatically operated by means of thermostats.

The hot water furnace resembles the hot air type, except that the jacket is filled with water instead of air. Such furnaces, too, are usually oil or gas fired and are automatic.

SOLAR HEATING SYSTEM

Solar heating is getting a lot of attention these days. In a **solar heating system,** the radiant energy in sunlight is used to heat air or water in solar collectors, and then the heated air or water is distributed throughout the building. A solar collector is simply an insulated box with a glass or plastic cover and a metal heat-absorbing plate inside. Sunlight passes through the transparent cover and heats the metal plate. Air, or water in pipes, is passed over the metal plate to absorb the heat and transport it to various locations in the building.

A typical system, using water, is shown in Figure 10.5. Water is pumped up to the solar collector, which is mounted on the roof. The water in the collector is heated (the temperature of the collector plate can rise to over 200°F), and the heated water then flows down to a large storage tank. The water in the radiator system is separate from the water in the solar collector system. The radiator water is pumped through a coil of pipe in the storage tank. This coil is called a *heat exchanger*. The heated radiator water then flows from the heat exchanger to the radiators.

PHOTO 10.1   This solar heating and cooling system for a school in Atlanta, Georgia, uses 576 panels to provide about 10 000 square feet of solar collector area. (Courtesy of Westinghouse Electric Corp.)

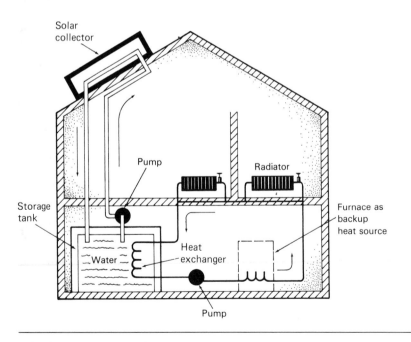

FIGURE 10.5   A Typical Solar Heating System.

10.2 Fluid Flow

In order to design heating systems, a technician needs more training than can be gained in an introductory physics course. As preparation for further courses, then, this text covers the basics needed for understanding fluid flow. We will simulate real conditions in our examples and problems, but keep in mind that a number of factors, such as pipe friction and pressure changes, are being ignored, so that we can concentrate on the basics.

In our first example—in Figure 10.6—water is flowing in a pipe at a velocity of 8 feet per minute. If we place a small floating object in the pipe at point A, the object will pass point B (8 feet away) 1 minute later. Similarly, all the water molecules passing A at any instant pass B 1 minute later. However, we want to know the *volume* of water flowing past a point in 1 minute, because liquid pumps are rated in gallons per minute (volume flow), while fans and blowers are generally rated in cubic

PHOTO 10.2   These pipes, varying in diameter from 10 inches to 36 inches, carry steam from a geothermal power plant in California. Fiber wool insulation is wrapped around the pipes to maintain the steam's 350 degrees Fahrenheit temperature. (Courtesy of Pacific Gas and Electric Co.)

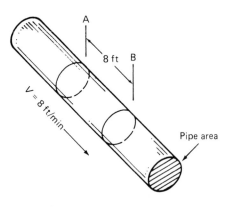

FIGURE 10.6   Illustration of Fluid Flow.

feet per minute (volume flow). If we are going to heat water in a furnace and pipe this water to a radiator, we want to know the volume of water flowing per minute so we can get some idea of the amount of heat being transported to a radiator and, in turn, to a room.

Now, let's go back to Figure 10.6 and consider the cylinder of water between points A and B. It may help to visualize this cylinder of water as a solid "chunk" of material with nothing in front or behind it. If we can determine the volume of the cylinder, we will have obtained the volume flowing past B each minute. In our discussion, R_f is used to represent the **flow rate** of a fluid. The units of flow rate generally found in handbooks are gallons per minute (GPM) or cubic feet per minute (CFM or ft³/min). Scientists use cubic feet per second and cubic meters per second, but we will not concern ourselves with these units.

* flow rate

The volume of fluid passing a given point in a specified time. It is the product of fluid velocity and the cross-sectional area of the pipe.

From geometry, we know that the volume of a cylinder is the product of its length and base area:

volume of cylinder = length × area of base

The base of a cylinder is a circle, and the area of a circle is:

$$A = \frac{\pi d^2}{4}$$

where d is the diameter of the circle. This basic formula for the volume of a cylinder is adjusted for flow rate by replacing length with velocity (the length for a 1 minute interval). We can now write:

---

* FLOW-RATE FORMULA

flow rate = velocity × area of pipe

$R_f = V \times A$

where R_f = flow rate in ft³/min
        V = velocity in ft/min
        A = area of pipe in ft²

To convert cubic feet per minute (CFM) to gallons per minute (GPM), multiply by 7.48. When working with a water flow system, we can convert cubic feet of water per minute to pounds of water per minute (lb/min) by multiplying by 62.4, since water weighs 62.4 pounds per cubic foot:

CFM × 7.48 = GPM
GPM × 0.1337 = CFM
CFM × 62.4 = lb/min   (water only)

We will not concern ourselves with fluids other than water, but the multiplying factors for other fluids can be determined from the table of densities given in Chapter 4, or they may be obtained from a handbook that gives densities in pounds per cubic foot. The following sample problems illustrate the procedures used to determine flow rate.

SAMPLE PROBLEM 10.1

Refer to Figure 10.6. Given the velocity of flow as 8.00 ft/min and the diameter of the pipe as 3.00 in., determine the flow rate in CFM and GPM.

SOLUTION

Wanted

$R_f = ?$

Given

$V = 8.00$ ft/min. Since the cross-sectional area (A) of the pipe must be in square feet, we must first change the diameter given in inches to feet:

$$d = 3.00 \text{ in.} \times \frac{1 \text{ ft}}{12 \text{ in.}} = 0.250 \text{ ft}$$

Next we determine area:

$$A = \frac{\pi d^2}{4} = \frac{3.14 \times (0.250 \text{ ft})^2}{4} = 0.0491 \text{ ft}^2$$

Formula

We can now use the formula for flow rate:

$$R_f = V \times A$$

Calculation

$R_f = 8.00$ ft/min $\times$ 0.0491 ft^2 = 0.393 ft^3/min
$R_f = \overline{0}.393$ CFM $\times$ 7.48 = 2.94 GPM

In the previous chapter, we saw how to determine the amount of heat flowing into or out of a substance. If a certain amount of heat is added by the furnace to a gallon of water, then this heat can be transported to the rooms in the home. The flow rate tells us how fast the heat can be transported in terms of Btu per minute or hour. Such information can also tell us how large a furnace (rated in Btu/hr) is required. This is demonstrated in Sample Problem 10.2.

SAMPLE PROBLEM 10.2

Let's say the fluid in Sample Problem 10.1 is water and the pipe leads to offices in an office building. The temperature of the water entering the offices is 185°F and coming out of the offices is 145°F. How much heat (H_r), in units of Btu/hr, is being delivered to the offices?

SOLUTION

Wanted

H_r = ?

Given

R_f = 0.393 CFM, T_1 = 185°F, and T_2 = 145°F. First, we must calculate the temperature difference: $\Delta T = T_1 - T_2$ = 185°F − 145°F = 40°F. Next, we determine the flow rate of the water in lb/min: 0.393 × 62.4 = 24.5 lb/min. We want the heat rate in Btu/hr, so we need the flow rate in lb/hr: R_f = 24.5 lb/min × 60 min = 1470 lb/hr.

Formula

We now use the heat formula from the previous chapter:

$$Q = w \times c \times \Delta T$$

Calculation

If we calculate the heat lost by 1470 lb of water, we will know the heat given up each hour:

$$Q = w \times c \times \Delta T = 1470 \text{ lb} \times 1 \text{ Btu/lb·°F} \times 40\text{°F}$$
$$= 58\ 800 \text{ Btu}$$

Therefore, 59 000 Btu/hr is being delivered to the offices. The furnace must be capable of supplying at least this amount of heat each hour.

10.3 Air Conditioning

There is more to making a house livable than merely raising the temperature of the air in the various rooms. For example, there is the necessity for cooling the air during some seasons and for keeping the amount of water vapor in the air at a proper level. **Air conditioning** is the process by which air can be cooled and by which the water vapor can be controlled.

We can easily tell that there is water vapor in the air by placing a saucer of water on the table. After an hour or so, some of the water will disappear: It evaporates into the air to form water vapor. Another test of water vapor in the air is to take a cold bottle from the refrigerator and permit it to stand in a warm room. The outside of the bottle becomes covered with water droplets. Where does this water come from? Obviously, not through the glass of the bottle. The drops of water are produced by condensation of the water vapor in the air as it comes in contact

PHOTO 10.3   Clouds are formed from water vapor condensing in the air. This satellite photo shows a hurricane east of Florida and Cuba, with winds of over 100 miles per hour. Of course, during a rainstorm, the relative humidity of the air is 100 percent. (Courtesy of NASA)

with the bottle's cold surface. Another example is clouds and mist, which are formed as water vapor condenses out of the air. If this condensation is heavy enough, drops of rain fall. If the temperature is cold enough, the water freezes and falls as snow or sleet. Dew and frost, often found on the ground early in the morning, are produced by condensation of the water vapor in the air.

The amount of water vapor in a given quantity of air is called the **humidity**. Generally, humidity is measured in grains of water vapor per cubic foot of air. (There are 437.5 grains to an ounce.) The total amount of water vapor a given quantity of air can absorb varies with the temperature of the air: The higher the temperature, the more water vapor the air can hold. Thus, at 0 degrees Fahrenheit, a cubic foot of air can absorb a maximum of about 0.5 grain. At 50 degrees Fahrenheit, it can hold up to 4 grains; at 70 degrees, 8 grains; and at 100 degrees, about 20 grains.

* **humidity**

The amount of water vapor present in the air.

Air at a given temperature does not always contain a certain amount of water vapor. It may have less than the maximum it can absorb. Thus, a cubic foot of air at 100 degrees Fahrenheit may have only 10 grains of water vapor. Under such conditions, we say that it has 50 percent of the water vapor it can absorb—that is, the **relative humidity** of the air is 50 percent.

* **relative humidity**

The percentage of water vapor actually present in the air compared to the maximum amount that could be present at a given temperature.

If the air contains less water vapor than the maximum amount it can hold at a given temperature—that is, if the relative humidity is less than 100 percent—it is able to absorb more. The lower the relative humidity of the air, the more quickly it can absorb water vapor. If the air contains the maximum amount of water vapor for a given temperature—that is, if the relative humidity is 100 percent—it can absorb no more, and any attempt to force more water vapor into the air will cause the excess to condense out as liquid water.

Now what has all this to do with the problem of making our homes livable? Simply this: The human body is a chemical machine that burns food as fuel. In this process, heat is produced. Some of this heat goes to keep the body at a constant temperature of 98.6 degrees Fahrenheit. However, more heat is produced than the body needs. The excess, therefore, must be gotten rid of.

Some of this excess heat is transmitted by convection and radiation to the air and surrounding objects. The bulk of it, however, is dissipated by the evaporation of perspiration that the sweat glands pour over the skin. Now, if the relative humidity of the surrounding air is too high, too little of this perspiration evaporates, and we feel hot and uncomfortable. On the other hand, if the relative humidity of the air is too low, evaporation takes place too rapidly. We may become chilled, and the skin and mucous membranes in our noses and throats may dry up.

It has been found that we are most comfortable when surrounded by air whose temperature is about 70 degrees Fahrenheit and whose relative humidity is about 50 percent. In order to be

PHOTO 10.4   This large heat pump is designed to operate the central air conditioning system of a commercial building. It has been operating under test at 0 degrees Fahrenheit for 24 hours a day for a year to test its durability. (Courtesy of Carrier Corp.)

comfortable, then, we must try to keep the air in the room constantly at these levels.

In winter, because the air is too cold for comfort, we raise its temperature by the various heating systems described earlier. However, as we raise the temperature of the air, we lower its relative humidity. The air becomes too dry, and the mucous membranes of our noses and throats tend to dry up. To be comfortable, we need more moisture in the air. One method for adding this moisture is to spray steam or water vapor directly into the warmed air before it is circulated in the various rooms of the house. Another method is to pass the warmed air through sprays or over pans of warm water and then blow it through the ducts leading into the rooms. In this way, the air picks up additional moisture. Still another method is to mix the warmed air with other air that has a high relative humidity. The mixture then has the proper amount of moisture, and this mixture is circulated.

In summer, it generally is the other way around. There is too much moisture in the air, because the hot summer air is able to absorb a good deal of moisture, especially in areas near large bodies of water. If we try to cool this hot air to a more comfortable level, we raise the relative humidity even higher, since the amount of water remains the same but the temperature is lower. The evaporation of our perspiration is slowed down in this air of high relative humidity, and as a result, we feel sticky and damp. You probably have heard people say, "It's not the heat, it's the humidity," when they feel uncomfortable in the summer. The problem, then, is to speed up the evaporation of perspiration so that our bodies feel cool and comfortable.

One way to increase comfort is to increase the circulation of the air in the room. If the air is stationary, the portion near our bodies absorbs as much evaporated perspiration as it can hold. Then it will absorb no more and evaporation stops. If we cause the air to circulate, we constantly change the air next to our bodies and thus prevent it from becoming completely saturated with evaporated perspiration. As a result, the evaporation continues, and we feel comfortable.

The fan has long been used to cause this circulation of air. Note that the fan does not cool the air. In fact, it actually increases the temperature of the air because of its mechanical energy, which is converted to heat. But we feel more cool and comfortable because the circulation of air speeds up the evaporation of perspiration. However, if the relative humidity of the air is high, even circulation will not help, because the fresh air coming in contact with our bodies already has nearly all the water vapor it can hold, and thus it will not absorb any appreciable amount of perspiration.

The best way to achieve comfort is to cool and remove some moisture from the air before it is circulated in the room. By passing the air over very cold surfaces or through sprays of cold water, the air becomes very cold, and its relative humidity becomes so high that some of its moisture condenses on the cold surfaces or in the sprays of cold water. The cold surfaces over which the air is passed may be cakes of ice or the coils of a mechanical refrigerator, as in an air conditioner. After losing some of its moisture through condensation, the air now has less water vapor. The air is then circulated in the room, where it mixes with the warm air. This circulation causes the cooler air to become warmer. Since the quantity of water vapor in the air hasn't changed, the relative humidity drops to a comfortable level.

There is still more to air conditioning than temperature and moisture control. The air contains dust and plant pollen that we wish to avoid inhaling. We may remove this dust and pollen in a number of different ways. One way is to force the air through filters before it enters the room. These filters consist of sheets of porous material, generally made from cellulose fibers. The clean air passes through the pores, but the dust and pollen are trapped and remain behind.

Another method is to "wash" the air by passing it through sprays of water. The water washes the dust and pollen out, and the clean air then is blown into the room. An alternate method for removing dust and pollen is to pass the air between electrically charged plates. The dust and pollen are attracted to the plates, but the air passes through.

Air conditioning is used not only in homes and offices; units also are installed in industrial plants to regulate the temperature and humidity conditions under which certain products, such as drugs, food products, and textiles, are manufactured. The manufacture of miniature ball bearings, precision instruments, and silicon chips requires dust-free atmospheres. Also, computer installations require air conditioning to prevent the computer from overheating.

We will not go into air flow problems because they are somewhat more complicated than liquid flow problems. Air, and of course other gases, are compressible, as we know, which affects the velocity and flow-rate relationship. We will leave air flow problems to more advanced courses.

10.4 Building Insulation

In industry, insulation of buildings, piping, and storage vessels is carefully determined. Too much insulation might cost more than the fuel that might be saved over the useful life of the plant or equipment, typically 20 years.

In most of our discussion, we will be talking about heat being conducted through the walls from a warm interior to a cold exterior. Of course, in the summer, the heat conduction is from outside to inside. In such a situation, insulation helps reduce the load on the air conditioner.

Can an object be heated to some temperature above its surroundings and then be stored in some container so that the object never loses its heat? The answer is no. The vacuum bottle is one of the most effective insulating devices made, but eventually, the temperature of the food in the bottle will become the same temperature as the surroundings. The reason is that heat can be

PHOTO 10.5 The thickness of insulation installed in a house ceiling must be determined by the type of insulation, the type of building structure, and the cost of insulation, versus the cost of heat saved. (Courtesy of Owens-Corning Fiberglas)

transmitted through solids, liquids, and gases by conduction and convection and through vacuums by radiation. Thus, the transfer of heat cannot be stopped. However, by proper choice of materials, the transfer of heat can be slowed down.

The problem of insulating buildings is similar to insulating food in a vacuum bottle. The heat inside a building is transmitted, or conducted, through the walls to the outside atmosphere and is lost to the building. Of course, in the summer, the reverse is true. The heating technician must be able to calculate the heat loss from a building (during the cold months) because the furnace must make up this loss. Different types of building construction and different kinds of insulating material affect this heat loss.

If we had a table telling us how much heat passes through a wall area of 1 square foot for each degree difference in temperature (between the inside and outside of the wall) and for various wall material and thicknesses, then all we would need to do is multiply this information by the total wall area, and the temperature difference, in order to find the total heat loss. Well, we do have such a table. Table 10.1 lists the overall **heat transfer**

TABLE 10.1 Thermal Resistance Values and Heat Transfer Coefficients for Various Walls and Building Materials

Type of Wall or Partition	Thermal Resistance (R)	Heat Transfer Coefficient (U)
Outside wall (wood siding, 2 × 4 stud, gypsum wallboard)	4.44	0.225
Solid brick wall (gypsum wallboard inside) 8 in. thick	3.91	0.256
Cinder block with brick face (gypsum wallboard inside)	6.33	0.158
Sloped roof, asphalt shingle (gypsum wallboard interior)	4.69	0.213
Solid wood door 2 in. thick	2.30	0.430
Concrete wall 8 in. thick	2.00	0.490
Concrete wall 8 in. thick with insulation (average)	6.70	0.150
Single pane window glass	0.91	1.100
Mineral fiber insulation from rock, slag, or glass:		
3 in. thick	11.00	0.091
6 in. thick	22.00	0.045
12 in. thick	44.00	0.022

Note: The thermal resistance (R) units are °F/Btu·hr·ft². The heat transfer coefficient (U) units are Btu/hr·ft²·°F. The values in the table are for the thicknesses given, not per inch thick.

Source: Table of data used with permission of American Society of Heating, Refrigerating, and Air Conditioning Engineers, Inc. (ASHRAE, Inc.).

coefficient for a number of walls and materials. The heat transfer coefficient (U) is in units of Btu per hour per square foot per degree Fahrenheit (Btu/hr·ft²·°F). The values listed in Table 10.1 apply to a specific thickness of a substance. Most building insulation handbooks do not have SI insulation data, so we will not bother with metric conversion here. Figure 10.7 shows the heat transfer through a 1 square foot area.

---

* heat transfer coefficient

The amount of heat that can be transmitted through a substance in 1 hour, for a 1 degree temperature difference, and for an area facing the heat flow of 1 square foot.

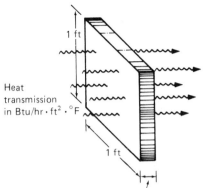

Heat
transmission
in Btu/hr · ft² · °F

1 ft

1 ft

Thickness, as specified in the handbook data tables

FIGURE 10.7   Illustration of Heat Transfer Coefficient.

The formula for the total heat transmitted each hour—that is, heat flowing through the walls each hour—is:

* HEAT TRANSMISSION FORMULA

total heat transmitted per hour = heat transfer coefficient
× area × temperature difference

$$H_{tr} = Q/hr = U \times A \times \Delta T$$

where H_{tr} = the total heat transmission per hour in Btu/hr

U = the heat transfer coefficient in Btu/hr·ft²·°F

A = the area in ft²

ΔT = the temperature difference in °F between each side of the material

We will now see how this formula is used to determine the heat transmitted through a window.

SAMPLE PROBLEM 10.3

A large glass window 10.0 ft high by 8.00 ft wide faces an outside temperature of 20.0°F and an inside temperature of 70.0°F. The architect wants to know how much heat will be lost per day (24 hours). This information will be combined with other data to determine the size of the heating system.

SOLUTION

Wanted

$Q = ?$

Given

$H = 10.0$ ft, $W = 8.00$ ft, $T_1 = 20.0°$F, and $T_2 = 70.0°$F. From Table 10.1, we find $U = 1.10$. We must calculate the area of the window, $A = 10.0$ ft $\times$ 8.00 ft $= 80$ ft^2, and the temperature difference, $\Delta T = 70.0° - 20.0° = 50.0°$F.

Formula

We will use the heat transmission formula:

$$H_{tr} = U \times A \times \Delta T$$

Calculation

$$H_{tr} = 1.10 \text{ Btu/hr·ft}^2·°F \times 80.0 \text{ ft}^2 \times 50.0°F = 4400 \text{ Btu/hr}$$

The heat lost in 24 hours is:

$$4400 \text{ Btu/hr} \times 24 \text{ hr} = 105\ 600 \text{ Btu}$$
$$\text{(round off to 106 000 Btu)}$$

Another important concept in heating calculations is the **thermal resistance** to heat flow. It is commonly referred to as the "R value." R values are listed in Table 10.1 in units of degrees Fahrenheit per Btu per hour per square foot (°F/Btu·hr·ft^2). The values listed in Table 10.1 apply to a specific thickness of a substance.

---

* thermal resistance

The resistance of a substance to heat flow in 1 hour, for a 1 degree temperature difference, and for an area facing the heat flow of 1 square foot.

The R value has several advantages. If you go into a store to buy insulation, the product information lists its R value. It is convenient to use the R value when comparing different insulating materials, because the greater the R value, the greater the resistance to heat flow. Also, on occasion, we may want to combine some of the wall and insulation materials that are listed in Table 10.1. Then we must obtain a combined U value to de-

PHOTO 10.6   Architects and builders must take into account a great many factors in building design and construction that affect heating systems, such as placement of windows, thickness and type of materials, and high-altitude effects of wind and temperature. Chicago's Sears Tower, shown here, is the tallest building in the world, at 1454 feet. (Courtesy of Chicago Convention and Tourism Bureau)

termine the heat transmission. A convenient method of doing so is to add up the R values of the material being used and then take the reciprocal to find the combined U value. We can do this calculation because the R value is inversely proportional to the U value. This method is shown here in algebraic symbols and is used in Sample Problem 10.4:

$$U = \frac{1}{R_1 + R_2 + R_3 \cdots}$$

SAMPLE PROBLEM 10.4

You work for a contractor who is building a small home with a total roof size of 24.0 ft by 40.0 ft. Your boss wants to determine whether it will be beneficial to put in a 6 in. layer of insulation under the roof. Using an outside temperature of 30.0°F and an inside temperature of 70.0°F:

A. Calculate the heat loss per hour through the roof without the insulation.

B. Calculate the heat loss per hour with the insulation.

SOLUTION A

Wanted

H_{tr} without insulation = ?

Given

$L = 24.0$ ft, $W = 40.0$ ft, $T_1 = 30.0$°F, and $T_2 = 70.0$°F. From Table 10.1, we find the U value for the roof, $U = 0.213$. We must calculate the area of the roof, $A = 24.0$ ft $\times$ 40.0 ft $= 960$ ft^2, and the temperature difference, $\Delta T = 70.0$°F $- 30.0$°F $= 40.0$°F.

Formula

$H_{tr} = U \times A \times \Delta T$

Calculation

Without insulation, the heat loss per hour is:

$$H_{tr} = U \times A \times \Delta T$$
$$= 0.213 \text{ Btu/hr·ft}^2\text{·°F} \times 960 \text{ ft}^2 \times 40.0\text{°F}$$
$$= 8180 \text{ Btu/hr}$$

SOLUTION B

Wanted

H_{tr} with insulation = ?

Given

From Table 10.1, we find R for the roof, $R = 4.69$, and for insulation, $R = 22$, in order to calculate total R:

$$R_t = 4.69 + 22 = 26.69$$

We now find the combined U value:

$$U = 1/R_t = \frac{1}{26.69} = 0.037$$

Formula

$$H_{tr} = U \times A \times \Delta T$$

Calculation

With insulation, the heat loss per hour is:

$$H_{tr} = 0.037 \text{ Btu/hr·ft}^2\text{·°F} \times 960 \text{ ft}^2 \times 40.0°F$$
$$= 1420 \text{ Btu/hr (round off to 1400 Btu/hr)}$$

With this information, the contractor can compare the cost of the heat being lost with and without the extra insulation. A decision can then be made on whether the savings in the fuel bill will pay for the extra insulation in a reasonable length of time.

Generally speaking, money can be saved by insulating an uninsulated house. A mistake some people make when doing their own insulating is to try and squeeze too much insulation into the available space. For example, if the framing for your walls is made of 2×4 studs, then a 3 in. thick batt of fiberglass insulation is appropriate. Trying to squeeze a 6 in. thick batt into the wall space will not provide the insulation specified for 6 inches.

Summary

There are two major types of central heating systems: hot air heating and hot water heating. The circulating hot air or hot water is heated by a furnace usually fueled by oil or gas. In solar heating systems, the sun's rays are used to heat the air or water. Blowers and fans force the hot air through the air ducts, and water pumps circulate the water through radiators and back to the furnace. The hot water system may use baseboard radiators or may have the heating pipes embedded in the entire ceiling, floor, or wall. This latter system is referred to as radiant heating

PHOTO 10.7   Temperature and humidity must be carefully regulated in places where magnetic tapes are stored for computer use. This automated library makes data available to a computer in less than a minute. (Courtesy of IBM)

because most of the heat is radiated to objects in the room and there is less heating of the air, thus reducing convection air currents.

Since heat is often transported throughout buildings by liquids, the technician must be able to calculate the flow rate (R_f) of liquids in pipes. Flow rates are in units of cubic feet per minute (CFM) or gallons per minute (GPM).

Air conditioning not only involves maintaining comfortable temperatures in hot weather, but includes other factors such as relative humidity. Humidity, the amount of water vapor in the air, can be controlled in a building. The most comfortable level is when the air contains about 50 percent of the maximum amount of water vapor it can absorb. For complete air conditioning, temperature and humidity must be controlled and dust and pollen must be removed from the air.

If we wish to have efficient operation of heating and air conditioning systems, the buildings must be properly designed and insulated. Insulation of buildings and equipment that operate above or below normal air temperatures is of prime importance in conserving energy. Heat is transmitted, or conducted, through various materials at different rates, depending on the thickness and area of the material, the type of material, and the temperature difference from one side of the material to the other. This transmitted heat is considered lost heat if it is leaving the building and an added load to the air conditioner if it is entering the building.

Key Terms

* hot air heating system
* hot water heating system
* radiant heating system
* solar heating system
* flow rate
* flow-rate formula

* air conditioning
* humidity
* relative humidity
* heat transfer coefficient
* heat transmission formula
* thermal resistance

Questions

1. Describe and explain the hot air heating system.
2. Describe and explain the hot water heating system.
3. Describe the hot air furnace.
4. What are the advantages of a radiant heating system?
5. What is the difference between flow rate and velocity of a fluid?
6. What is meant by humidity? By relative humidity?
7. Explain what is meant by the saying, "It is not the heat that bothers us, but the humidity."
8. Give two methods for reducing the relative humidity of air.
9. What does the term "air conditioning" mean?
10. Can the flow of heat be stopped?
11. Refer to Table 10.1. Does a high value for U indicate a high rate of heat transfer or a low rate? What do high or low R values indicate?
12. What factors determine the total amount of heat conducted through a wall?
13. How is the R value related to the heat transfer U?

Problems

FLUID FLOW

1. Find the flow rate in CFM and GPM for a sump pump with a discharge hose area of 0.00850 ft² (1-1/4 in. diameter) and a water velocity of 943 ft/min.

2. Refer to Problem 1. In an emergency, the motor on the pump was replaced with a smaller one, and the water velocity was reduced to 586 ft/min. Find the flow rate in CFM and GPM.

3. Solve Sample Problem 10.1 if the pipe diameter is 1.5 in.

4. An irrigation pump is rated at 34 GPM. What is the velocity of water in (a) a 5.0 in. diameter pipe? (b) a 2.5 in. diameter pipe?

5. What size pipe is required for a boiler feed pump rated at 22 GPM if we want a velocity of $1\overline{0}$ ft/min? Give the diameter in inches.

6. A solar heating installation has been designed to have the water coming from the collector and going to the storage tank at a temperature of 205°F. The water from the tank flowing to the collector is at 145°F. The connecting pipes are 1 in. in diameter, and the velocity of flow is $4\overline{0}$ ft/min. Determine the flow rate in CFM and GPM.

7. Determine the quantity of heat delivered in one hour to the storage tank in Problem 6. Give the answer in Btu.

BUILDING INSULATION

8. A typical window has a glass area of 15.0 ft². The temperature difference between the inside and outside of the window is 45.0°F. What is the heat loss in Btu/hr?

9. A concrete wall, 8 in. thick, has an area of 165 ft². The temperature difference between the inside and outside of the wall is 38.0°F. What is the heat loss in Btu/hr?

10. Find the heat loss (in Btu/hr) through an 8.00 in. solid brick wall that is 8.00 ft high and 30.0 ft long. The outside temperature is 30.0°F, and the inside temperature is 70.0°F.

11. A wood-siding building 25.0 ft × 55.0 ft × 10.0 ft high has a standard roof as specified in Table 10.1. What furnace capacity is required to supply the heat loss while maintaining $7\overline{0}$°F inside when the outside temperature is $-1\overline{0}$°F?

Give the answer in Btu/hr. Assume heat loss through the floor is negligible, and ignore heat loss through the windows and doors.

12. What is the total thermal resistance of a solid wood door with 3 in. of glass wool added?

13. A wood-siding outside wall has 3 in. of glass wool insulation. What is the combined heat transfer coefficient (U)?

14. Solve Problem 11 with 6 in. of glass wool insulation in the roof and 3 in. of glass wool in the walls.

15. A manufacturer of water heaters wants to reduce his heat loss to 8.1 Btu per square foot of heater surface per hour. If the metal wall of the heater reaches 160°F and room temperature is $7\overline{0}$°F, what thickness of glass wool insulation is required?

16. A commercial structure has poured concrete walls and ceiling (8 in. thick). The dimensions are $7\overline{0}$ ft × 45 ft × 15 ft high. It has three large glass windows, each 15 ft wide by 10 ft high.

    a. What is the heat loss per hour when the inside temperature is $7\overline{0}$°F and the outside temperature is 0.0°F? (*Hint:* Place the windows in one wall. Solve for each of the four walls separately and the ceiling separately.)

    b. What air conditioning capacity (in Btu/hr) is required if the inside temperature is maintained at 65°F and the outside temperature is $1\overline{0}0$°F?

Computer Program

This program will determine the heat loss (transmission) per hour from the formula $H_{tr} = U \times A \times \Delta T$. Use this program to check your answers to some of the chapter problems.

PROGRAM

```
10   REM   CH TEN PROGRAM
20   PRINT "THIS PROGRAM WILL CALCULATE THE HEAT LOSS"
21   PRINT "(OR HEAT TRANSMISSION) IN BTU/HR."
25   PRINT
30   PRINT "IF AREA IS GIVEN-TYPE 1.   IF LENGTH AND "
31   PRINT "WIDTH ARE GIVEN-TYPE 2."
35   INPUT X
40   PRINT "YOU HAVE CHOSEN NO. - ";X
```

```
50    PRINT
60    IF X = 1 THEN   GOTO 100
70    IF X = 2 THEN   GOTO 200
80    END
100   PRINT "TYPE IN THE U VALUE, THEN THE AREA IN SQ.FT.,"
101   PRINT "AND THEN THE TEMPERATURE DIFFERENCE IN DEG. F."
110   INPUT U,A,T
120   PRINT
130   PRINT "U = ";U" .  AREA = ";A;" SQ.FT."
131   PRINT "TEMPERATURE DIFFERENCE = ";T;" DEG. F."
140   PRINT
150   LET QR = U * A * T
160   PRINT "HEAT LOSS = ";QR;" BTU/HR."
170   GOTO 80
200   PRINT "TYPE IN THE U VALUE, THEN THE LENGTH IN FEET,"
201   PRINT "THE WIDTH IN FEET, AND THEN THE TEMP. DIFF. IN DEG. F."
210   INPUT U,L,W,T
220   PRINT
230   PRINT "U = ";U;" LENGTH = ";L;" FT."
231   PRINT "WIDTH = ";W;" FT.  TEMP. DIFF. = ";T;" DEG. F."
240   PRINT
250   LET QR = U * L * W * T
260   PRINT "HEAT LOSS = ";QR;" BTU/HR."
270   GOTO 80
```

NOTE

Lines 150 and 250 contain the heat transmission formula.

A person can do more work with the aid of heat engines. Here, a large farm tractor, driven by a diesel engine, pulls a chisel plow. (Courtesy of John Deere and Company)

Chapter 11

Objectives

When you finish this chapter, you will be able to:

- [ ] Define a heat engine as a device designed to obtain useful work from heat energy, and recognize the Carnot engine as an idealized imaginary engine.
- [ ] Recognize the modified engine efficiency formula: e = work out/work in = $(Q_{in} - Q_{out})/Q_{in} = (T_{in} - T_{out})/T_{in}$. The efficiency of an engine equals output work (W_{out}) divided by input work (W_{in}), or input heat energy (Q_{in}) minus heat out (Q_{out}) divided by input heat (Q_{in}), or input temperature (T_{in}) minus output temperature (T_{out}) divided by input temperature (T_{in}).
- [ ] Diagram a steam turbine, its heat source and condenser, and discuss its merits.
- [ ] Diagram a gasoline engine, explain the operation of valves and flywheel, and discuss compression ratios and efficiencies.
- [ ] Compare a diesel engine with a gasoline engine.
- [ ] Diagram a refrigerator (air conditioner or heat pump), explain its operation, and compare with engines.
- [ ] Explain the operation of a jet engine and compare with other engines.

Heat Engines

Machines that convert heat energy into mechanical energy (or vice versa) are called heat engines. We can think of a heat engine as a device that performs work and that does the following:

Takes in a substance—usually a gas at a high temperature.
Extracts some of the heat of the substance and converts this heat to mechanical energy (work).
Then discharges the substance at a lower temperature.

Note that a refrigerator reverses this process. Refrigeration will be discussed later in the chapter. We begin this chapter with a discussion of an ideal heat engine and continue with a number of practical engines, such as the steam turbine, the gasoline and diesel engines, the refrigerator, and the jet engine. The practical engines are built to utilize different types of available fuel, as well as to try to approach ideal conditions. Each type of engine has its own advantages and disadvantages.

The technician should be familiar with the more common types of heat engines, and understand what thermal efficiency means and how to calculate it. As a technician, you may have to apply some heat engine to your product or to a production operation, in which case, you may be involved in measuring engine performance or engine emissions.

11.1 The Ideal Heat Engine

Suppose we could put an ideal gasoline engine into an automobile. The **ideal engine** would extract energy from the burning gasoline, and all this extracted energy would be used for moving the vehicle. No energy would be needed to overcome friction in the engine. None would be needed to operate valves, water pumps, or fans. And no heat would be lost through the engine casing. Of course, no actual, or practical, engine can reach these ideal conditions, because energy is required to operate needed accessories and to overcome friction. Also, loss of heat through the engine casing cannot be stopped. Even so, in our efforts to develop more efficient engines, it is important to know the thermal efficiency of an ideal engine that operates with the same working substance and in the same temperature range as the practical engine. This set of data from the ideal engine, when compared with the data from the practical engine, tells us how close to "ideal" the practical engine is operating. French physicist and engineer Sadi Carnot (1796–1832) developed a simple formula to determine the

PHOTO 11.1 Hydroelectric turbines help convert the energy of falling water into electric energy, but some energy is lost in the process due to heat losses. No real engines can operate at 100 percent efficiency. (Courtesy of the Bonneville Power Administration)

percent of heat energy that an ideal engine could convert to work. This ideal heat engine is frequently referred to as the Carnot engine.

∗ ideal engine

One in which all the heat energy extracted from the working substance produces useful work.

First, though, we must become acquainted with the formula for **efficiency** for any machine, whether it is a heat engine, a set of gears, a lever, or some other device. The formula for efficiency is:

$$e = \frac{\text{work out}}{\text{work in}} \times 100$$

Since efficiency is usually expressed as a percentage, we must multiply by 100 to convert the answer into percent.

---

* efficiency

The ratio of useful work gotten out of a machine compared to the work put into a machine.

To understand the efficiency formula as it applies to a heat engine, we must form a mental picture of internal energy flowing into the engine at a high temperature and, after work is done, flowing out at a lower temperature. Remember, heat is the amount of internal energy flowing between two objects. Carnot likened the flow of internal energy (heat) through a heat engine to the flow of water over a waterwheel. In order for a waterwheel to operate and produce work, the water has to flow to the wheel from a reservoir higher than the reservoir into which the water is expelled.

Let's go over this once more. We are saying two things. First, in order for a waterwheel to operate, water must flow through, or past, the wheel. We can't just place the wheel in a pond, or lake, and expect it to operate. Similarly, with a heat engine, internal energy of a substance must flow "through" a heat engine. The heat can't just stop at the engine.

The second point is that in order for water to flow, there must be a reservoir, or source, of water at a higher elevation than the one into which the water empties. The same point applies to heat. In order for internal energy to flow, there must be a source of energy at a *higher temperature* than the reservoir into which the internal energy flows. The heat from the engine must be dumped into some reservoir, usually the atmosphere or a body of water. Since these reservoirs are at everyday temperatures, the heat coming into the engine must have a higher temperature.

The work done by a heat engine (called *work out*) is equal to the difference in the heat coming in and going out of the engine. See Figure 11.1. The efficiency of the heat engine is called the **thermal efficiency**.

*** thermal
efficiency**

The ratio of the amount of internal energy converted to useful work compared to the amount of internal energy put into the engine.

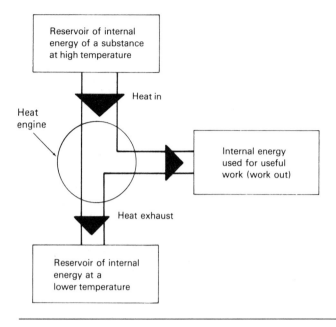

FIGURE 11.1   The Flow of Internal Energy through a Heat Engine.

The formula for thermal efficiency is:

$$e = \frac{\text{work out}}{\text{work in}} \times 100 = \frac{(\text{heat in} - \text{heat out})}{\text{heat in}} \times 100$$

or
$$e = \frac{(Q_{\text{in}} - Q_{\text{out}})}{Q_{\text{in}}} \times 100$$

This formula is not in its most practical form. The derivation of the final formula is beyond the scope of this text, but we can understand the concept from the heat formulas we worked with in Chapter 9. For example, in the formula $Q = w \times c \times \Delta T$, the heat ($Q$) changes when the temperature difference ΔT changes, assuming that the quantity of substance and the specific heat remain constant.

Based on this relationship, Carnot proved that the thermal efficiency of an ideal heat engine could be determined by the absolute temperatures of the substance flowing through the engine. The resulting formula is:

★ THERMAL EFFICIENCY FORMULA

efficiency = (inlet temperature − exhaust temperature) ÷ inlet temperature × 100

$$e = \frac{T_{in} - T_{out}}{T_{in}} \times 100$$

where e = the thermal efficiency in % (thus we multiply by 100)

T = the absolute temperature in K

The formula indicates that the larger the difference between the inlet temperature (T_{in}) and the exhaust temperature (T_{out}), the more efficient the engine. As mentioned a few paragraphs back, in most cases the heat is exhausted to the atmosphere, over which we have no control. So T_{out} is usually at, or above, atmospheric temperature. Since it is not practical to lower exhaust temperatures below the normal everyday air (or water) temperatures, engineers try to raise the incoming temperature of gases. The high incoming temperatures are limited by the materials used in the heat engines. Let's see how the efficiency changes if we are able to double the inlet temperature of a heat engine, as in Sample Problem 11.1.

SAMPLE PROBLEM 11.1

A particular engine we are working with has heat coming into it at 100.0°C and exhausting to the atmosphere at 20.0°C.
  A. Determine the thermal efficiency of the engine.
  B. Determine the thermal efficiency if the inlet temperature is doubled to $20\overline{0}$°C.

SOLUTION A

Wanted

e = ?

Given

T_{in} = 100.0°C and T_{out} = 20.0°C

Formula

$$e = \frac{T_{in} - T_{out}}{T_{in}} \times 100$$

Calculation

First, we must convert the given temperatures to K:

$$T_{in} = 100.0°C + 273°C = 373 \text{ K}$$
$$T_{out} = 20.0°C + 273°C = 293 \text{ K}$$

Now we can use the thermal efficiency formula:

$$e = \frac{T_{in} - T_{out}}{T_{in}} \times 100 = \frac{373 \text{ K} - 293 \text{ K}}{373 \text{ K}} \times 100$$
$$= \frac{80}{373} \times 100 = 21.4\%$$

SOLUTION B

Wanted

$$e = ?$$

Given

$$T_{in} = 200°C \text{ and } T_{out} = 20°C$$

Formula

$$e = \frac{T_{in} - T_{out}}{T_{in}} \times 100$$

Calculation

We first convert to K:

$$T_{in} = 200°C + 273°C = 473 \text{ K}$$
$$T_{out} = 20.0°C + 273 = 293 \text{ K}$$

Then we substitute into the formula:

$$e = \frac{473 \text{ K} - 293 \text{ K}}{473} \times 100 = \frac{180}{473} \times 100 = 38\%$$

We can use this information to determine whether the required changes to the engine will be economical. Some of the changes that may have to be made are installing new heating elements to obtain the higher temperature, replacing engine parts with new materials to withstand the higher temperatures, and improving the engine cooling system.

11.2 The Steam Turbine

At one time, you have probably played with a water hose by pushing a rubber ball around the yard with the force of the stream of water. In a similar way, the force of water against the blades of a waterwheel will cause it to turn. Steam under high pressure can also be used to turn a wheel. Such a device, called a **steam turbine,** is illustrated in Figure 11.2.

The turbine has a number of advantages over most other heat engines. Since it has only one rotating assembly, it is simpler and operates more smoothly. Maintenance and lubrication requirements are low, and the rotating blades produce less vibration than the back-and-forth motion of pistons. The turbine can operate at much higher speeds than most other engines. However, it is less efficient when operating at low speeds or partial loads.

PHOTO 11.2   Steam turbines rotate at such high speeds that the slightest vibration could quickly create great damage. Tolerances in the manufacture of turbines must be as small as 0.001 inch, and maintenance inspections must be thorough. (Courtesy of Penn Power and Light)

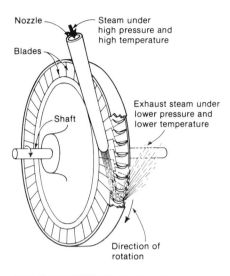

Nozzle — Steam under high pressure and high temperature

Blades

Shaft

Exhaust steam under lower pressure and lower temperature

Direction of rotation

FIGURE 11.2   A Steam Turbine with Just One Set of Movable Blades and with the Casing Removed.

These characteristics make the steam turbine most suitable for large central-station electric plants and oceangoing vessels.

To operate the steam turbine, water is heated in a closed boiler until it forms steam at a pressure of several hundred pounds per square inch. This steam is then directed through pipes and emitted through a nozzle that strikes against the curved blades of the wheel. The impact of the steam against the blades causes the wheel to revolve at great speed.

Modern steam turbines may have as many as 1500 or more blades, mounted in as many as 20 sets on a single shaft. As Figure 11.3 indicates, each set of movable blades is mounted on a wheel whose diameter is larger than that of the preceding set, just as the pistons of multiple steam engines are progressively larger in diameter to compensate for the increasing loss in pressure.

In modern steam turbines, steam at pressures as high as 2300 pounds per square inch travels through the turbine at speeds as great as 1200 miles per hour. The outer rim of the wheel may travel as fast as 600 miles per hour, and 100 000 horsepower or more may be generated.

The turbine is called an *external combustion engine,* because the combustion that converts the chemical energy of the fuel to heat energy takes place in a boiler external to the engine itself. If we calculate the energy of the burning fuel and compare it

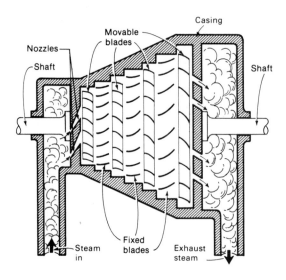

Casing

Movable
blades

Nozzles

Shaft

Shaft

Steam
in

Fixed
blades

Exhaust
steam

FIGURE 11.3   Cross-Sectional View of a Large Steam Turbine with Four Sets of
Movable Blades.

with the mechanical-energy output of the turbine, we find that
all but about 27 to 35 percent of the energy has been wasted.
First of all, about half the energy is wasted in the boiler and in
the pipes leading the steam to the engine or turbine. Waste occurs
partly because of incomplete combustion of the fuel, particles of
which may be carried up the chimney as black smoke. Again,
the hot gases that escape up the chimney carry with them a good
deal of the heat energy. Heat is lost, too, by radiation from the
hot boiler and pipes going to the engine or turbine.

Of the remaining heat that reaches the engine or turbine,
more is dissipated by radiation from the hot machine. More energy is lost overcoming friction, and still more is wasted as the
still hot steam comes out of the exhaust port. If the turbine is
not equipped with a condenser, energy is lost pushing the exhaust
steam out against the atmospheric pressure. If a condenser is
used, this loss is reduced considerably.

What is a condenser? Well, if the steam from the turbine is
exhausted directly into the surrounding atmosphere, the steam
must overcome atmospheric pressure of about 15 pounds per
square inch (psi) to push its way out of the turbine. Thus, the
mechanical-energy output of the turbine is cut down by the amount
necessary to push the exhaust steam into the air. However, if
the exhaust steam is discharged into a series of pipes—that is,
a condenser—where it can be cooled by air or water, the steam

condenses. Because of this condensation, the pressure in the pipes drops down to a pound or so per square inch. Thus, less work is needed to discharge the exhaust steam into these low-pressure pipes than would be needed if the steam were discharged directly into the air, and more useful energy is delivered to the turbine wheel.

Now is a good time to combine some of the heat and power information that we have learned in previous chapters. We can determine the amount of power a turbine can deliver for a given flow of steam by modifying the heat–temperature formula in Chapter 9:

$$Q = w \times c \times \Delta T$$

Remember that heat energy (Q) is measured in Btu in the U.S. Customary System. If this heat energy is used to perform work, then Q represents the amount of work. Also, power is work divided by time:

$$P = \frac{Q}{t}$$

For turbine applications, this power is referred to as heat rate (R_h). The heat rate is the heat given up by the steam each hour as it flows through the turbine. To find the heat rate, we need to know the steam flow (W_s) in pounds per hour; that is, $W_s = w/t$, where w is the weight of steam in pounds. The heat rate can be found by modifying the heat–temperature formula to include time:

$$Q/\text{hr} = w/\text{hr} \times c \times \Delta T$$

We can rewrite this formula for turbine applications as follows:

*** HEAT RATE
FORMULA**

heat rate = steam flow × specific heat capacity
× temperature difference

$$R_h = W_s \times c \times \Delta T$$

where R_h = the heat rate in Btu/hr
W_s = the steam flow in lb/hr
c = the specific heat capacity in Btu/lb·°F
ΔT = the change in temperature in °F

The heat rate can be converted into horsepower by multiplying by 0.000 393:

horsepower = heat rate (Btu/hr) × 0.000 393

Bear in mind that the answers we obtain will be approximate. In order to illustrate the topic more clearly, we will disregard losses due to: (1) heat loss to the surroundings, (2) heat required to overcome friction, and (3) a turbine (or other heat engine) not letting the steam (or gas) expand under ideal conditions. Also, remember that we have only the data in Table 9.1 for the specific heat of steam, and that this has not been corrected for pressure differences.

Even so, technicians and engineers regularly make these approximations as a first step in the design of a turbine for a particular application. Sample Problem 11.2 shows how to determine the power and efficiency of a turbine.

SAMPLE PROBLEM 11.2

Steam enters a turbine at 312°F and leaves at 212°F at atmospheric pressure. Six hundred pounds of the steam flow through the turbine each hour.
  A. Determine the heat rate.
  B. Find the horsepower output.
  C. Determine the thermal efficiency of the turbine.

SOLUTION A

Wanted

R_h = ?

Given

W_s = 60$\bar{0}$ lb/hr, T_{in} = 312°F, and T_{out} = 212°F. From Table 9.1, we find the specific heat capacity for steam, c = 0.48. The temperature difference is ΔT = 312°F − 212°F = 100.0°F.

Formula

We use the heat rate formula:

$$R_h = W_s \times c \times \Delta T$$

Calculation

$$R_h = 600 \times 0.48 \times 100°F = 29\ 000 \text{ Btu/hr}$$

This answer is the amount of heat turned into useful work each hour.

SOLUTION B

Wanted

$HP = ?$

Given

$R_h = 29\ 000$ Btu/hr

Formula

The horsepower output can be found by multiplying the heat rate by 0.000 393.

Calculation

$HP = 29\ 000$ Btu/hr $\times$ 0.000 393 $= 11.4$ hp

SOLUTION C

Wanted

$e = ?$

Given

$T_{in} = 312°F$ and $T_{out} = 212°F$

Formula

$$e = \frac{T_{in} - T_{out}}{T_{in}}$$

Calculation

First we must convert the temperatures into °C and then K using the formulas from Chapter 9:

$$T_C = \frac{5}{9}(T_F - 32°F) \qquad \text{and} \qquad T_K = T_C + 273°C$$

First, T_{in}:

$$T_C = \frac{5}{9}(312°F - 32°F) = \frac{5}{9} \times 280 = 156°C$$

$$T_K = 156°C + 273°C = 429 \text{ K}$$

Next, T_{out}:

$$T_C = \frac{5}{9}(212°F - 32°F) = \frac{5}{9} \times 180 = 100°C$$

$$T_K = 100°C + 273°C = 373 \text{ K}$$

Now we can use the thermal efficiency formula:

$$e = \frac{T_{in} - T_{out}}{T_{in}} \times 100 = \frac{429 \text{ K} - 373 \text{ K}}{429 \text{ K}} \times 100$$

$$= \frac{56}{429} \times 100 = 13\%$$

This information can help us decide on the size of pipes and other equipment needed. Also, if we want to adjust the power up or down, we can use this information to decide whether to adjust the flow rate of the steam or the inlet and outlet temperatures of the steam.

11.3 The Gasoline Engine

The various engines discussed in the remainder of this chapter are all the *internal combustion* type—that is, the conversion from chemical to heat energy takes place within the engine itself, rather than in some external device such as a boiler. Most common is the familiar **gasoline engine**, perfected in 1876 by the German scientist Dr. N. A. Otto. The cylinder of a gasoline engine is shown in Figure 11.4. Let us see what happens to a mixture of air and gasoline as it is sucked into the engine.

The liquid gasoline is changed to a gas and mixed with 15 parts of air (by weight) in the carburetor. This combustible mixture then is fed into the cylinder of the gasoline engine. There

PHOTO 11.3   Car engines are put together with their control mechanisms in an assembly line and then sent to where the car bodies have been put together. Larger or faster cars often have 8-cylinder engines; smaller cars usually have more economical 4- or 6-cylinder engines. (Courtesy of GM Assembly Division)

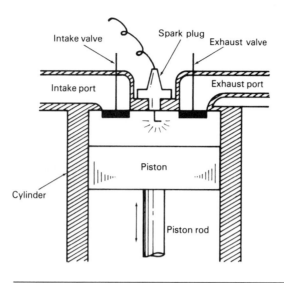

FIGURE 11.4 Cross-Sectional View of the Cylinder of a Gasoline Engine.

it is compressed and ignited by a spark. As it burns, the gas expands and pushes down the piston, thus producing the mechanical energy that turns the wheels.

On the input stroke of the gasoline engine (see Figure 11.5), the piston descends, and the gasoline vapor-and-air mixture is drawn into the cylinder through the open intake valve. During the compression stroke, both intake and exhaust valves are closed. The ascending piston compresses the gas mixture in the top portion of the cylinder. As the piston reaches the top of its stroke and starts descending, a spark from the spark plug ignites the mixture. The heat generated by the burning gas causes it to expand, and the piston is pushed down in its power stroke.

The gas mixture was compressed before it was ignited in order to raise its temperature before ignition. The hotter the gas mixture before ignition, the more it will expand when it burns, and the pressure against the piston will therefore be greater. Thus, a greater force will be exerted during the power stroke. In modern automobile engines, the gas is compressed to about 1/9 of its original volume before it is ignited. We say the engine has a 9-to-1 compression ratio.

During the compression stroke, the pressure in the cylinder rises to about 100 pounds per square inch. When the mixture is ignited and burns, a temperature as high as 4000 degrees Fahrenheit may be reached. As a result of this great heat, the pressure

PHOTO 11.4   The transmission of a car enables the driver to change speed or go in reverse by using a different set of gears in the driveshaft. The workers shown are connecting a 4-cylinder engine to the car's transmission. (Courtesy of GM Assembly Division)

in the cylinder increases to about 400 pounds per square inch. It is this pressure that gives the automobile engine its power.

Why, then, do we not increase the compression ratio to, say, 10-to-1, or higher? The answer lies in the fuel we use. The mixture in the cylinder burns in a fraction of a second, but, even in this short period of time, it must burn smoothly and evenly. It must not burn all at once or in several different spots at the same time. Should it do so, we get a characteristic "knocking," or pinging, noise in the engine, which indicates that the fuel is not burning efficiently. Present-day gasoline for use in automobiles will not permit a compression ratio of much more than 9-to-1 without knocking. However, it is much improved over the gasoline in use not so many years ago, when a compression ratio of 3-to-1 was the highest practical.

At this point, a question may arise. We can see how the exploding vapor forces the piston to move down during the power stroke, but what causes the piston to move during the other parts of the cycle? The answer is that not one but four or more cylinders are used in modern cars. Each cylinder produces a power stroke in turn. The pistons are connected to the same crankshaft in such a way that when one piston is being forced down by the explosion in its cylinder, some of its power is used to force the other pistons through the other portions of their cycles. Figure 11.5 portrays the cycle of a 4-cylinder gasoline engine.

In addition, a heavy iron wheel, called a *flywheel,* is mounted on the crankshaft. The revolving crankshaft causes the flywheel to rotate. Because of its great mass, the flywheel tends to keep rotating and, in turn, tends to keep the crankshaft revolving, and the crankshaft then forces the pistons through the nonpower portions of their cycles.

Now let us examine the action of the valves. There are two for each cylinder: an intake valve controlling the flow of the gasoline-and-air vapor into the cylinder and an exhaust valve controlling the flow of the burnt gases out of the cylinder after the explosion has taken place. During the intake stroke, the intake valve is open, and the descending piston sucks in the explosive vapor. The exhaust valve is closed, or the incoming gas would escape through the exhaust port. During the compression stroke, both valves are closed, and the ascending piston compresses the gas in the cylinder. During the power stroke, both valves remain closed. The spark ignites the gas, and the piston is forced down by the explosion. Finally, during the exhaust stroke,

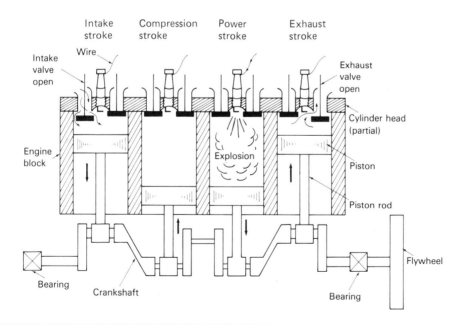

FIGURE 11.5   The 4-Cylinder Gasoline Engine.

the exhaust valve is open and the intake valve is closed. The ascending piston forces the burnt gases out through the exhaust port, past the open exhaust valve, through the exhaust manifold (a pipe leading from the exhaust port), and ultimately into the air.

How are these valves operated so that they will open or close at the right time? A gear wheel is attached to the crankshaft and revolves with it. This gear meshes with a second gear that causes another shaft fastened to it to revolve. On this shaft (called the *camshaft*), a set of two cams for each cylinder is fastened, as Figure 11.6 shows. These cams move the valve rods, and thus the valves travel up and down as the camshaft rotates. By proper timing of the cams, the valves may be made to open or close at the proper time. Thus, we can see why the gear wheels are called the *timing gears*.

Only about 30 percent of the chemical energy of the gasoline-vapor-and-air mixture is converted into mechanical energy in the gasoline engine. This percent figure is referred to as the overall thermal efficiency. For any engine, it is simply the amount of useful work obtained from the engine compared to the heat energy supplied by the fuel, as stated in the section on the ideal engine. The list in Table 11.1 compares the overall thermal efficiency for various types of engines.

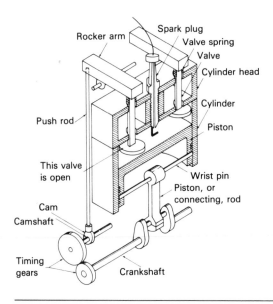

FIGURE 11.6   How the Cams Operate the Valves of a Gasoline Engine.

TABLE 11.1   Comparison of Thermal
Efficiencies for Various Engines

Two-cycle gasoline engines	15–24%
Four-cycle gasoline engines	23–32%
Aircraft, supercharged gasoline engines	31–36%
Automobile diesels	36–42%
Central-station powerplant with steam turbine	27–35%

From *Automobile Engines,* 8th ed., by A.W.
Judge. Robert Bentley, Inc., Cambridge, Mass.

Part of the energy in the fuel is lost because of incomplete burning of the fuel. Another part is lost as heat that is absorbed and carried away by the water of the cooling system and as heat that is radiated into the surrounding atmosphere by the hot engine. And a large portion of the energy is lost in the heat of the exhaust gases discharged into the air. Incidentally, still more mechanical energy is lost in pushing these exhaust gases out against the atmospheric pressure.

A breakdown of the losses in a typical automobile engine is shown in Table 11.2. Nevertheless, the gasoline engine remains one of our best sources of motor power. It is popular because it requires little or no warm-up time, whereas considerable time is

TABLE 11.2   Heat Losses in a Typical
Gasoline Engine

Heat required to overcome friction in the engine and turn the cooling fan	4%
Heat needed to pump gases in and out of the engine	4%
Heat carried away by cooling water	20%
Heat lost in the exhaust gases	45%
Heat available for "useful output"	<u>27%</u>
	100%

From *Automobile Engines,* 8th ed., by A.W.
Judge. Robert Bentley, Inc., Cambridge, Mass.

required to heat up boilers to the proper temperature to operate steam turbines. The gasoline engine is more compact—that is, no large boilers or condensers are required. It can change speed quickly and accept large load changes. It is ideal for automobile and airplane use. It can be made small and compact for use on power mowers, chain saws, motorcycles, and so forth.

11.4 The Diesel Engine

Let us examine the problem of compression ratio in the gasoline engine more closely. Suppose we compressed the fuel-and-air mixture in the cylinder, not to one-ninth of its original volume, as is done in the present-day automobile, but even more, say to one-sixteenth of its volume. We would not need a complicated ignition system, since the heat of compression would be great enough to ignite the fuel without requiring an electric spark.

What about pre-ignition? Would not the fuel start burning at the wrong portion of the cycle? If air alone is sufficiently

PHOTO 11.5 Diesel engines have more weight per horsepower output than gasoline engines, which makes diesel more appropriate for large mobile equipment where engine weight is not a major consideration. This freight train is powered by a 3000 horsepower diesel engine. (Courtesy of Union Pacific Railroad)

compressed, it will not burn, no matter how high the heat. It is only an air-and-fuel mixture that can burn. Suppose, then, that we compress only the air. The pressure and the heat will rise, but there will be no pre-ignition because there is nothing to burn. Now, at the very end of the compression stroke, we inject the fuel into the cylinder in the form of a fine spray. The fuel will instantly mix with the hot air and start burning. Thus, there can be no pre-ignition. The burning can take place only at the end of the compression stroke, which is the proper portion of the cycle.

How can we prevent engine knock? Knocking occurs because of the uneven burning of fuel in the cylinder. If the fuel is injected in a fine, even spray, and the fuel burns as it enters the cylinder, then it will burn evenly and there will be no knocking.

The principle of compressing air so that it becomes hot enough to ignite the fuel is the principle of the **diesel engine,** which was invented by the German engineer Rudolph Diesel in 1892. The diesel engine closely resembles the gasoline engine. (Look at the simplified diagram in Figure 11.7.) There are differences between the diesel and gasoline engines, however. In place of an ignition system, there is a *fuel injector,* which, at the proper time, injects the fuel into the cylinder in the form of a fine, mistlike spray. Since the fuel need not be mixed with air outside the cylinder, there is no carburetor.

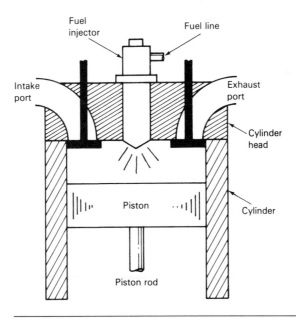

FIGURE 11.7   The Cylinder of the Diesel Engine.

What fuel do diesel engines use? Early types could burn almost any combustible liquid. Even powdered coal has been used successfully. Modern diesel engines, however, use a petroleum fuel that is carefully designed to meet their needs. This fuel must start burning the instant it reaches the hot air in the cylinder. If it does not, too much fuel may accumulate in the cylinder, and when this accumulation starts burning, it produces a knock similar to the knocking produced by uneven burning in the gasoline engine.

Now let us compare the diesel engine with the gasoline engine. The diesel engine is more efficient, especially when not running at full speed. This is due chiefly to the higher compression ratio, which may be as high as 25-to-1. Approximately 40 percent of the chemical energy of the burning fuel may be converted to mechanical energy. Another advantage is that the diesel engine runs cooler than the gasoline engine. This advantage is especially pronounced at lower speeds. Moreover, since diesel fuel will not evaporate so readily, it is safer to store. Finally, the diesel engine requires neither an ignition system nor a carburetor. However, this advantage is offset by the need for an injector, a costly device that must be carefully constructed.

The chief advantage of the gasoline engine is its lower weight per horsepower. Whereas the diesel engine may weigh from 6 to 15 pounds per horsepower, the gasoline engine used in automobiles weighs about 5 pounds per horsepower, and gasoline engines used in airplanes may weigh as little as 1-1/4 pounds per horsepower. This advantage has prevented the diesel from replacing the gasoline engine in airplanes. Also, gasoline engines require less cranking effort because of their lower compression ratio.

Like the gasoline engine, the diesel engine generally is constructed with several cylinders. Most common are 2- to 4-cylinder engines. Diesels used for ships and locomotives may have even more cylinders. Large diesel-electric locomotives are powered by two 16-cylinder diesel engines, each rated at 3300 horsepower. Although the diesel is becoming popular in automobiles, its use is in heavier mobile equipment such as tractors, trucks, buses, locomotives, and small oceangoing vessels.

11.5 The Refrigerator

Would you believe that a **refrigerator** is a heat engine? Well, it is. As diagrammed in Figure 11.8, it uses mechanical energy from a pump to extract heat energy from food, thus lowering the

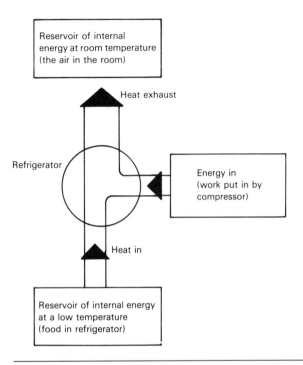

FIGURE 11.8 The Flow of Internal Energy in a Refrigerator.

temperature of the food. To understand the concept, let's start with a very simple refrigerator.

Recall from Chapter 9 that when a liquid evaporates, it takes in heat. Here, then, is an answer to our problem. Place the food in a tin can, then set the can in a pan of water and cover it with a cloth, the ends of which dip in the water. Figure 11.9 shows this arrangement. The water is drawn up the cloth by capillary action. Then it evaporates into the air, taking the necessary heat to do so from the tin can. The cooled can takes heat from within it, and this cool air keeps the food cold.

As a matter of fact, this principle is used in a certain type of water bottle designed to keep drinking water cool. The bottle is made of porous clay, and a little of the water it contains seeps through the pores to the outer surface of the bottle. There the water evaporates into the air, cooling the bottle as it does so. In this way, the water inside is kept cool by contact with the cold bottle.

For the refrigeration of food, however, such a device is not satisfactory. The water does not evaporate fast enough to produce sufficient cooling to prevent the food from spoiling. Of course, we

FIGURE 11.9   A Simple Refrigerator Using the Cooling Effect of Evaporating Water.

could use some other liquid, such as Freon, which will evaporate much more rapidly and thus produce the required amount of cooling, and is not poisonous or flammable. However, a couple of obstacles stand in the way: Freon evaporates so rapidly that the pan would require constant refilling, and Freon is quite expensive. What we need is a closed chamber, such as the pipe shown in Figure 11.10, in which the Freon could evaporate without coming in contact with the outside air. The Freon vapor then could be recovered and used over and over again.

There is a flaw in our refrigerator, however. True enough, the Freon will evaporate and take in heat, as latent heat of evaporation. But after the Freon has changed to a gas, how is it converted back again to a liquid to repeat the process?

In Chapter 9, we saw that pressure raises the boiling point of a substance. We must compress the Freon vapor to change it back to a liquid. However, since the temperature also increases with pressure, we still have Freon gas. Our next step is to cool the Freon gas to about room temperature. As the Freon cools (at

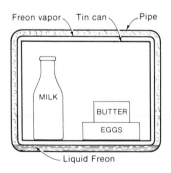

FIGURE 11.10   A Simple Refrigerator Using an Evaporating Liquid in a Closed Chamber.

the higher pressure), it condenses to a liquid. Therefore, at some point in the closed chamber, we must insert a compressor and some cooling device (Figure 11.11) to convert the Freon vapor back to liquid Freon. The compressor can be operated by any convenient device, such as a gasoline engine or more commonly, an electric motor.

There is yet another obstacle to be overcome. As has already been mentioned, when we compress a gas, we also increase its temperature. It would hardly do to have hot Freon vapor circulating through the pipes of our refrigerator. Some method must be found to cool the Freon after it has been compressed. A device similar to the radiator used in an automobile, and similar to the condenser used in some turbines, is employed for this purpose. The hot, compressed Freon circulates through a series of small pipes that are exposed to the outside air. These pipes give off their heat to the air, and as the Freon assumes room temperature, it condenses to a liquid. It is now ready to evaporate again. Metal fins attached to these pipes increase the radiating surface and thus help in cooling. Often a fan, operated by the same motor that drives the compressor, blows cool air over the pipes.

And now, believe it or not, we have designed a *mechanical refrigerator* that operates on the same principles basic to most home refrigerators. Figure 11.12 shows such a refrigerator in simplified form. The piston compresses the freon gas, which is cooled as it passes through the finned tubes comprising the condenser and radiator. The cooled liquid is then forced through the needle valve. The needle valve chokes up the flow through the closed system so that a partial vacuum exists in the pipes between the valve and the compressor. As a result, the liquid refrigerant

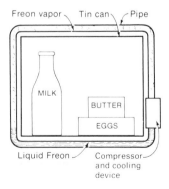

FIGURE 11.11   A Simple Refrigerator with Compressor and Cooling Device Added to Change the Freon Vapor to a Liquid.

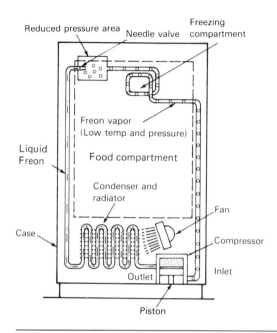

Reduced pressure area Needle valve Freezing compartment

Freon vapor (Low temp and pressure)

Liquid Freon

Food compartment

Condenser and radiator

Fan

Case

Compressor

Outlet Inlet

Piston

FIGURE 11.12   The Mechanical Refrigerator.

is sprayed into an area of reduced pressure, and this speeds up evaporation, thus increasing the rate of cooling. Remember, when a liquid boils, or vaporizes (even at low temperature), it absorbs heat.

Since the greatest cold is produced at the point where the evaporation first takes place, the cold gas is led through a coil of pipes that surrounds the freezing compartment. Here the temperature is low enough to freeze water, thus producing ice cubes. Then the cold gas is led through a series of pipes surrounding the food compartment and, finally, back to the compressor.

The entire unit is placed in a metal case whose walls are heavily insulated to prevent outside heat from entering. A thermostat (not shown) starts the motor running when the temperature inside the refrigerator becomes dangerously high (about 40 degrees Fahrenheit), and turns the motor off when the temperature becomes safe once more. Consequently, a fairly even temperature is maintained within the refrigerator at all times. The thermostat is adjustable so that the temperature of the interior of the refrigerator can be maintained at any desired level.

Another type of refrigerating device is the *freezer,* a device in which fresh meats and vegetables may be stored for weeks or even months without spoiling. The freezer resembles the mechanical refrigerator, except that it is adjusted to produce an internal temperature below the freezing point of water. The insulation of its walls is thicker, generally from 4 to 5 inches thick, to help keep out external heat. The fact that the freezer is opened far less frequently than the ordinary household refrigerator helps maintain the cold of its interior. Most household refrigerators on the market have built-in freezer compartments operating from the same compressor that cools the refrigerator.

The commercial apparatus used for making ice operates on the same principles we have just discussed. Of course, it must be very much larger to be able to produce tons of ice daily.

11.6   The Jet Engine

To understand how a jet engine works, blow up a toy rubber balloon. Pinch the neck of the balloon between your fingers to prevent the compressed air from escaping. Hold it at arm's length

PHOTO 11.6   The jet engine shown here is rated at 5000 pounds of thrust and is used in lightweight military aircraft. Its highest efficiency comes at high speeds. (Courtesy of Northrop Corp.)

and then release it. The balloon will dart through the air, its compressed air streaming out of the opening at the rear. When all the air has escaped, the balloon will fall to the ground.

What made the balloon fly around? Well, if we recall Newton's Third Law of Motion (Chapter 5), we remember that for every action, there is an equal and opposite reaction. The compressed air, rushing out of the mouth of the balloon in one direction, gives rise to a reaction force causing the balloon to move in the opposite direction. This principle applies to the **jet engine,** a simplified version of which is shown in Figure 11.13.

The air is pulled in through the intake duct by the rotation of a compressor, which rotates at a tremendous speed. The air is whirled around by the blades, and is thrown out and backward at a higher pressure toward the combustion chamber. In this way, the compressor acts like a tremendous vacuum cleaner.

As the compressed air enters the combustion chamber, it mixes with sprays of fuel from the fuel inlets. This fuel may be gasoline, kerosene, or some other suitable liquid. The hot wires ignite the mixture and the resulting heat increases the pressure enormously. The hot gases then pass through the tailpipe and out through the rear nozzle in a fiery jet. It is this jet that propels the engine forward.

However, before it enters the tailpipe, the hot gas first strikes against the blades of a turbine. This turbine resembles the steam turbine discussed earlier in this chapter, and operates in the same way. The fiery gas spins the turbine at high speed. This turbine is connected to the compressor by means of a shaft, and thus furnishes the power to operate it. Simple though this engine appears to be, it is capable of generating a tremendous force. Up to 20 000 pounds of thrust can be developed by large jet engines.

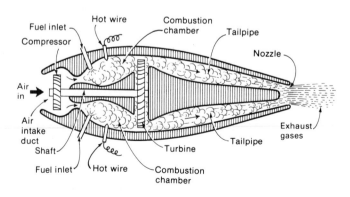

FIGURE 11.13   The Jet Engine.

Since the thrust of a jet engine is relatively constant, the engine can develop more power in high-speed airplanes than in low-speed trains or automobiles. Why is this so? Remember, power is work divided by time: $P = W/t$. Normally, we think that to do more work in a given amount of time, we have to exert more force. Generally, this is correct. However, the formula for work (work = force × distance) includes both force and distance. If the distance could be increased in a given time period while maintaining the force constant, more work could be done in that given time. This means more power. A high-speed airplane can travel the greater distance in a given time period; thus, jet engines are more suitable for airplanes than for other forms of transportation.

Summary

Heat engines are devices for converting heat energy to mechanical energy, and vice versa. The heat engines discussed are the steam turbine, the gasoline and diesel engines, the refrigerator, and the jet engine.

Carnot developed a simple formula to determine the thermal efficiency of ideal engines—engines in which all the heat energy extracted from the working substance produces work. In an actual engine, heat must be used to overcome friction and to operate the necessary valves, pumps, and fans. Heat is also lost through the engine casing. So, the thermal efficiency of an actual engine is always less than that of an ideal engine.

The steam turbine is an external combustion engine that operates on the principle that steam under high pressure and at high temperature can be directed against the blades of a rotor, causing the rotor to turn. The steam turbine is ideal for stationary operation where large power output is required, and for propulsion of large ships. The rotor is the only moving part, other than the incoming steam valve, so maintenance and lubrication requirements are low when compared to the requirements of other engines. The turbine does not operate efficiently at part loads, and cannot change speed rapidly.

The gasoline engine is an internal combustion engine that runs on gasoline and is the major type of engine used in automobiles. A mixture of gasoline vapor and air is fed into the cylinder and, at the proper moment, is ignited by a spark plug. The increased heat due to the burning fuel increases the pressure in the cylinder, pushing the piston down. The piston, which is con-

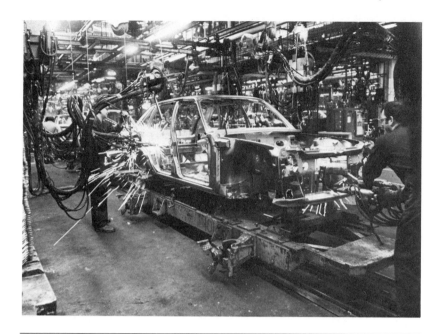

PHOTO 11.7   The advantages of the gasoline engine are most apparent in small and lightweight applications, such as cars, as opposed to larger vehicles, such as trains or large airplanes. Modern car design has been tending to smaller and more economical car bodies. (Courtesy of GM Assembly Division)

nected to a crankshaft, causes the crankshaft to turn. The gasoline engine is dependable, operates well at partial loads, and can change speed rapidly.

The diesel engine is also an internal combustion engine, but it differs from the gasoline engine in that the fuel is injected into the cylinder after the air in the cylinder has been compressed sufficiently by the piston. The compressed air becomes hot enough to ignite the fuel as it is injected into the cylinder, thus creating the pressure to drive the piston down again. The diesel engine has a higher thermal efficiency than the gasoline engine, but the engine runs slower and is heavier. Its major use is in trucks, locomotives, and large construction vehicles.

The refrigerator is a heat engine, too. It works in reverse, so to speak. Mechanical energy is used to remove heat. Mechanical energy (supplied by a compressor) compresses Freon vapor circulating around the refrigerator. Since the compression increases the temperature of the vapor above room temperature, it is cooled with the air of the room. This causes the Freon to condense to a liquid at room temperature. When the pressure is removed from the liquid (at the needle valve), its temperature drops, and, on absorbing heat from the food compartment, it turns to a vapor again.

The jet engine uses part of the energy of expanding gas to turn a turbine wheel, which drives an air compressor that compresses the air prior to mixing with the fuel. Useful work is obtained when the hot gas is exhausted at high speed, thus creating a reaction force that drives the airplane forward. The jet engine is most efficient at high speeds, and therefore is most useful for powering airplanes.

Key Terms

* ideal engine
* efficiency
* thermal efficiency
* thermal efficiency formula
* steam turbine

* heat rate formula
* gasoline engine
* diesel engine
* refrigerator
* jet engine

Questions

1. What is a heat engine?
2. What is an ideal heat engine?
3. Define thermal efficiency.
4. Can a thermal efficiency of 100 percent be attained?
5. Describe the construction and operation of a steam engine.
6. In a gasoline engine, what is meant by "compression ratio"?
7. Compare the gasoline engine with the steam engine.
8. Describe and explain the diesel engine.
9. What are the basic differences between the gasoline and diesel engines? Explain the advantages and disadvantages of each.
10. Describe the construction and operation of a refrigerator.
11. Describe and explain the operation of the jet engine.

Problems

1. Do Sample Problem 11.1 with the initial temperature raised to $30\overline{0}°C$.

2. An engine adapted for operation in space uses a gas raised to a temperature of 677°C and has an exhaust temperature of 66.0°C. What is the thermal efficiency?

3. The heat from a geothermal well is used to provide steam to a turbine that is connected to an electric generator. If the steam temperature into the turbine is $40\overline{0}°C$ and the exhaust temperature is 20.0°C, what is the thermal efficiency?

4. The maximum steam temperature used for turbine operation is about $60\overline{0}°C$. If the exhaust temperature is $12\overline{0}°C$, what is the thermal efficiency?

5. You have helped develop a new material for turbine blades that will allow steam temperatures up to $130\overline{0}°F$. If the exhaust temperature is $22\overline{0}°F$, what is the thermal efficiency?

6. Solve Sample Problem 11.2 if the inlet temperature of the steam is $40\overline{0}°F$ and the outlet temperature is $10\overline{0}°F$. The steam flow is the same, $60\overline{0}$ lb/hr.

7. Solve for the heat rate and horsepower output of Sample Problem 11.2 if steam enters the turbine at 612°F.

8. A gas turbine is the same as a steam turbine except that the hot gases of combustion (usually from burning natural gas) are passed directly through the turbine. A particular gas turbine has an inlet temperature of $180\overline{0}°F$ and an outlet temperature of $30\overline{0}°F$. The gas flow is 50 $\overline{0}$00 lb/hr. The mixture of burned gas and air has a specific heat capacity of 0.61. Determine (a) the heat rate and (b) the horsepower output.

9. A hot-air engine has a flow rate of 1150 lb/hr. Air has a specific heat capacity of 0.24, and the temperature difference for the air going in and coming out of the engine is 480°F. Find the heat rate and the horsepower output.

10. A steam turbine has a heat rate of $10\overline{0}$ 000 Btu/hr and a temperature difference of $40\overline{0}°F$. What is the steam flow in pounds per hour?

11. When the fuel burns in a particular gasoline engine, the temperature of the air reaches $100\overline{0}°F$ and the exhaust temperature is $60\overline{0}°F$. What is the thermal efficiency?

12. For the engine in Problem 11, the cooling system is removing heat at the rate of $20\overline{0}$ Btu/min. How much horsepower is being removed from the engine by the cooling system?

13. Refer to Problem 11. Diesel engines usually have lower exhaust temperatures than gasoline engines. If this engine were a diesel with an exhaust temperature of $40\overline{0}°F$, what would be the thermal efficiency?

14. What is the thermal efficiency of a refrigerator if the Freon operates between the temperatures of $-25°C$ and $+1\overline{0}°C$? Consider $1\overline{0}°C$ to be T_{in} for use in the formula.

Computer Program

This program will determine the thermal efficiency of an ideal engine. You may use it to check the answers to some of the chapter problems. Also, you may want to use the program to assist in plotting a graph of efficiency (e) versus temperature in (T_{in}) [keep the temperature out (T_{out}) constant]. Place the efficiency scale on the vertical y-axis and the temperature in on the horizontal x-axis. Increase the temperature in by 10 or 20 degree increments.

PROGRAM

```
10   REM  CH ELEVEN PROGRAM
20   PRINT "PROGRAM FOR THERMAL EFFICIENCY. IF TEMP. IS IN DEG. F - "
21   PRINT "TYPE 1, IF IN DEG. C - TYPE 2."
30   INPUT A
40   PRINT "YOU SELECTED -- ";A
50   IF A = 1 THEN  GOTO 100
60   IF A = 2 THEN  GOTO 200
70   END
100   PRINT
110   PRINT "TYPE TEMP. IN THEN TEMP. OUT."
120   INPUT T1,T2
130   PRINT "TEMP. IN = ";T1;"DEG. F."
140   PRINT "TEMP. OUT = ";T2;"DEG. F."
150   LET I = (5 / 9) * (T1 - 32) + 273
160   LET O = (5 / 9) * (T2 - 32) + 273
170   LET E = ((I - O) / I) * 100
180   PRINT
190   PRINT "THERMAL EFFICIENCY = ";E;" %."
195   GOTO 70
200   PRINT
```

```
210    PRINT "TYPE TEMP. IN THEN TEMP. OUT."
220    INPUT T1,T2
230    PRINT "TEMP. IN = ";T1;" DEG. C."
240    PRINT "TEMP. OUT = ";T2;" DEG. C."
250    LET I = T1 + 273
260    LET O = T2 + 273
270    GOTO 170
```

NOTES

Line 40 displays your temp. scale.

Lines 150, 160, 250, and 260 convert temps. to Kelvin scale.

Line 170 contains the efficiency formula.

Part IV

Sound and Light

Sound is the lifeblood of radio stations. Disc jockeys and other radio personnel pay careful attention to such characteristics of sound as pitch, loudness, and tone. (Courtesy of Station KPNW, Eugene, Oregon)

Chapter 12

Objectives

When you finish this chapter, you will be able to:

- [ ] Diagram sound waves, explain compression and rarefaction, and define periodic motion.
- [ ] Explain characteristics of sound: loudness, quality, fundamental frequency, and overtones.
- [ ] Explain the relationship between velocity, frequency, and wavelength of sound waves, and how different substances transmit sound at different velocities.
- [ ] Recognize the wave motion formula $V = f\lambda$: velocity of sound (V) equals frequency of vibration (f) times wavelength (λ).
- [ ] Discuss the difficulties in measuring and controlling the loudness of sound (noise pollution).
- [ ] Define a transducer.
- [ ] Discuss sound reflection, echoes, and echo control.
- [ ] Distinguish among natural, forced, and sympathetic (resonant) vibrations.
- [ ] Describe and explain the phenomenon of beats.
- [ ] Describe and explain the Doppler effect.

Introduction to Sound

So far we have discussed the areas of mechanical energy and heat energy. Sound is also a form of energy, and most animal life can detect and react to it. An understanding of sound is important to the technician, for a number of reasons. Obviously, it is basic to the communications industry. Less obviously, some of the concepts of sound can be applied to that area of physics dealing with electromagnetic waves, which we will discuss later in the text. In addition, sound is used in burglar alarms, depth finders on ships, submarine detection, machining operations for brittle materials such as glass and ceramics, thickness measuring, and materials testing. Sound has also found new applications in medical technology.

We will not be able to discuss all these specialized applications in this text, but we will discuss a number of basic facts about sound. In this chapter, we will consider such topics as velocity, frequency, wavelength, loudness, reflection, resonance, beat frequency, and the Doppler effect.

12.1 How Sound Is Produced

Sound is produced when bodies are set in motion. However, not just any kind of motion will produce sound; there are rather rigid requirements for this motion. First of all, it must be **periodic motion**—that is, the pattern of motion must be repeated at constant intervals. For example, if a pendulum is set in motion, it will swing first in one direction and then in the opposite direction. The back-and-forth swing then will be repeated. Each succeeding back-and-forth swing will be somewhat smaller than the one before, but each complete swing will be made in the same length of time.

∗ periodic motion   The motion of some vibrating body that occurs at regular time intervals.

The length of time for a complete back-and-forth swing is called a *period*. For example, a pendulum with a period of one-half second completes a back-and-forth swing in one-half second. As another example of periodic motion, consider a guitar string rigidly fastened at both ends. As the string is plucked, it is first pulled to one side. When it is released, its elasticity causes it to swing to the other side. Like the pendulum, it swings back and forth, each back-and-forth motion (called a *vibration*) being completed in a certain length of time.

Any elastic substance will vibrate if it is struck. The number of vibrations per second, or **frequency** (*f*), depends on the size, shape, and material of the thing vibrating. The unit of frequency is the hertz (Hz), named after the German physicist Heinrich Rudolph Hertz (1857–1894). For example, an object vibrating 20 times each second is said to be vibrating at a frequency of 20 hertz. It may also be said to move at a frequency of 20 vibrations per second or 20 cycles per second (cps).

∗ frequency   The number of vibrations per second.

The second requirement for the motion with which we are going to deal is a restriction on the frequency of vibration. When dealing with sound, we are concerned with motion whose frequency is between approximately 20 and 20 000 hertz. We set this limitation because the human ear is sensitive to vibrations that lie approximately between these two figures, depending on individual differences. Whether there can be a sound if there are no human ears present to hear it will be left to the philosophers to decide. In this section, we will concern ourselves only with vibrations in this frequency range and their effects.

Now let us return to the vibrating guitar string. As it comes forward (Figure 12.1A), the string crowds the molecules of air in front of it, producing a *compression*. This compression of molecules travels away from the string, passing its energy from molecule to molecule of the air in its path.

You might understand this transfer of energy better if you were to place a number of blocks, such as dominoes, on end next to each other. You tip the first one. It falls against its neighbor, upsetting it. This one, in turn, upsets the next one, and so forth. Thus the motion travels the full row of dominoes, even though each one moves only a slight distance. In a similar manner, the compression may travel great distances through the air, even though each molecule of air moves only slightly.

As the string moves back (Figure 12.1B), it leaves a space into which the air molecules can rush. Having more room, the molecules are spaced farther apart. We say that a *rarefaction*— that is, a reduction in pressure (the opposite of compression)— has been produced. The rarefaction follows the compression through the air. As the string continues to vibrate, alternate

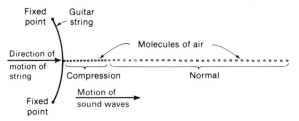

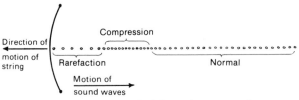

A. Vibrating string moves to the right, compressing the air molecules next to it.

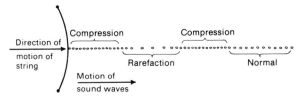

B. Vibrating string moves to the left, producing a rarefaction that follows the compression through the air.

C. Vibrating string moves to the right again, producing another compression of the air molecules.

FIGURE 12.1   Production of a Sound Wave.

---

compressions and rarefactions are produced in the air next to it. These alternate compressions and rarefactions travel through the air and are called **sound waves** (Figure 12.1C).

∗ sound waves   Alternate compressions and rarefactions of an elastic substance, such as air, that travel through a substance.

In our illustration, we have shown the compressions and rarefactions produced only by one point on the string. Obviously, all the air next to the full length of the string is set in similar motion. This applies to the air on the other side of the string as well. Thus, as Figure 12.2 shows, the vibrating guitar string sends out sound waves in all directions, somewhat like ripples on a pond into which a stone has been thrown.

12.2 Characteristics of Sound

There are a number of characteristics of sound that permit our ears to distinguish one sound from another. For example, we can easily tell the difference between the sound of a whistle and the rumble of thunder. We say the whistle is high-pitched, whereas the thunder is low-pitched. The *pitch* of a sound depends on the frequency of vibration of the object producing it. If the frequency is high, the sound is high-pitched; if the frequency is low, the sound is low-pitched. You will recall that the human ear can hear frequencies that lie only between about 20 and 20 000 vibrations per second (20 and 20 000 hertz).

Another characteristic of sound is its *loudness*. This characteristic depends on the force, or intensity, with which the object producing the sound vibrates. Thus, if the guitar string of our

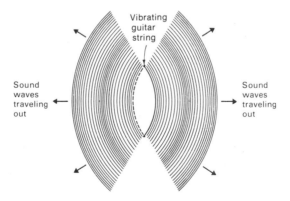

FIGURE 12.2   Sound Waves Traveling Out in All Directions from the Source of Sound. The lines close together represent compressions, and those farther apart represent rarefactions.

illustration is plucked gently, it will set up feeble compressions and rarefactions in the air, and these will strike softly against our ears. On the other hand, should the string be plucked vigorously, the compressions and rarefactions will be stronger and will strike harder against our ears. The sound that we hear will be louder.

However, there are other factors to consider. If the string vibrates strongly but at a great distance from our ears, the compressions and rarefactions of the sound wave would lose much of their energy before they reached us. The compressions and rarefactions would strike gently against our eardrums, and the sound that we hear would be soft.

There is still another factor. As we will see, sound waves can travel through a number of substances besides air. As a matter of fact, sound waves travel better through most of these substances than they do through the air. If the sound waves are traveling through a substance that can transmit them better than air, they will sound louder than they would if they were traveling through air. Thus, the loudness of a sound depends on the substance through which it passes.

It is possible for an object to produce sound waves that are too feeble for our ears to detect. We say that such sounds lie below the level of audibility.

Finally, there is the *quality,* or *tone,* of a sound. We can tell the difference between the tones of a violin and a trumpet, even though they both play the same note. Frequently, we can recognize the identity of a person merely from the quality of his or her speech.

The reason that two instruments or two voices sound different is the presence of overtones. Rarely in nature do we obtain a sound of only one frequency. Generally, when an object is struck or a string is plucked, sound waves of a number of different frequencies are produced. The frequency of the dominant sound wave is known as the *fundamental frequency*. The other sound waves that are produced at the same time are the *overtones,* or *harmonics*. The quality of a sound, then, depends on the frequencies and the relative strengths of the overtones produced at the same time as the fundamental frequency.

A violin string may produce the same fundamental frequency as a trumpet, but the frequencies of the overtones and their relative strengths will be different for the violin and for the trumpet. Thus middle C (whose fundamental frequency is 256 hertz) played on a violin will not sound the same as the same note played on the trumpet.

12.3 Velocity, Frequency, and Wavelength

Now let us return to our sound wave. You will recall that the vibrating guitar string sets the molecules of air in motion, producing a series of compressions and rarefactions that travel through the air, creating sound waves. Obviously, if there is to be a sound wave, there must be some substance whose molecules are able to pass on these compressions and rarefactions. We call this substance through which the sound wave travels the *medium*.

The medium used in our illustration is air. However, there are other media that may also transmit sound waves. As a matter of fact, any substance that has elasticity will transmit sound. Generally, solids are best. Steel, for example, will transmit a sound wave about 15 times as fast as air. Liquids are the next best media. Water, for example, transmits sound waves about 4 times as fast as air. Gases, generally, are the poorest media.

The rate at which sound waves travel through a medium depends on the elasticity and density of the medium. In the case of gases, the nature of the gas and its temperature affect its elasticity. Table 12.1 gives the velocity of sound in various media.

Now, let's continue our discussion of the guitar string. Each vibration of the string in Figures 12.1 and 12.2 produces 1 compression and 1 rarefaction in the air. The frequency of vibration will determine the number of compressions and rarefactions of the sound wave that will pass a given point in 1 second. If the string used in our illustration has a frequency of 1000 vibrations per second, then 1000 compressions and 1000 rarefactions will strike our eardrums each second. Our brain will

TABLE 12.1 Velocity of Sound in Various Media

Medium	Velocity	
	Feet/Second	Meters/Second
Aluminum	16 740	5100
Copper	11 670	3560
Steel (low carbon)	16 410	5000
Wood	10 000–15 000	3050–4570
Water	4794	1460
Air (at 68°F)	1126*	343

*The velocity of sound in air is frequently rounded to 1100 ft/s.

receive these impulses and give us the sensation of hearing a sound of 1000 vibrations per second (1000 hertz).

Associated with the speed of travel and the frequency is the concept of **wavelength** of the sound wave. Although a wavelength may be measured from any point on a wave to the corresponding point on the next wave, we generally say that it is the distance from the peak of one compression to the peak of the adjacent compression. A wavelength is illustrated in Figure 12.3. The symbol for wavelength is the Greek letter "lambda" (λ).

---

* wavelength

The distance between the points of maximum compression of two adjacent compressions of a sound wave.

Note also in Figure 12.3 that a sound wave is often represented graphically by a curve called a **sine wave.** The sine wave we have shown indicates the variation in air pressure along the wave. Each wave "peak" represents a point of maximum compression (or point of maximum rarefaction).

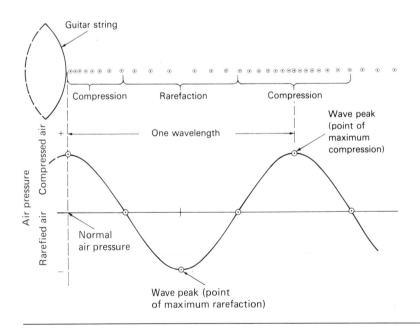

FIGURE 12.3   Illustration of Wavelength.

| * **sine wave** | The shape of a wave pattern that has characteristics related to the sine of an angle. |

If our string is vibrating at 1000 vibrations per second, it will send out 1000 compressions and 1000 rarefactions every second. If these travel at the rate of 1126 feet per second, the distance between any two successive compressions (or rarefactions) will be approximately 1 foot. We say, then, that the sound has a wavelength of about 1 foot. If the string vibrates 500 times per second, the wavelength will be about 2 feet.

To help us illustrate the relationship between wavelength, velocity, and frequency, consider the four soldiers marching in single file shown in Figure 12.4. They are evenly spaced 5 feet apart and walk past you in 10 seconds. How fast are they walking? If we can determine how far the last man walks to reach you in 10 seconds, we can solve the problem. The general formula for velocity is $V = \Delta D/t$. For this application, the change in distance (ΔD) is the number of soldiers (N) times the distance between the soldiers (d):

$$V = \frac{Nd}{t} = \frac{4 \text{ soldiers} \times 5 \text{ ft}}{10} = 2 \text{ ft/s}$$

Now, if peaks of sound waves are passing us at the same rate as the soldiers in this example, we can find the velocity of

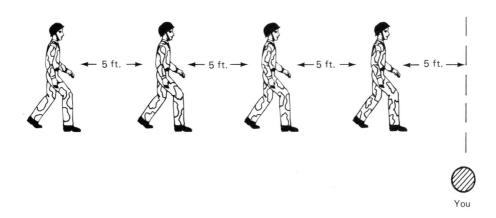

FIGURE 12.4   Illustration of Frequency: Marching Soldiers.

the sound wave in a similar way. But we must first adjust the formula a bit so it will be easier to use with waves. We will begin this "adjustment" by asking a related question: "What is the distance between wave peaks for a sound having a frequency of 550 hertz?" A frequency of 550 hertz means that 550 compression peaks (N) pass a given point in 1 second. We know that these peaks are evenly spaced, and we take the speed of sound in air as 1100 feet per second. Using the formula just derived from the soldiers and substituting, we have:

$$V = \frac{Nd}{t} = 1100 = \frac{550d}{1}$$

Now, we can find the distance:

$$d = \frac{1100}{550} = 2 \text{ ft}$$

This formula is simplified for wave theory as follows: Since frequency is the number of peaks (or vibrations) passing us each second, then N/t actually represents frequency. We know the symbol for frequency is f. Also, d is not used to indicate wavelength (peak-to-peak distance), but the Greek letter lambda (λ), as mentioned earlier. After making these changes, we have:

---

* WAVE MOTION FORMULA

velocity = frequency × wavelength

$$V = f \times \lambda$$

where V = the velocity in ft/s or m/s (depending on the units used for wavelength)
f = the frequency in Hz
λ = the wavelength in ft or m

SAMPLE PROBLEM 12.1

What is the wavelength (in meters) of a sound with a frequency of $80\overline{0}0$ Hz? Use 343 m/s for the speed of sound in air.

SOLUTION

Wanted $\lambda = ?$

Given $f = 80\overline{0}0$ Hz and $V = 343$ m/s

Formula First we must rearrange the formula:

$$V = f \times \lambda \quad \text{to} \quad \lambda = \frac{V}{f}$$

Calculation

$$\lambda = \frac{343 \text{ m/s}}{80\overline{0}0 \text{ Hz}} = 0.0429 \text{ m}$$

Such a small value is usually expressed in millimeters. To find millimeters, we multiply meters by 1000:

$$\lambda = 0.0429 \times 1000 = 42.9 \text{ mm}$$

Knowledge of the wavelength of a particular sound frequency is necessary when designing equipment to produce the sound, and for calibrating instruments used to determine the thickness of materials or liquid levels.

There is one other characteristic of sound to consider at this point. When sound travels from one medium to another, as through a wall and into a room, the velocity and wavelength change in each medium, but the frequency remains constant. Thus, when making calculations for sound traveling through some other medium, you must know how to apply the formula. Problem 18 at the end of this chapter involves a device for detecting microscopic cracks in a metal part. This sounding device is called a **transducer**. In the problem, it converts an electrical signal into sound at a specified frequency. The crack reflects the sound back, thus alerting the technician to the flaw.

---

* transducer   A general term for devices that convert some sort of signal from one form of energy to another. A loudspeaker and a microphone are examples of transducers.

12.4 Difficulties in Measurement and Control of Loudness

As stated earlier, loudness depends on the pressure, or intensity, of the sound wave. In order for the ear to hear a sound, the sound must exert a pressure of at least one ten-millionth of a pound per square inch on the ear drum. At the other extreme, if the pressure is over approximately one-hundredth of a pound per square inch, the sound becomes uncomfortable.

Such knowledge of sound is extremely important to the electronic sound communication industry and to environmentalists concerned with noise pollution. In order to produce sound, the communications industry must supply sufficient power for the listener to hear, but not so much as to be deafening. The power should not vary to such an extent that the listener must continually adjust the volume control knob. (Sometimes when you have your television turned to low sound volume, commercials seem to come through louder—because the technician at the station

PHOTO 12.1   Sound levels were carefully monitored during the first landing of a supersonic airliner (the Concorde) in Los Angeles International Airport. Noise studies have been carried out at airports to protect nearby residents from "noise pollution." (Courtesy of Wyle Laboratories)

has turned up the power just a bit.) Environmentalists want to keep industrial noise and airplane noise at airports below undesirable levels.

One problem in measurement and control of sound is that the ear is more sensitive to frequencies in the 3000–4000 hertz range than to the extremely low or extremely high frequency ranges. Another problem is that musical sounds, speech, and noise are generally made up of a multitude of frequencies. Trying to determine the loudness level of a given sound is not a direct measurement; the sound must be compared to the equivalent loudness of a sound at a single frequency. A frequency of 1000 hertz generally is used for comparison. Also, sound levels cannot be measured directly because some sounds block out others at the ears.

Still another problem is that the ear hears changes in loudness roughly equivalent to a logarithmic scale. We will not go into a mathematical explanation here, but essentially, this means that if the power of a sound source were increased by 10 times, the ear would not hear a sound 10 times as loud. Instead, the ear would indicate that the sound was only twice as loud.

12.5 The Reflection of Sound Waves

What happens when sound waves strike an obstacle? Well, part of their energy is transmitted as sound waves through the medium of the obstacle itself. Another portion of the sound wave, however, may be reflected from the obstacle in the same way that light is reflected from a mirror.

Look at Figure 12.5 and suppose that a woman is standing at one end of the lake, which is about a quarter of a mile in diameter. If she shouts at a clump of trees located on the opposite shore, the sound waves will travel through the air across the lake and will be reflected back from the trees. A listener stationed on the shore at a point about halfway between the shouter and the trees will hear two calls. One shout is from the sound wave that travels directly from the shouter to the listener. The other results from the sound wave that travels across the lake and is reflected back from the trees. Since this second sound wave must travel a greater distance, it will reach the listener a little later than will the first sound wave (in this example, a little more than a second later). The human ear can distinguish between sounds that are about one-fifteenth of a second, or longer, apart. Thus, the listener hears first the shout and, about a second later, the reflected sound. We call the reflected sound an *echo*.

PHOTO 12.2   This radar/radio antenna in Arecibo, Puerto Rico, has a diameter
of 1000 feet. It can detect radar waves or radio waves with wavelengths as short as
6 centimeters reflected from objects as far away as the planet Saturn. In this way, it
has been used to explore the surfaces of planets in the Solar System. (Courtesy of
NASA)

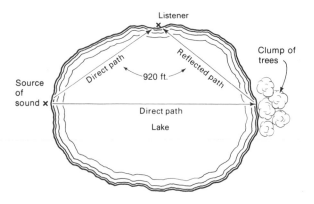

FIGURE 12.5   Production of an Echo.

Echoes may be beautiful in open country, but when they occur in a room or a concert hall, they are nuisances. They are not troublesome in rooms of ordinary size, since the sound waves reflected from the walls reach the hearer's ears in too short a time to be discerned. But if the room is 40 feet or more in length, the time it takes the sound wave to reach an opposite wall and return may be more than one-fifteenth of a second and the echo may be detected.

Under such circumstances, the room must be made sound-proof. The walls may be constructed of special, inelastic, porous material, or heavy cloth drapes may be hung over the walls. Sound waves are absorbed when they encounter materials that contain pores or crevices. The waves are reflected back and forth from one side of the crevice to the other, until most of their energy is dissipated as heat. Thus very little of the sound wave remains to be reflected back into the room. Often, floors and ceilings are similarly soundproofed.

Echoes can also be helpful to us. For instance, depth-sounding devices, radar, sonar, fish-finding devices, and flaw-detecting devices for metals all use the echo principle. The formula is simply our old velocity formula, $V = \Delta D/t$, that we are already familiar with. Sample Problem 12.2 demonstrates an echo problem. Of course, instruments simply measure the time interval required for a sound to reach an object and return, but their scales are usually calibrated in distance units for direct reading.

SAMPLE PROBLEM 12.2	A sonar device on a ship emits a high-frequency sound, and the echo is reflected back from a submarine in a time of exactly 4.00 s. How far away is the sub?
SOLUTION	
Wanted	$\Delta D = ?$
Given	$t = 4.00$ s. From Table 12.1, we find the velocity of sound in water, $V = 4794$ ft/s.
Formula	We rearrange: $$V = \frac{\Delta D}{t} \quad \text{to} \quad \Delta D = V \times t$$

Calculation
$$\Delta D = 4794 \text{ ft/s} \times 4 \text{ s} = 19\,176 \text{ ft}$$

This answer is the total distance the sound has traveled; the sub, of course, is only half that distance away. Therefore, the sub is 9590 ft away.

Normally, sound waves travel out in all directions from the source of sound. Consequently, the sound energy that travels in any one direction is only a fraction of the energy of the whole. If we wish to concentrate all the sound energy in one direction, instead of permitting it to spread in all directions, we may use reflecting devices that trap the sound waves and cause them to travel in the desired direction. One such device, as shown in Figure 12.6, is the megaphone, which is a funnel-shaped tube. The sound waves enter the small end, and when they attempt to spread out, they are reflected back from the inner surfaces of the megaphone. As a result, when the sound waves emerged from the large opening, they are all traveling in the same direction. Because the entire sound energy is directed in a single direction, the resulting sound is louder than if no megaphone were employed.

12.6 Natural Frequency and Forced Vibrations

When we discussed the loudness of sound, we said that three factors involved were the intensity of vibration of the source, the distance between the source and our ears, and the medium through which the sound wave traveled. There is still another factor—namely, the area of the vibrating surface.

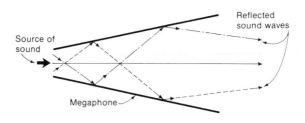

FIGURE 12.6   The Megaphone.

Let us see why. If we consider the guitar string of our previous illustration, the amount of air in contact with the thin vibrating string is quite small, and only a small amount of air is set in motion. Suppose, however, that a large board is set vibrating. The amount of air in contact with its surface is large. Thus, a larger amount of air is set in motion and, if the other factors (such as intensity, distance, and medium) remain constant, the sound we hear will be louder.

Use is made of this fact to *amplify* a sound, or make it louder. In the case of a violin, for example, the sound of the vibrating strings would be too weak to be heard in a large room. Accordingly, the energy of the vibrating strings is transmitted to a sounding board (which is the wooden body of the violin) through a thin strip of wood, called the bridge. When this sounding board vibrates, a larger amount of air is set in motion, and the sound we hear is louder.

If an object is permitted to vibrate freely—that is, free from any external influences except gravity and friction—it will vibrate at its **natural frequency,** or natural period, which is determined by the nature of the object, its size, and its shape. It is possible for one vibrating object to transmit its energy to a second object, and to make this second object vibrate at the same frequency, regardless of the natural frequency of this second object. This effect is called **forced vibration.** The vibrating violin string, which transmits its energy through the bridge to the sounding board and forces the latter to vibrate at the same frequency, is an example. Similarly, the sounding board of a piano amplifies the sound of the strings by forced vibration.

---

* natural frequency	The frequency at which an object, because of its shape, size, and mass, vibrates more easily than at other frequencies.
* forced vibration	The vibration of an object at a frequency other than its natural frequency. This vibration must be forced on the object.

12.7 Sympathetic Vibration and Resonance

The actual pressure exerted by sound waves in air is very small. As stated earlier, the sound wave that produces the faintest sound that can be detected by the average human ear exerts a pressure of about one ten-millionth of a pound per square inch against the eardrum. The pressure of the sound wave producing the loudest sound the ear can stand may be only about one-hundredth of a pound per square inch. Thus, we see that if a sound wave encounters an obstacle, the amount of energy it can impart to that obstacle will be very small. Normally, this slight amount of energy is dissipated by the obstacle. However, if the natural frequency of the obstacle is the same frequency as the sound wave striking it, then the object may be set vibrating quite vigorously and thus emit a sound wave of its own.

We can compare what happens to the sound waves to a child sitting on a swing. As in the case of the pendulum, the length of the ropes determines the number of times per second the swing will be able to travel back and forth. This is the natural frequency of that swing. To set the swing in motion, it is not necessary to push very hard. A slight push will set it swinging, and a continuation of these slight pushes soon will have the swing moving vigorously. However, these slight pushes cannot be delivered at random. If you push as the swing is traveling toward you, you will slow it down. If you try to push the swing after it has passed you and is going away, you will find that you have missed it. The proper time to push the swing is at the instant when it has stopped traveling toward you and is starting to travel away from you. Since there is only one such instant during each complete back-and-forth swing, you must push at that instant. In other words, your pushes must be *in step* with the natural frequency of the swing. Then each slight push will increase the motion of the swing and soon it will be moving vigorously.

In the same way, sound waves will cause an object to vibrate if the frequency of the sound waves is the same as the natural frequency of the object. The first push will make the object vibrate, although these vibrations may not be strong enough to cause an audible sound. However, the cumulative effect of succeeding pushes *in step* with the vibrations of the object soon has the object vibrating vigorously. We call this effect **sympathetic vibration.** We say that the object vibrating and the source of the sound wave that causes it to vibrate are in **resonance.**

---

* **sympathetic vibration**	The effect that occurs when an object is caused to vibrate at its natural frequency by a sound wave having the same frequency.
* **resonance**	The effect that occurs when a sound wave causes sympathetic vibration in an object.

As an example, you may find that striking a certain note on a piano will cause a nearby object to emit that same note. The vibrating piano string and the nearby object are in resonance. A favorite trick of singers who have powerful voices is to shatter a thin glass by singing at it. First, the glass is struck to enable it to emit a sound at its natural frequency. Then the singer repeats that sound. The glass starts to vibrate because of sympathetic vibrations. Then the singer repeats the note, but with more volume. The glass vibrates more vigorously. This process is continued until the glass vibrates so strongly that it shatters.

A column of air may also be set in sympathetic vibration (may be made to resonate) if the column is of the proper length. Like a vibrating string, a vibrating air column may produce a musical sound. Suppose that you blow across the open end of a tube. A musical whistle will be produced. The pitch of the sound depends on the length of tube, or, more accurately, the length of the vibrating air column within the tube. The shorter the air column is, the higher the pitch will be; the longer the air column is, the lower the pitch will be.

The original vibrations or puffs may have little energy. But if they are in resonance with the vibrating air column, the air column may be made to vibrate quite strongly and give out a loud sound, just as the swing discussed previously may be made to move quite vigorously by the application of a series of feeble pushes. As an example, we can place a vibrating string just over the open mouth of a tube that is closed at the far end. In Figure 12.7A, the string has reached the end of its motion away from the tube and is starting back. The air below the string is compressed, and a compression wave peak starts down the tube. In Figure 12.7B, the string has reached the midpoint of its swing. If the tube is of the proper length, the wave peak will have reached the bottom. There it is reflected and starts back to the opening. In Figure 12.7C, the string has completed one-half a vibration and is starting back. The wave peak has reached the top of the tube and is ready to aid the string in its backward motion, because the tube is exactly the right length for the wave to travel down and up in the time required for the string to make one-half a vibration. In other words, for resonance to occur, the wave traveling down and up the tube must cover a distance equal to one-half the wavelength of the vibrating string; therefore, the tube itself must be equal in length to one-quarter of the wavelength.

When this condition occurs, the energy of the reflected compression wave will aid the vibrations of the string. As a result, the string (and the air column) will vibrate more vigorously and

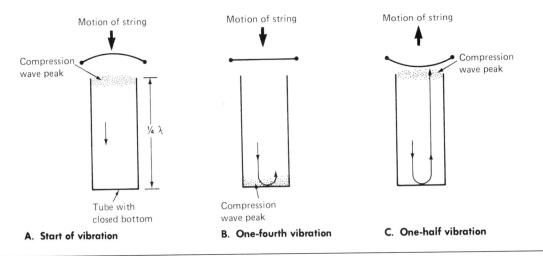

FIGURE 12.7   Illustration of Resonance.

a louder sound will result. If a condition other than resonance prevails, the vibrations of the string and those of the air column will interfere with each other and their energies will be dissipated. A little thought will show that resonance may be achieved if the wave travels down and up the tube while the string is making 1/2 vibration, 1-1/2 vibrations, 2-1/2 vibrations, 3-1/2 vibrations, and so forth. In other words, the length of the tube may be 1/4 wavelength, 3/4 wavelength, 5/4 wavelength, 7/4 wavelength, or any other odd quarter-wavelength.

Now, looking at Figure 12.8, suppose that the tube is open at both ends. When the compression wave reaches the bottom of the tube, instead of finding a hard surface to reflect it, it finds an open space. The compressed air molecules rush into that space. As they do so, the pressure in the bottom of the tube is reduced, and a rarefaction results. The air molecules in the tube above this rarefaction rush down into the area of reduced pressure, and this action reduces the pressure farther up the tube. Thus, if the tube is open at both ends, a compression wave traveling down the tube produces, when it reaches the bottom, a rarefaction that travels up the tube.

Since a rarefaction occurs half a vibration after a compression, if resonance is to be achieved, the string must have completed a full vibration by the time the rarefaction reaches the top of the tube. Thus, the length of the open tube must be equal to half the wavelength of the vibrating string. Similarly, reso-

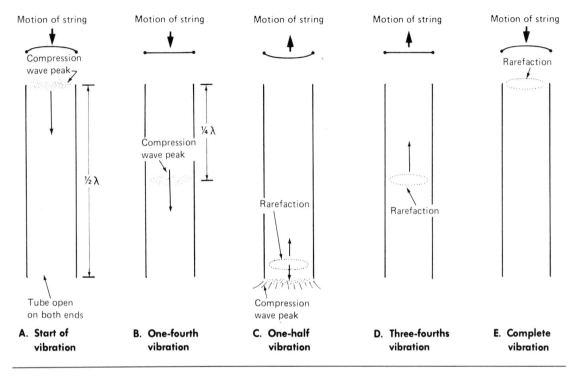

Motion of string

Compression
wave peak

Tube open
on both ends

**A. Start of
vibration**

Motion of string

Compression
wave peak

¼ λ

**B. One-fourth
vibration**

Motion of string

Rarefaction

Compression
wave peak

**C. One-half
vibration**

Motion of string

Rarefaction

**D. Three-fourths
vibration**

Motion of string

Rarefaction

**E. Complete
vibration**

½ λ

FIGURE 12.8   Illustration of Fundamental Frequency.

nance may also be achieved if the length of the open tube is equal to 2/2 wavelength, 3/2 wavelength, or any other half-wavelength.

Therefore, if we have a tube that is closed at one end, the air column within it will resonate with any sound wave, provided that the frequency of the sound wave is such that the length of the air column is equal to 1/4 wavelength, 3/4 wavelength, or any other odd multiple of the quarter-wavelength of that sound wave. If the tube is open at both ends, the air column will resonate with any sound wave if the length of the column is equal to 1/2 wavelength, or any multiple of the half-wavelength. The lowest frequency at which a given air column will resonate is called the *fundamental frequency*. All other resonating frequencies are called *overtones*, or *harmonics*.

**SAMPLE
PROBLEM
12.3**

Determine the frequency of a 6.00 in. organ pipe that is (A) closed at one end and (B) open at both ends.

SOLUTION A

Wanted

$f = ?$

Given

The pipe is 6.00 in. long. From the previous discussion and from Figure 12.7, we note that the pipe is one-quarter of the wavelength of its natural frequency. Therefore, we can calculate the wavelength as being four times the length of the pipe, $\lambda = 4 \times 6.00$ in. $= 24.0$ in. (2 ft). From Table 12.1, we find the velocity of sound in air, $V = 1126$ ft/s.

Formula

We must rearrange the wave motion formula:

$$V = f \times \lambda \qquad \text{to} \qquad f = \frac{V}{\lambda}$$

Calculation

$$f = \frac{1126 \text{ ft/s}}{2 \text{ ft}} = 563 \text{ Hz}$$

SOLUTION B

Wanted

$f = ?$

Given

From Figure 12.8, we see that the open pipe is one-half the wavelength of the natural frequency. Thus, we can calculate the wavelength, $\lambda = 2 \times 6.00$ in. $= 12.0$ in. (1 ft).

Formula

$$f = \frac{V}{\lambda}$$

Calculation

$$f = \frac{1126 \text{ ft/s}}{1 \text{ ft}} = 1130 \text{ Hz}$$

This information is, of course, important to musicians. However, the technician, who must help design and service not only musical instruments but industrial instruments, also makes use of the resonant characteristics of sound waves. This basic knowledge of how a column of air may be made to vibrate at its natural frequency (resonate) is valuable when investigating machine vibrations and noise.

12.8 Beats

What happens when two sound waves meet? If the two waves are at the same frequency, they reinforce each other, and the sound we hear is louder. All the compressions and rarefactions are in step, or as technicians say, *in phase*.

If the two sound waves are not at the same frequency, the compressions and rarefactions are not all in phase. At one instant, the sounds may be in phase and reinforce each other. At another instant, they may be out of step *(out of phase)*. In this case, a compression, which crowds the molecules of the medium, may encounter a rarefaction, which spreads the molecules. The two tend to neutralize each other.

The ear, then, will hear a sound caused by a blending of the two sound waves. When they are in phase, the sound will be louder. When they are out of phase, the sound will be softer. The sound will rise and fall a number of times per second, which is equal to the difference in frequency between the two sound waves.

PHOTO 12.3   Musicians tune their instruments to the same pitch by listening to the beat frequency heard when the instruments play the same note. The more beats heard per second, the farther apart the frequencies are. When no beats can be heard, the instruments are in tune. (Courtesy of the New York Chamber Soloists)

This rising and falling off of the sound is called **beats**. For example, if a sound wave whose frequency is 256 vibrations per second encounters a sound wave whose frequency is 260 vibrations per second, 4 beats per second will be produced—that is, the sound intensity will rise and fall four times per second.

∗ beats

A rising and falling off of sound that is heard when two different frequencies strike the ear.

To demonstrate beats, let the lines in Figure 12.9 represent the compression wave peaks of two frequencies, one of 9 hertz and one of 6 hertz. This is somewhat like looking down on ocean waves from an airplane and watching lines of waves approach the shore. The total interval in the figure represents a 1 second time span. Notice that the lines coincide three times. The difference in the two frequencies results in 3 beats per second.

This phenomenon is used in all modern radios and television sets. The radio frequency from a station is mixed with a different radio frequency generated by the receiver. The difference between these frequencies, the **beat frequency**, is then amplified and converted to sound. To illustrate, when a 1000 hertz sound is mixed with a 1500 hertz sound, we hear sounds at 1000 Hz, 1500 Hz, 2500 Hz (1000 + 1500 = 2500 Hz), and 500 Hz (1500 − 1000 = 500 Hz). The 500 hertz sound is called the beat frequency.

∗ beat frequency

The difference in frequency between two original frequencies.

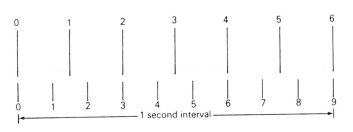

FIGURE 12.9 Illustration of Beat Frequency.

12.9 Doppler Effect

Have you ever noticed a car horn, a train whistle, or a fire siren change pitch (frequency) as the vehicle passes you? The name for this apparent frequency change is the **Doppler effect,** named in honor of Christian J. Doppler (1803–1853), an Austrian mathematician and physicist who first correctly explained this apparent change in frequency.

---

* Doppler effect

The change in frequency of a sound brought about by relative motion between the source and the listener.

In order to understand Doppler's explanation, we must again make use of the wave motion formula:

$$V = f \times \lambda$$

When applying this formula, subscripts must be used to indicate specific conditions. However, we will not apply a subscript to V when it indicates the speed of sound in the medium (air, for example). All other symbols will have subscripts: f_t is the true frequency generated by the source, and λ_t is the true length of the wave.

Let's see what happens when you are approaching a source emitting a constant sound—for instance, a bird in a tree. Look at Figure 12.10. The bird is creating a sound wave according to the formula:

$$V = f_t \lambda_t$$

By rearranging the formula, we have:

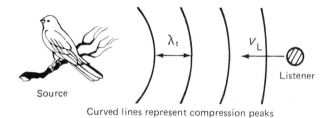

Curved lines represent compression peaks

FIGURE 12.10   The Doppler Effect: Listener Approaching the Sound Source.

* DOPPLER FORMULA 1

$$\lambda_t = \frac{V}{f_t}$$

Now, if the listener were stationary, he or she would hear this true frequency. If the frequency is 500 hertz, then 500 compression peaks are passing the listener in 1 second. However, if the listener moves toward the source, then more than 500 compression peaks will pass the listener in 1 second. (Remember the marching soldiers? If you start marching in the direction opposite to the soldiers, more soldiers will pass you in a given time interval.) The sound waves are traveling by the listener at a faster *relative* velocity. This relative velocity is equal to the velocity of sound (V) plus the velocity of the listener (V_L): ($V + V_L$). Therefore, the listener hears a different frequency. This apparent frequency (f_{ap}) is determined by the formula:

$$V + V_L = f_{ap}\lambda_t$$

Again, we can rearrange:

* DOPPLER FORMULA 2

$$f_{ap} = \frac{(V + V_L)}{\lambda_t}$$

Note that the wavelength doesn't change (see Figure 12.10).

A formula without the wavelength would be desirable. Referring back to formula 1, we can substitute V/f_t for λ_t in formula 2:

$$f_{ap} = \frac{(V + V_L)}{V/f_t}$$

Thus, for situations in which the listener is moving toward the sound source, we have:

* DOPPLER FORMULA 3

$$f_{ap} = \frac{f_t(V + V_L)}{V}$$

If the listener is moving away from the sound, then we subtract the listener's velocity from the sound's velocity:

* DOPPLER FORMULA 4

$$f_{ap} = \frac{f_t(V - V_L)}{V}$$

Note that if you are moving away from the sound source at the speed of sound, the relative velocity is zero, and you hear no sound.

Surprisingly enough, if you are stationary and the sound source moves toward you, a different formula is required. (Additional Doppler formulas will be left for other texts.) Sample Problem 12.4 illustrates the Doppler effect.

SAMPLE PROBLEM 12.4

We are in a car traveling at 60.0 mi/hr (88.0 ft/s) toward a firehouse siren producing a single frequency of $50\overline{0}0$ Hz. Let the speed of sound in air be 1130 ft/s. What is the apparent frequency of the sound?

SOLUTION

Wanted

$f_{ap} = ?$

Given

$V = 1130$ ft/s, $V_L = 88.0$ ft/s, and $f_t = 50\overline{0}0$ Hz

Formula

Since the listener is moving toward the sound, we must use formula 3:

$$f_{ap} = \frac{f_t (V + V_L)}{V}$$

Calculation

$$f_{ap} = \frac{5000 \text{ Hz} (1130 \text{ ft/s} + 88 \text{ ft/s})}{1130 \text{ ft/s}} = 5390 \text{ Hz}$$

The answer in Sample Problem 12.4 and the use of these formulas for the Doppler effect have practical applications with the radio frequencies used in radar, high-frequency sound in sonar, and light frequencies in astronomy. In radar and sonar, the distance to a target is measured by the time for the signal to bounce back, or echo; and a frequency shift in the return signal means that the target is approaching or retreating. By comparing the frequency of the outgoing and return signals, the direction and speed of target movement can be determined.

Similarly, astronomers speak of the "red shift" or "blue shift" of a star. The apparent frequencies of the light from a star shift from the true values depending on whether the star is approaching or receding from the earth.

Summary

An understanding of sound is not only important in its own area of physics, but the principles and concepts also apply to other areas of physics where energy is transmitted in waves with periodic vibrations.

PHOTO 12.4   A symphony orchestra contains different types of musical instruments, including strings, winds, and percussion. The acoustics of the hall they play in must be designed to prevent echoes and let the overall sound retain its clarity. (Courtesy of Boston Symphony Orchestra)

Sound is transmitted through elastic materials by a series of compressions and rarefactions of the molecules of the material. Although the molecules themselves move very little, they affect their neighbors so that a wave of compression and rarefaction passes through the material. Note that a complete wave includes a compression and a rarefaction of the molecules.

Sound is usually produced by some vibrating object. The number of vibrations each second, and therefore the number of compression peaks (or rarefaction peaks) we receive each second, is called the frequency. The unit of frequency is the hertz (Hz). A high-frequency sound is called a high-pitched sound, and the human ear can hear sounds generally in the range from 20–20 000 hertz. A loud sound has more energy than one not so loud. This energy is in the form of a higher than usual pressure for the compression and a lower than usual pressure for the rarefaction. The quality of a sound depends on the frequencies making up the sound. Also, sound travels at different speeds in different substances.

Each frequency has its own wavelength, as determined by the medium it is traveling in. A wavelength is easiest to understand if we say that it is the distance from the peak of one compression to the peak of the adjacent compression. (Actually, the wavelength may be measured from any point on a wave to the corresponding point on the next wave.)

The velocity of sound, the frequency, and the wavelength are related by the formula: $V = f \times \lambda$. When using this formula for different media, remember that the frequency of the sound doesn't change. A 1000 hertz sound will have the same pitch (frequency) in air, water, or other liquids, gases, or solids.

It is difficult to measure loudness because the ear hears some frequencies better than others, some frequencies block others, and the ear hears sounds in a logarithmic proportion.

Sound waves can be reflected from obstacles (an echo), and this fact is used to detect submarines and other underwater objects, such as schools of fish. Sound can be used to determine the depth of the ocean floor or to detect microscopic cracks in metal parts.

Sound can force an object to vibrate at the same frequency (forced vibration). Sounding boards on musical instruments are a good example. If the object's natural frequency of vibration is the same as the sound, we say the object's vibration and the sound are in resonance. In order to cause a column of air to vibrate in resonance with the applied vibration, the column of air must be in a tube with a length that is related to the wavelength of the sound. For closed tubes, the length must be 1/4, 3/4, or any other odd quarter-wavelength. For open tubes, the length must be 1/2, 1, 1-1/2, or any other half-wavelength.

When two sounds meet, their combined effect produces an additional sound with another frequency. This sound, having a frequency equal to the difference between the original two frequencies, is called the beat frequency. When two compression peaks reach the ear at the same instant, the sound is amplified. When a compression peak of one sound and a rarefaction peak of the other sound reach the ear at the same instant, they nullify each other. This sequence of sound produces the beat frequency.

The Doppler effect is caused when the sound source, or listener, or both, are moving. The motions cause the listener to hear an apparent frequency change. If the source and/or listener are approaching, the frequency (or pitch) sounds higher than it actually is. If the two are receding, the pitch sounds lower.

Key Terms

* sound
* periodic motion
* frequency
* sound waves
* wavelength
* sine wave
* wave motion formula
* transducer
* natural frequency
* forced vibration

* sympathetic vibration
* resonance
* beats
* beat frequency
* Doppler effect
* Doppler formula 1
* Doppler formula 2
* Doppler formula 3
* Doppler formula 4

Questions

1. Define "frequency." What are its units?
2. What is meant by a sound wave? How is it produced?
3. What is meant by the pitch of a sound? By its loudness? By its quality?
4. What determines wavelength? That is, where is the measurement made?
5. When sound goes from one medium to another, what changes—velocity, frequency, or wavelength?
6. Can the loudness of a sound be measured directly? Explain.

7. Explain the term "echo."

8. What is meant by "natural frequency"?

9. What is meant by resonance? How does this differ from a forced vibration?

10. Explain how a column of air can be made to resonate.

11. What are beats? How are they produced?

12. What is meant by "beat frequency"?

13. What is the Doppler effect?

Problems

CHARACTERISTICS OF SOUND

In the following problems, use 1100 ft/s for the velocity of sound in air.

1. Solve Sample Problem 12.1, but change the speed to 1460 m/s for sound in water. The frequency remains the same.

2. Solve Sample Problem 12.1, but change the frequency to $56\overline{0}$ Hz. The speed remains the same.

3. What is the wavelength, in feet, of a $20\,\overline{0}00$ Hz sound in air?

4. Thunder has a frequency of approximately 35 Hz. What is the wavelength in feet?

5. The highest frequency transmitted by a telephone is about 2740 Hz. Using the velocity of sound in air, find the wavelength in feet.

6. If the sound in Problem 3 is transmitted through aluminum, determine its wavelength in feet.

7. What is the wavelength of a $10\overline{0}$ Hz sound in water? Give the answer in meters.

8. Two sounds in air have wavelengths of 5.5 ft and 2.5 ft. What are their frequencies?

9. It is discovered that a sound wave in an unknown medium has a frequency of $50\overline{0}$ Hz and a wavelength of 10.0 m. Determine the velocity in meters per second and look at Table 12.1 for the material.

10. A sound in air has a wavelength of 1.6 m. What is its frequency?

11. Bats emit short ultrasonic squeaks at frequencies up to $15\overline{0}\,000$ Hz. What wavelength does this represent?

12. A sonic device is used to measure the thickness of aluminum foil. For best operation, the wavelength should be equal to four times the thickness of aluminum foil, which in this case, is 0.012 in.
    a. What frequency should be used?
    b. What frequency should be used if the material is copper foil? (*Remember:* Wavelength should be measured in feet.)

13. A sound wave travels 22 400 ft in 2.00 s. What is the velocity?

14. A sound wave travels 532 ft in 0.040 s. What is the velocity?

REFLECTION OF SOUND

15. A ship's depth finder receives its return signal 3.00 s after sending it. Bearing in mind the speed of sound in water, find the water depth (a) in feet and (b) in meters.

16. A hunter shouts across a pond and his echo returns in 2.50 s. How far is it across the pond? (Trees at the edge of the pond reflect the sound.)

17. A bat sends out a signal and receives an echo from a wall 0.005 s later. How far is the bat from the wall?

18. In order to detect microscopic flaws in a metal part, a high-frequency sound signal is sent out from a transducer at one end of the steel bar shown in Figure 12.11. What is the time interval for the transducer to receive the echo from (a) the end of the bar 30 in. away and (b) the flaw 12 in. away?

19. Refer to Figure 12.5. What are the time intervals for a listener (stationed as shown) to hear a sound by the shortest path and by the echo from the trees?

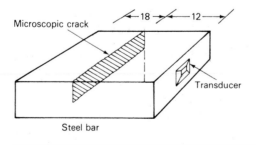

FIGURE 12.11   Steel Bar with a Flaw, for Problem 18.

RESONANCE AND BEATS

In the following problems, use 1126 ft/s for the velocity of sound in air.

20. The largest open pipe in a particular church's organ is 8.00 ft long. Determine the frequency.

21. Solve Sample Problem 12.3 with a pipe 18.0 in. long.

22. Determine the shortest length of a closed tube needed to produce a 256 Hz sound (middle C).

23. How long would an open pipe have to be to produce a 550 Hz sound?

24. What is the fundamental frequency that a 16 ft long open-end organ pipe can produce?

25. What is the beat frequency between a 3.8 kHz and a 4.0 kHz sound?

DOPPLER EFFECT

26. We are in a plane traveling at $20\overline{0}$ mi/hr (294 ft/s) away from a tornado that produces a very loud sound at $70\overline{0}$ Hz. What is the apparent frequency we hear?

27. A stationary submarine sends out a 35 $\overline{0}$00 Hz sonar signal and receives the echo 4.00 s later. The target ship is moving toward the submarine at 30.0 ft/s. (*Remember:* The sound is in water.)
    a. How far away is the target?
    b. What frequency does a listener on the target ship hear?

28. You are on a train traveling at 70.0 mi/hr (103 ft/s). A crossing bell has a primary frequency of 8.00 kHz.
    a. What is the apparent frequency of the bell as you approach the crossing?
    b. What is the apparent frequency of the bell as you leave the crossing?

Computer Program

This program is for the Doppler formulas. You may use this program to check your answers to the appropriate chapter problems. Also, you can experiment to see how the apparent frequency changes with changes in velocity of the listener. This data can be used to plot a graph of apparent frequency versus listener velocity.

PROGRAM

```
10   REM   CH TWELVE PROGRAM
20   PRINT "DOPPLER FORMULAS.   ALL VELOCITIES MUST HAVE THE SAME"
21   PRINT "UNITS AND BE EITHER FT/S OR M/S."
30   PRINT "TYPE IN THE TRUE FREQUENCY IN HERTZ, THE VELOCITY OF"
31   PRINT "SOUND IN THE MEDIUM AND THE VELOCITY OF THE LISTNER."
40   INPUT F,V,VL
50   PRINT
60   PRINT "FREQUENCY = ";F;" HZ. VELOCITY = ";V;" FT/S (M/S)."
61   PRINT "VELOCITY OF LISTNER = ";VL;" FT/S (M/S)."
70   PRINT "IF LISTNER IS APPROACHING SOUND SOURCE-TYPE 1, IF "
71   PRINT "MOVING AWAY FROM SOUND SOURCE-TYPE 2."
80   INPUT A
90   PRINT "YOU HAVE SELECTED -- ";A
100  PRINT
110  IF A = 1 THEN   GOTO 200
120  IF A = 2 THEN   GOTO 300
130  PRINT "THE APPARENT FREQUENCY IS ";FAP;" HZ."
140  END
200  LET FAP = F * (V + VL) / V
210  GOTO 130
300  LET FAP = F * (V - VL) / V
310  GOTO 130
```

NOTES

Lines 60 and 61 confirm your data input.

Lines 200 and 300 are the two Doppler formulas.

A hot steel billet at "red heat" is being formed in a steel mill. The steel is hot enough to radiate energy in the form of light. (Courtesy of Phoenix Steel Corp.)

Chapter 13

Objectives

When you finish this chapter, you will be able to:

- [ ] Explain how atoms may become excited and how these excited atoms give off tiny bundles of light energy called photons.
- [ ] Compare longitudinal (sound) and transverse (light) waves.
- [ ] Diagram and discuss the electromagnetic wave spectrum—in particular, how visible light fits into the total spectrum.
- [ ] Compare transparent, translucent, and opaque substances.
- [ ] Diagram and explain the polarization of light and how polarization filters operate.
- [ ] Describe the difference between laser light and ordinary, or normal, light.

Introduction to Light

As a technician, you must understand the basic information in this chapter before you can work with optical equipment or electro-optical devices. If you already know something about light, then you know something about electromagnetic waves, which we will discuss in this and later chapters. In this chapter, we discuss the theory of how light is produced and the basic characteristics of light, such as speed, wavelength, and frequency. Sections on polarized light and laser light and their uses are also included.

13.1 Theories of Light

When we heat a bar of iron, we know that the heat energy applied causes the molecules of iron to move faster. We say the bar becomes "hotter." As it becomes hotter, the bar emits more and more radiant heat energy. As we continue to heat the bar of iron, we notice a peculiar effect: In addition to emitting radiant heat energy, the bar becomes red-hot and emits a new kind of energy—light energy. This new energy travels from the hot iron to our eyes. The light stimulates our eyes, and as a result, an impulse is sent to our brain, which interprets this impulse as the sensation of sight.

How the hot iron emits light energy, and how this energy reaches our eyes, is one of nature's deepest secrets. For many years, scientists pondered these problems, and only recently have they developed an explanation. Their explanation is theoretical—a guess, at best—and even now the theories are being subjected to searching tests and constant revision as new facts are discovered.

Although a thorough explanation of light energy is beyond the scope of an introductory book such as this, we may state that, as a result of the work of men such as the two German physicists and mathematicians Max Planck (1858–1947) and Albert Einstein (1879–1955), scientists today believe that light is emitted from a body in small bundles of energy, called quanta, or **photons.** Just as a water molecule is the smallest complete particle of water, so is a photon the smallest complete particle of light energy. Photons exhibit characteristics of both waves and minute particles, or bundles, of energy. Each bundle may be considered as a train of waves. Photons, which are emitted by light-giving or **luminous bodies,** travel through space and, striking the eye, give up their energy to produce the sensation of sight.

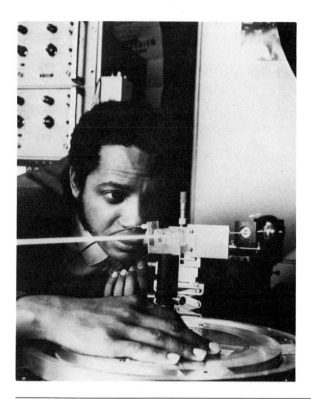

PHOTO 13.1   The laser produces an intense beam of light of a single frequency, enabling it to transmit great amounts of energy. Lasers are finding application in many different areas of industry; here, Dr. Gerard Alphonse of RCA Laboratories uses a laser beam (the white rod extending from right to left) in an experimental television system. (Courtesy of RCA)

∗ photon   The smallest complete particle of light energy.

∗ luminous body   A body that emits energy in the form of light.

What makes luminous bodies emit these photons? In an attempt to solve this problem, scientists turn to the electron theory concerning the structure of matter. From Chapter 1, we recall that Niels Bohr, the Danish scientist, gave us a picture of the atom where a central nucleus is surrounded by concentric shells of planetary electrons. Now, if a certain amount of energy, such as heat energy, is applied to the atom, one or more electrons contained in an inner shell may be forced to leave that shell and jump to an outer shell. We say the atom is *excited*. The displaced electron has acquired extra energy. However, the attraction of the nucleus soon pulls the displaced electron back to its normal position. As the electron falls back, it loses its extra energy. It is this extra energy that is shot out as a photon.

The concept of photons is helpful when studying such topics as light-sensitive devices (for example, photocells), lasers, and phenomena at the atomic level. As a technician or engineer, though, you will find that most of your work with light is based on the concept that light travels in waves.

When we considered a sound wave, we visualized a wave where the particles, or molecules, of the medium vibrated back and forth a short distance in line with the direction in which the wave was traveling. We call such a wave a **longitudinal wave**. Figure 13.1A demonstrates a longitudinal wave.

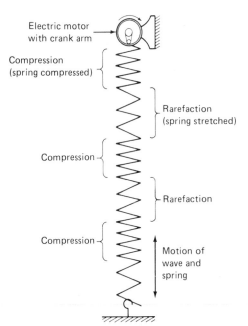

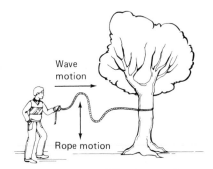

A. **Longitudinal wave:** Vibrations are in the same direction in which the wave is traveling.

B. **Transverse wave:** Vibrations are in a direction perpendicular to the direction in which the wave is traveling.

FIGURE 13.1   Two Types of Waves.

| * **longitudinal wave** | A wave whose vibrations are in a direction parallel to the direction in which the wave is traveling. |

However, there is another type of wave by means of which energy can be transmitted. As an example, assume that you have a length of rope, one end of which is tied to a tree and the other end is in your hand, as Figure 13.1B shows. If you flip your hand sharply up and down, a hump forms in the rope and travels as a wave from your hand to the tree. In this way, the energy of motion is transmitted through the rope. The particles of the rope do not travel in line with the direction that the wave travels. Your hand, which still holds the end of the rope, is no nearer the tree than before. Rather, the particles of the rope move up and down, at *right angles* to the direction that the wave travels. We call such a wave a **transverse wave**. Light waves are transverse waves, and, like sound waves, they have speed, wavelength, and frequency.

| * **transverse wave** | A wave whose vibrations are in a direction perpendicular to the direction in which the wave is traveling. |

13.2   The Speed, Wavelength, and Frequency of Light

How fast do light waves travel—that is, what is the speed of light? For many years, it was thought that the speed of light was infinite, that light traveled instantly from its source to the eye, even though these were miles apart. Eventually, some scientists suspected that the speed of light was not infinite, but rather was so great that it could cover any distance on the earth in too short a time to be measured. Recent experiments have set the value for the speed of light (rounded off) at 186 000 miles per second (300 megameters per second). This speed is accepted as the speed of light either in a vacuum or in air, since the difference is insignificant.

To understand wavelength and frequency as applied to light waves, we need to examine the transverse wave. Recall the example of wave motion in the rope tied to the tree (Figure 13.1). If you continue to flip your hand up and down, a series of waves moves toward the tree. In a similar fashion, small waves, or

PHOTO 13.2   The wavelength of light emitted by an excited atom is always the same. Since 1960, the international standard unit of length, the meter, has been defined as 1 650 763.73 wavelengths of orange-red light from the atomic isotope krypton-86. This light may be seen with the type of setup shown here. (Courtesy of the National Bureau of Standards)

ripples, are formed when you drop a pebble in a pond of still water. As Figure 13.2 indicates, ripples travel away from the center of disturbance in expanding, concentric circles. Each ripple is a transverse wave traveling away from the center of disturbance. A leaf or small wood chip floating nearby will move up and down as each wave passes, but will not be carried across the pond with the waves. This fact indicates that transverse waves cause each molecule of the medium (the water) to move up and down at right angles to the direction of travel of the wave. Note that no molecules get any closer to shore, because they are not carried along with the wave, just as the particles of rope in our previous example got no closer to the tree. As these molecules of water move up and down, they form crests above the surface of the pond, and depressions, or valleys, below the surface. If we

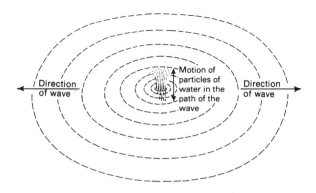

FIGURE 13.2   Transverse Wave Caused by Throwing a Pebble into a Still Pond.

were to take a cross-sectional view of the pond, it would appear as in Figure 13.3. Note that the motion is periodic—that is, the pattern is repeated at regular intervals.

Recall the discussion in the previous chapter on sound. The distance between two adjacent crests (or between two adjacent valleys) is called the wavelength (λ). The number of crests, or peaks, that pass a fixed point in a unit of time is called the frequency (f). The unit of frequency, the hertz (Hz), is the number of crests passing a fixed point in 1 second. If the speed of the wave is constant, then the longer the wavelength is, the lower will be the frequency; and the shorter the wavelength is, the higher will be the frequency.

The wave motion formula ($V = f \times \lambda$) in the previous chapter can also be used for light. The only difference is that scientists have agreed that the letter c should be used exclusively to represent the speed of light. Therefore, the formula $V = f \times \lambda$ becomes:

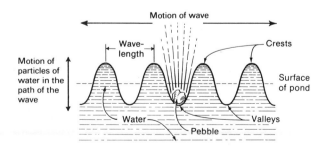

FIGURE 13.3   Cross-Sectional View of a Pond after a Pebble Is Dropped into It, Showing the Crests and Valleys Produced by the Motion of the Wave.

$$c = f \times \lambda$$

While this formula is a useful one, light wave frequencies are extremely large, and wavelengths are extremely small. Thus, calculations are a nuisance without the use of advanced mathematical notations, which are beyond this book. So, we will not become involved with speed of light problems in this text.

13.3 Electromagnetic Waves

When an excited atom returns to its normal state, a wave of energy is shot out. The general name for this wave, whose existence was predicted in 1867 by the brilliant English mathematician James Clerk Maxwell, is the **electromagnetic wave**. The speed of travel of an electromagnetic wave is always the same, 186 000 miles per second in a vacuum or in air. However, its frequency depends on the amount of energy emitted by the atom. The greater this energy is, the higher will be the frequency.

---

* electromagnetic wave

A wave that has electrical and magnetic characteristics, and is emitted by an excited electron as it returns to its normal position in an atom. Light waves are electromagnetic waves.

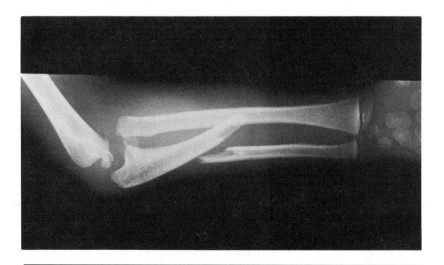

PHOTO 13.3   X-rays are also part of the electromagnetic spectrum. Their ability to penetrate body tissues and affect a photographic plate has enabled medical technicians to detect broken bones and other ailments. (Courtesy of Ruth McCarroll)

The entire frequency range of electromagnetic waves is called the electromagnetic spectrum. Visible light—that is, energy that is able to stimulate our eyes to produce the sensation of vision— forms only a small part of this spectrum. Included in this spectrum are the radiant heat waves, or rays, we encountered in the section on heat. These waves are produced in a manner similar to the production of visible light waves, differing only in frequency and wavelength. The heat rays have a lower frequency (and, accordingly, a longer wavelength) than the visible light waves. Also included in the spectrum of electromagnetic waves are radio waves, ultraviolet rays, X-rays, and gamma rays. Table 13.1 shows the electromagnetic spectrum arranged in order of ascending frequencies.

Note that the frequencies of the various types of waves, or rays, tend to overlap. For example, the highest-frequency radio waves and the lowest-frequency heat rays overlap. Actually, the waves in that overlapping region show the characteristics of both

TABLE 13.1 The Electromagnetic Spectrum

Type of Wave or Ray	Frequency, Hertz (Hz)	Wavelength	
		Meters (m)	Angstrom Units (Å)
	300 000 (300 kHz)	1000 (1 km)	
AM radio	3 000 000 (3 MHz)	100	
Radio waves	30 000 000 (30 MHz)	10	
TV, FM radio	300 000 000 (300 MHz)	1	
	3 000 000 000 (3 GHz)	0.100 (100 mm)	
Microwaves	30 000 000 000 (30 GHz)	0.010 (10 mm)	
	300 000 000 000 (300 GHz)	0.001 (1 mm)	
	3 000 000 000 000 (3 THz)	0.000 100 (100 μm)	
Infrared or heat rays	30 000 000 000 000 (30 THz)	0.000 010 (10 μm)	100 000
	300 000 000 000 000 (300 THz)	0.000 001 (1 μm)	10 000
Visible light rays			
Ultraviolet rays X-rays Gamma rays	3 000 000 000 000 000 (3000 THz)	0.000 000 100 (100 nm)	1 000

radio and heat waves, and are characterized by the method used to generate the waves.

The frequency limits of the various types of waves or rays shown in Table 13.1 are by no means fixed or certain. Constant revisions take place as new experimentation adds further knowledge. Note, too, the arrangement by wavelength in descending order of length. Since the meter becomes an awkward unit to use, especially in the region of visible light, the **angstrom unit** (Å) is also used to measure wavelength. One meter is equal to 10 000 000 000 angstrom units, so named after the Swedish scientist Anders Angström (1814–1874). Some of the various types of waves or rays listed in Table 13.1 will be discussed in appropriate sections of this book.

---

* angstrom unit   A unit of length equal to one ten-billionth of a meter.

13.4 Visible Light

Let us turn once again to visible light—that is, electromagnetic waves whose wavelengths are approximately between 4000 and 8000 angstrom units. Different wavelengths of light appear as different colors. For instance, blue light has wavelengths in the 4300–4900 angstrom range, and yellow light is in the 5500–5900 anstrom range. Figure 13.4 shows the full visible light spectrum.

Suppose that we heat our bar of iron once more until it becomes luminous—that is, until it emits visible light. The light rays are sent out in all directions, but each ray travels in a straight line. Now, what happens when the light rays encounter some object? The result depends on the nature of the object. Certain substances, such as air or window glass, permit light to go through them with very little change. We say that these substances *transmit* light and that they are **transparent.** Other substances, such as oiled paper or ground glass, will partially transmit

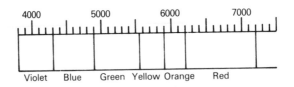

FIGURE 13.4   Visible Light Spectrum (Wavelength in Angstrom Units).

PHOTO 13.4   Most visible objects are seen because they reflect light from a light source. The bright part of the planet Venus shown here is reflecting light from the sun to the space probe camera. Since no light reaches the far side of the planet, it remains invisible.

light. Such substances are called **translucent**. Still other substances, such as iron, wood, or stone, block the passage of light. We say that such substances are **opaque**.

* transparent substance	A substance that allows light waves to pass through it.
* translucent substance	A substance that will partially transmit light. Objects viewed through a translucent substance will appear indistinct or as a shadow.
* opaque substance	A substance that completely blocks the transmission of light.

When an opaque object blocks the passage of light, what happens to the light energy? Part of the light energy is taken in, or *absorbed,* by the object. Generally, this absorbed energy is changed to heat energy, although some of it may be changed to other types of energy, for example, chemical energy. If the surface of the opaque object is rough and dark, it will usually absorb light better. Another part of the light energy will bounce off, or be *reflected* by, an opaque object. If the surface of the opaque object is smooth and shiny, it will be a better reflector.

Reflection plays an important part in our lives. We see the world around us chiefly by reflected light. Light originates in luminous bodies. Our chief sources of natural light are the sun and, to a much smaller degree, the distant stars. We create light artificially by burning gas, wood, oils, or fats, and we create light electrically by heating substances until they become luminous or by electrical discharges through gases. We see all nonluminous bodies by the light they reflect. For example, the moon is seen by the reflected light of the sun. You see this page by the light it reflects. If no light were present, you could not see it.

13.5 Polarized Light

As we have seen, light is a transverse wave—that is, the vibrations are at right angles to the beam's line of travel. The waves of polarized light are modified. To understand what happens when light waves are modified, we must examine them more closely. The examples of the rope tied to a tree and the ripples in pond water help explain the meaning of a transverse wave. But light waves are more complicated. Light waves vibrate in all directions at right angles to the direction of the wave motion (the direction of the beam of light).

To visualize this concept more clearly, let's go back to Figure 13.1, showing the rope tied to a tree. Only this time, you and three or four friends are each holding a rope tied to the same spot, and all the ropes are held close together. Now, you start flipping your rope up and down. At the same time, your friends start flipping their ropes. However, one friend uses a horizontal motion, and the others flip their ropes at various angles to the horizontal. We have to imagine that this action can be done without tangling the ropes. Also, we must view each rope vibration not as a separate wave, but as part of the overall wave moving from you and your friends toward the tree. We must think of a light wave in this way also.

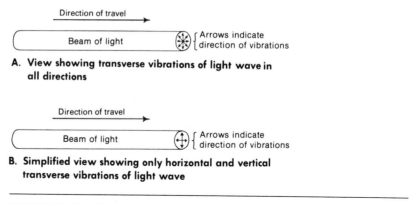

FIGURE 13.5   *Vibrations of a Light Wave.*

It is normal for a light wave from our various sources of illumination to vibrate in all directions at right angles to the beam of light, as illustrated in Figure 13.5A. However, to simplify our discussion, only the vertical and horizontal portions of the wave will be used to describe how polarized light is modified (Figure 13.5B).

Suppose, now, that this beam of light were to strike an opaque sheet that had a number of extremely fine horizontal slots, as in Figure 13.6A. The vertical vibrations would be blocked off, but the horizontal vibrations would pass through the slots. We say that now the light is **polarized**—that is, the transverse vibrations are all in one direction or plane. The intensity of the beam is reduced, but to the eye, polarized light appears the same as normal, unpolarized light.

∗ polarized light   Light whose waves vibrate in only a single direction, perpendicular to the direction in which the light is traveling.

Suppose that the opaque sheet is rotated so that the slots are vertical, as is shown in Figure 13.6B. Now the vertical vibrations will pass through, and the horizontal ones will be blocked off. Again, the light is polarized, but this time in a vertical direction. The eye still cannot detect the difference.

But suppose now that we use two such opaque sheets, one with horizontal slots and the other with vertical slots (Figure 13.6C). The light passing through the first sheet will be polarized in a horizontal direction. These horizontal vibrations cannot pass

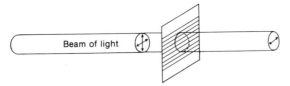

A. Vertical light wave vibrations are blocked, and light is horizontally polarized.

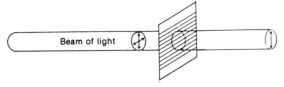

B. Horizontal light wave vibrations are blocked, and light is vertically polarized.

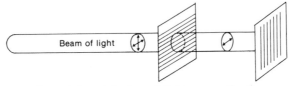

C. Both vertical and horizontal light wave vibrations are blocked, resulting in no light.

FIGURE 13.6   Illustration of How Light Is Polarized.

through the vertical slots in the second sheet. As a result, no light comes through at all.

Certain crystals, such as tourmaline, have the property of being able to polarize light. A sheet of polarizing material can be made by depositing certain fine crystals on a celluloid film in such a way that practically all the crystals are lined up in the same plane. If two sheets of polarizing material are placed one on top of the other so that the crystals of both are in the same plane, normal light will pass through with very little opposition. But if one sheet is rotated until its crystals are at right angles to those of the other, practically all the light will be blocked out.

When light strikes a smooth surface, such as a highway or a sheet of water, the vertical vibrations are absorbed, leaving the horizontal vibrations to be reflected. We say the reflected light is polarized, as Figure 13.7 indicates. Since the eye cannot

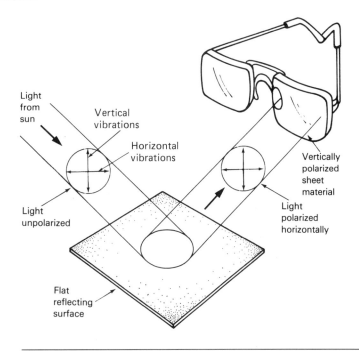

FIGURE 13.7 Illustration of Polarized Sunglasses Reducing Glare from Reflecting Surfaces.

distinguish between polarized and unpolarized light, the reflected rays may cause glare. However, if glasses of polarizing material are worn, and the lenses are properly aligned, a portion of the reflected polarized light will be blocked out, and the glare will be reduced.

The uses of polarizing materials continue to expand. In addition to polarized sunglasses, we will mention two others. Photographers use polarized filters to reduce glare from the sky or from reflecting surfaces such as water. Also, photographers who want to take a picture through a window, but are bothered by glare, use a polarizing filter.

Engineers and technicians use polarized light in stress analysis of beams and other structural parts. Figure 13.8 shows a device that utilizes a clear plastic model (light has to pass through it) of a part and that allows a small load to be applied to the part. The plastic part transmits polarized light differently in the sections that are highly stressed. Thus, the technician can view those locations on the part that have the highest stress.

FIGURE 13.8   Device for Analyzing the Stress Variations in a Beam by Using Polarized Light. (Courtesy of Measurements Group/Vishay Research & Education Division)

13.6   The Laser

In 1964, the American scientist C.H. Townes shared the Nobel Prize in physics for the discovery and development of the laser. The **laser** is a device for producing a beam of light with some unusual and valuable properties. The properties most interesting to us include the following:

> The laser beam of light contains a tremendous amount of energy in a small space. The beam is used for such purposes as drilling holes in diamonds to make wire drawing dies and making complex cuts in sheet metal parts.
> The light beam can be brought to a very sharp focus with lenses. Eye surgeons use this property to weld detached retinas back into place.

PHOTO 13.5 The sheet metal cutting ability of a laser beam is being demonstrated. Air may be fed through the plastic tubes to the laser beam. The air, under pressure, speeds the cutting process and blows away the molten and oxidized metal. (Courtesy of Coherent General, Inc., Palo Alto, CA)

In the communications industry, the light beam can carry much more information than can standard radio waves or copper telephone wires.

To understand how a laser works, we must discuss photons. Earlier we noted that a bundle of photons may be considered as a train of waves. Photons in normal, or ordinary, white light from the sun or from incandescent lamps are said to be incoherent. That is, the photons may have different wavelengths, and those that do have the same wavelength are out of phase (out of step) with each other. Figure 13.9 illustrates the difference between photons that are in phase and those that are out of phase. Photons in light from a laser, though, are coherent. That is, all the photons have the same wavelength, and all are in phase with each other.

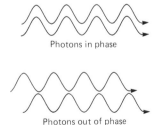

Photons in phase

Photons out of phase

FIGURE 13.9 Photons in and out of Phase.

We know that if a substance is heated, its atoms absorb energy. The atoms become excited as the extra energy causes the electrons to move up to orbits with a greater radius. When an electron returns to its original orbit, the energy is released by emitting a photon. The wavelength of the emitted photon depends on the amount of energy it contains.

However, another type of emission process is used by the laser. If a photon with a suitable wavelength crashes into an atom in an excited state, the atom may emit a photon with the same wavelength and the same phase as the photon striking the atom. The original photon continues on its path, and the emitted photon goes with it. Now there are two photons that crash into other excited atoms, thereby producing more photons. Not only does this effect amplify the number of photons, but the photons all have the same wavelength, and all are in phase with each other. The light produced is said to be coherent and to have been produced by stimulated emission.

The term "laser" is the acronym for Light Amplification by Stimulated Emission of Radiation. The first successful laser, still popular, is the ruby laser, shown in Figure 13.10. It uses a crystal of ruby, which is made of aluminum oxide with some chromium atoms added. The chromium atoms are raised to an excited en-

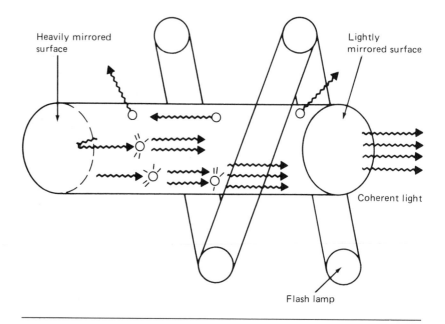

FIGURE 13.10   The Ruby Laser.

ergy state by absorbing light from an external flash lamp. Some chromium atoms give off photons of a specific wavelength. Some of the emitted photons are lost because they pass off through the sides of the crystal. But other emitted photons move in a path parallel to the long axis of the ruby crystal. These photons strike other chromium atoms, and more photons are released in the direction along the axis of the crystal. This effect is the amplification of light. The photons are reflected back and forth between the mirrored surfaces, striking more chromium atoms, which release more photons. As more and more photons are emitted, the light becomes more intense. Finally, the photons burst through the lightly mirrored surface to form a beam of coherent light.

Laser machining is rapidly becoming commonplace in industry, and technicians will be involved in the operation. Increasingly, manufacturers are turning to lasers to machine such materials as tungsten, titanium, and natural diamond. Also, lasers can be used to cut paper, leather, rubber, and plastics more quickly and cleanly than is possible with conventional methods.

Summary

Light has a dual personality. In some instances, as in the operation of a photocell, it appears to travel as small bundles of energy, called photons. In most technical applications, though, it appears to travel, and behave, as a wave of energy. Light waves are transverse waves and travel at a speed of 186 000 miles per second (300 megameters per second) in a vacuum or in air.

The formula for the speed of light is the same as for the speed of sound, except the symbol for velocity—that is, speed of light—is $c = f \times \lambda$. Light waves are part of the electromagnetic wave spectrum. Visible light has wavelengths between the limits of 4000 and 8000 angstrom units. Light of different wavelengths will appear in different colors. Some materials, such as glass, transmit light waves easily and are called transparent. Materials that transmit only a small amount of light are called translucent. Materials that block out light completely are called opaque. Usually, when light is absorbed, the energy is changed to heat.

Light can be polarized, which means that the transverse waves vibrating in one direction can be absorbed. The light waves that continue through the polarizing material vibrate at 90 degrees to the direction of the absorbed waves. For example, light from the sun striking a smooth reflecting surface, such as a pond, will have the vertical vibrations absorbed, and only the horizon-

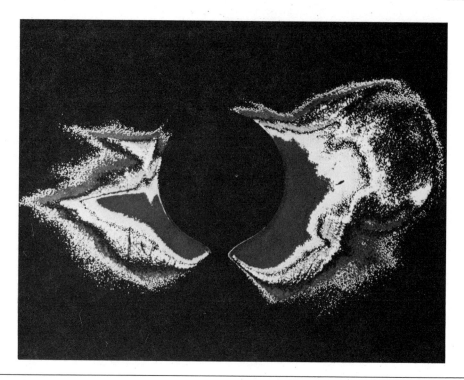

PHOTO 13.6   This computer-enhanced photo of the sun, taken from the Skylab space vehicle, blocks out the main disk of the sun to show the light emitted by the hot gases surrounding the sun, called the solar corona. The different shadings indicate different levels of brightness. (NASA photo)

tal vibrations of the light wave will be reflected. Polarized light is used in technical applications such as stress analysis.

Light from a laser is coherent light. The laser beam contains a large amount of energy, can be focused sharply, and can act as a carrier for information transmission.

Key Terms

* photon
* luminous body
* longitudinal wave
* transverse wave
* electromagnetic wave
* angstrom unit

* transparent substance
* translucent substance
* opaque substance
* polarized light
* laser

Questions

1. Discuss the quantum or photon theory of light. What causes a luminous body to emit photons?

2. What is meant by a longitudinal wave? A transverse wave? Which type is a light wave?

3. What is meant by the wavelength of a light wave? By its frequency? Explain the relationship between the wavelength and the frequency.

4. What is the electromagnetic spectrum? What is the relationship between the waves or rays of which it is composed?

5. What quantity does an angstrom unit measure?

6. An electromagnetic wave with a wavelength of 10 meters has a frequency of _____.

7. An electromagnetic wave with a frequency of 300 gigahertz has a wavelength of _____.

8. What happens to light rays as they meet a transparent body? An opaque body?

9. What is polarized light? How do glasses of polarizing material reduce the glare from a highway?

10. Describe a use, other than sunglasses, for a polarizing device.

11. How does a laser amplify light?

12. What is meant by coherent light?

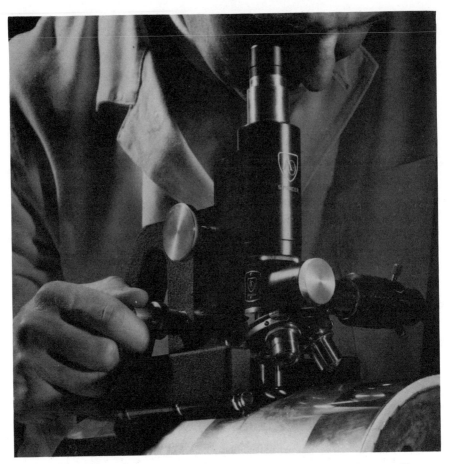

A microscope utilizes the refractive and reflective properties of light to magnify images. Here, a technician in the printing industry uses a microscope to check the etched image on a printing plate.

Chapter 14

Objectives

When you finish this chapter, you will be able to:

- ☐ State the law of reflection. Employ a diagram showing the angle of incidence and angle of reflection.
- ☐ Define virtual image and use a diagram of an image in a mirror to demonstrate that light rays only appear to originate at the image.
- ☐ Diagram and explain the phenomenon of refraction at the interface between two different transparent materials.
- ☐ Define the index of refraction as the ratio of light's velocity in a vacuum to its velocity in a material.
- ☐ Diagram and explain the phenomenon of total internal reflection.
- ☐ Diagram and explain how refraction is used in concave and convex lenses.
- ☐ Construct ray tracing diagrams to show how real and virtual images are formed, located, and magnified by convex and concave lenses.
- ☐ Recognize the lens magnification formula $m = D_i/D_o$: The magnification of an image (m) equals image distance (D_i) divided by the object distance from the lens.
- ☐ Recognize the lens formula $1/f = (1/D_o) + (1/D_i)$: The reciprocal of the focal length (f) equals the reciprocal of the object distance (D_o) plus the reciprocal of the image distance (D_i).

Reflection and Refraction

Light, as we have seen, originates with luminous bodies. Such luminous bodies may be natural sources of light, such as the sun and stars, or they may be artificial sources, such as candles or electric lamps. When light strikes an object, some of the light is absorbed by the object and converted into heat. Another portion of the light is reflected from the surface of the object. Finally, if the object is transparent, a portion of the light is transmitted through the object.

Since we are still in an introductory stage in our discussion on light, this chapter has very few technical applications to discuss. The information on reflection, refraction, and lenses relates to cameras, telescopes, microscopes, and electro-optical devices. We will see how lenses make use of the refractive properties of light and consider the meaning of such terms as angle of incidence, index of refraction, virtual image, convex lens, and concave lens.

14.1 Reflection

Let us investigate what happens to light that is reflected. In Figure 14.1, a light ray from a luminous body strikes a smooth reflecting surface. The angle formed by this light ray (called the incident ray) with a line drawn perpendicular to the reflecting surface at the point where the ray touches the surface is called the **angle of incidence**. The angle formed by this perpendicular line and the light ray that is reflected from the surface is called the **angle of reflection**.

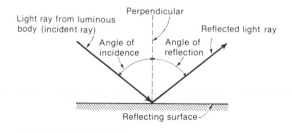

FIGURE 14.1 Reflected Light. When light is reflected, the angle of incidence is equal to the angle of reflection.

PHOTO 14.1   Most bright, shiny metals can serve as a mirror. This metallized disk produces a virtual image: It appears to be behind the disk and is not produced by any light source. The disk was developed to store television pictures for later retrieval and projection. (Courtesy of RCA)

* **angle of incidence**	The angle between the incident ray of light and a line drawn perpendicular to the reflecting surface at the point where the ray strikes the surface.
* **angle of reflection**	The angle between the reflected light ray and a line drawn perpendicular to the reflecting surface at the point where the reflected ray begins.

Now that the two angles are defined, we can state the **law of reflection:**

* Law of Reflection | The angle of incidence is always equal to the angle of reflection.

In Figure 14.2A, we can see what happens when rays of light strike a smooth surface. The reflected rays are orderly and parallel to each other. This effect is called *specular light,* which means "light produced by a mirror." For each ray, the angle of incidence is equal to the angle of reflection. In Figure 14.2B, however, we can see what happens when rays of light strike an uneven or rough surface. For each ray, the angle of incidence is still equal to the angle of reflection. But since the incident rays strike the surface at different angles, the reflected rays are scattered, or *diffused.* It is this diffusion—caused chiefly by rough surfaces and dust particles in the air—that permits light to illuminate the spaces behind opaque objects, spaces that otherwise would be in total darkness.

We are all familiar with the ordinary mirror. A mirror is made by coating the back of a sheet of flat glass with a thin layer of silver. The glass, being transparent, does not reflect much light. Instead, it transmits the light to the silver coating, which does the actual reflecting. The glass serves merely to protect the fragile silver layer and to prevent it from tarnishing in the air.

When you look in a mirror, light from a luminous source is reflected from you to the mirror and then back to your eye. You see an image of yourself. This image is full size and in the same vertical position as you are. However, the right hand of the image is at your *right,* while if a person were facing you, his or her right hand would be at your *left.* Thus, the mirror image is reversed. If you were to hold a line of writing in front of the mirror, the image would be erect, but reversed, reading from right to left instead of from left to right.

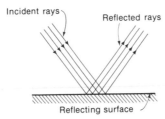

A. Specular light reflected from a smooth surface.

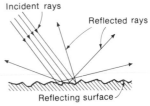

B. Diffused light reflected from a rough surface.

FIGURE 14.2   Specular and Diffused Light.

Note, too, as Figure 14.3 shows, that your image seems to be as far behind the mirror as you are in front of it. The light seems to be coming to your eye from behind the mirror, when actually it is being reflected from the surface. We call such an image a **virtual image.** If you were to try to project such an image on a screen, you would be unable to do so. You can see such an image only when it is formed by your eye.

---

* virtual image   An image that is not formed by light rays. It only appears (to the eye) to be formed by light rays.

14.2 Refraction

When a light ray strikes a transparent object, a certain amount of the light is reflected from the surface; a certain amount is absorbed by the object and transformed into heat; and the rest is transmitted through the object. Let us examine this transmission a little more closely.

We have stated that the speed of light through air or a vacuum is approximately 186 000 miles a second. But this speed does not hold for light passing through other media. The speed

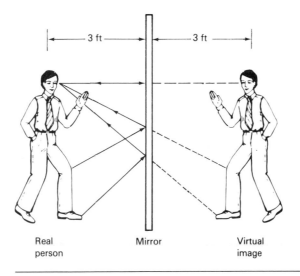

Real
person   Mirror   Virtual
                                          image

FIGURE 14.3   Virtual Image. An image that is not formed by light rays is a virtual image.

of light through pure water, for example, is about three-quarters of its speed in air, and in glass, its speed is only about two-thirds of its speed in air.

Scientists, when working with light in substances other than a vacuum, make use of the ratio of the speed of light in a vacuum to its speed in other substances. This ratio is called the **index of refraction.** For example, the index of refraction for water is 1.33. This figure is obtained by dividing 186 000 miles per second (speed of light in a vacuum) by 139 500 miles per second (speed of light in water).

---

* index of refraction   The ratio between the speed of light in a vacuum and the speed of light in a given transparent substance.

When a ray of light travels through air and strikes a transparent object whose index of refraction is greater than 1 (which is the index of refraction of air), it is slowed up. When it passes through this object and emerges into the air again, it resumes its normal speed. Besides the change of speed, something else happens to light going from one medium to another. Figure 14.4 shows a portion of a glass prism, in which the wave crests of light are illustrated by straight lines similar to the way in which the peaks of a sound wave were illustrated in Figure 12.9. When a light ray strikes the glass obliquely, rather than perpendicularly, the light ray bends. Note that as the bottom of a wave crest enters the glass, the upper portion of that crest is still in the air. Thus, the wave crest in the glass is traveling at a slower speed

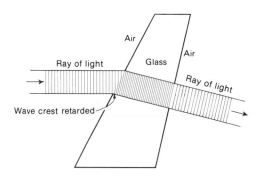

FIGURE 14.4   Diagram Showing What Happens When a Light Ray Strikes a Glass Surface Obliquely. (Reflected and absorbed parts of light are not shown.)

than the portion still in the air. The result is that the ray bends down toward the base of the prism.

You may understand this effect more clearly if you think of yourself running along the ground. Your feet and head move at the same speed and you remain erect. But suppose you were suddenly to encounter sand. The sand would slow up your feet. Your head and the upper portion of your body would tend to keep traveling at their original speed, and you would fall forward.

Once the entire wave crest is in the glass, all portions travel at the same speed, and the ray travels in a straight line in the new direction. Note that the wavelength changes, but the frequency does not. In Figure 14.4, we have assumed that the ray of light emerges into the air perpendicularly to the surface. Consequently, all portions of each wave crest emerge simultaneously. The speed increases, but there is no further bending of the ray. We call this bending of a ray of light as it obliquely enters another medium, **refraction.** Since it depends on the difference between the speeds of light in the two media, refraction is a function of the difference between the indexes of refraction of the two media. The greater this difference, the greater is the refraction.

---

* refraction   The bending of a ray of light when it passes from one transparent medium into another.

Now let us consider what happens to a ray of light when it both enters and leaves a glass prism at an oblique angle. Figure 14.5 shows that as a ray of light enters the glass prism, it is refracted down toward the base, as in the previous illustration. In the glass, the light ray travels in a straight line. As the ray leaves the glass, however, the upper portion of the wave crest enters the air first. Again, this upper portion travels faster than the lower portion; and again, the ray bends down toward the base of the prism.

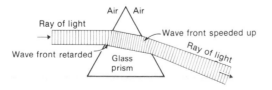

FIGURE 14.5   Diagram Showing How a Light Ray Is Refracted as It Enters and Leaves a Glass Prism.

While we are discussing the refraction of light, we must examine another of its aspects: A prism does not refract light of different wavelengths to the same degree. In 1666, Isaac Newton discovered that if a ray of white light (which is made up of light of all wavelengths, or colors) is passed through a prism, the emerging ray, as Figure 14.6 indicates, consists of a band of light of various colors. This band of colored light follows a definite pattern. At one end is red, and at the other end is violet. Between these two limits, the other colors appear in the order shown. Breaking up white light into its component colors is called *dispersion,* and the band of color is called the *visible spectrum.*

Note that red light (whose wavelength is longest) is refracted the least. Violet light (whose wavelength is shortest) is refracted the most. The other colors are refracted in varying degree, depending on their wavelengths. This dispersion of white light is what causes a rainbow—that is, sunlight is refracted by the raindrops in the air.

Let us perform another experiment with refraction. Suppose, as in Figure 14.7, that we have a source of light in a tank of water. If a ray of light from this source strikes the surface of the water perpendicularly, as at point A, the ray will emerge without refraction. If the ray strikes the surface obliquely, as at point B, it will be bent slightly. As the angle at which the ray strikes the surface is increased, as at point C, the ray is bent more as it enters the air. When a certain critical angle is reached, as point D shows, the ray will be bent so far that it travels along the surface of the water and does not enter the air. If the angle

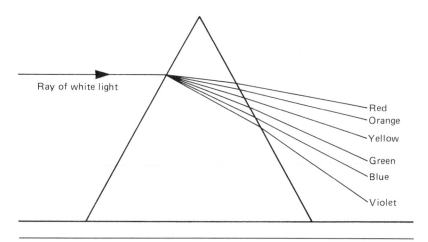

FIGURE 14.6 ˙Diagram Showing How a Prism Disperses White Light.

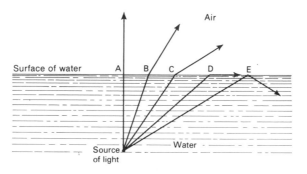

FIGURE 14.7   Illustration Showing the Total Internal Reflection of Light.

is increased still more, as at point E, the ray will be reflected back from the surface into the water. We call this phenomenon the **total internal reflection** of light.

---

* total internal reflection

A phenomenon that occurs when the angle of incidence is large enough and the light attempts to pass from an optically dense medium to an optically rarer medium.

Note that total reflection may occur only when light attempts to pass from a medium in which its speed is slower (optically denser) into a medium in which its speed is greater (optically rarer). If the situation were otherwise, the bending would take place the other way and the light ray would penetrate the surface, regardless of the angle at which it might strike.

We utilize total internal reflection when we cut diamonds and other precious stones in such a way that the internal reflection of the light rays from face to face causes the gem to sparkle when illuminated. Similarly, the luminous road markers along highways are cut in such a way that the light from the headlights of oncoming automobiles is reflected back and forth inside the transparent material of the markers, causing them to glow.

Total internal reflection is employed in many optical instruments through the use of total-reflection prisms. These prisms usually are made of glass and are cut so that one angle is a right angle (90 degrees) and the other two are 45 degree angles. Three positions in which these prisms may be used are illustrated in Figure 14.8; we shall consider them further in the next chapter.

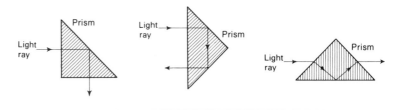

FIGURE 14.8   Diagram Showing How Total-Reflection Prisms May Be Used to Reflect Light Rays.

PIPING LIGHT THROUGH GLASS FIBERS

The use of light to transmit information is probably as old as civilization itself. In one of Longfellow's famous poems, the words "one if by land, two if by sea" were Paul Revere's instructions in 1775 concerning the number of lanterns to place in the church tower to warn of the British advance. The modern way of using light to communicate is to pipe the light through glass fibers. This technique is one of the most modern applications of the *total internal reflection phenomenon.* Laser light with a single wavelength is piped through optical glass fibers about the size of human hair (approximately 0.005 inch in diameter). At present, some 144 glass fibers can be bundled together in a cable about one-half inch in diameter. This lightguide cable can carry as many voice transmissions at one time as four regular cables, each having 900 copper wires, and each cable being almost 3 inches in diameter.

Guiding light through glass by total internal reflection has been useful technically since about 1950. The glass fibers are used in industry for cathode ray tube faceplates and for inspection of hard-to-view cavities in products. The fibers also aid doctors by allowing them to view internal body cavities, such as the stomach. The distances for these applications are short, and therefore light loss is not a problem. The types of optical glass available in the fifties were adequate for these applications, but not for long-distance light transmission.

We can understand that if the diameter of a glass fiber changes suddenly, or if air bubbles or other impurities are trapped in the glass, the light can be scattered through the sides of the fiber and lost. Also, when laying cable, sharp turns must be avoided or the light may not be reflected around the bend. It took many years to develop the glass materials and manufacturing processes needed to reduce light loss enough to transmit light over the long distances required by the communications industry. The glass fibers must have high purity, long length, and a proper index of refraction for the type of laser light being used. At present, repeaters (signal boosters) are placed at about 20 mile intervals along a transmission line.

14.3 Lenses

We make use of refraction of light in the **lens,** which consists of a transparent substance, such as glass, whose opposite surfaces are curved so that light rays enter or leave obliquely. In Figure 14.9, we have a cross-sectional view of a lens whose middle is thicker than its ends. We call such a lens a **convex lens.** Generally, the curvature of each surface is shaped so as to form part of a sphere.

---

* lens

A transparent device shaped to diverge or converge a beam of light rays.

* convex lens

A lens designed to converge light rays. If the rays are parallel before passing through the lens, they will converge at some fixed point.

If we draw a horizontal line running through the center of the lens and striking each surface perpendicularly, we have the *principal axis.* A ray of light following this principal axis will not be bent because it enters and leaves the lens perpendicularly to the surfaces. But any ray of light that is parallel to the principal axis and that does not pass through the center of the lens will strike the surfaces obliquely and, thus, will be refracted as it enters and as it leaves.

Suppose our source of light is so far from the lens that all the rays that strike the lens are parallel to one another. Theoretically, this source must be at an infinite distance from the lens, but practically, it may be fairly near the lens. Each ray, other than the central ray, will be refracted as it enters and as it leaves the lens. The amount of refraction of each ray will depend on the angle at which it strikes or leaves the lens. Regardless of this angle, however, each ray will be so refracted that it will converge with every other ray at a point (F) behind the

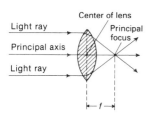

FIGURE 14.9 Convex Lens. When parallel light rays exit the lens, they are focused at the point of principal focus, which is determined by the focal length (f) of the lens.

PHOTO 14.2   The camera in an office copier has a convex lens that produces real, inverted images of the same size as the object. (Courtesy of Xerox Corporation)

lens. This converging of rays to meet in a point is called *focusing*, and the point at which they meet is called the **principal focus.** The greater the curvature of the surfaces of the lens, the shorter the distance and the shorter the **focal length** (*f*) of the lens. An image of the source can be seen if a screen is placed at the point of focus. We call such images **real images.**

---

* **principal focus**   The fixed point at which parallel light rays are converged by a convex lens.

* **focal length**   The distance between the center of a lens and its principal focus.

* **real image**   An image that is formed by light rays and that may be projected on a screen.

Now let us move the source of light to a point a little more than two focal lengths away from the lens (Figure 14.10A). The rays from the source are no longer parallel to one another, but diverge. As they strike the lens, they are refracted and focused

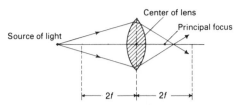

A. Light rays from a source more than two focal lengths in front of a convex lens are brought to a point of focus between one and two focal lengths behind the lens.

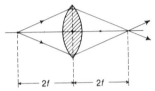

B. Light rays from a source two focal lengths in front of a convex lens are brought to a point of focus two focal lengths behind the lens.

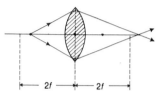

C. Light rays from a source between one and two focal lengths in front of a convex lens are brought to a point of focus more than two focal lengths behind the lens.

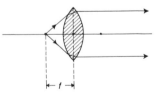

D. Light rays from a source one focal length in front of a convex lens become parallel as they leave the lens and cannot be brought to a point of focus.

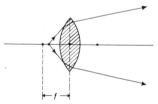

E. Light rays from a source less than one focal length in front of a convex lens diverge as they leave the lens and cannot be brought to a point of focus behind the lens.

FIGURE 14.10 Focal Points of a Convex Lens When Distance from Light Source Varies.

at a point behind the lens exactly as before. But this point now lies somewhere between one and two focal lengths away from the center of the lens.

As the source of light is moved closer to the lens, the point of focus moves farther away from the lens. When the source, as Figure 14.10B shows, is two focal lengths from the lens, the point of focus is also two focal lengths away. Moving the source still closer to the lens causes the point of focus to move even farther away. In Figure 14.10C, the source is between one and two focal lengths from the lens, and the point of focus is more than two focal lengths away.

In Figure 14.10D, the source is one focal length from the center of the lens. The rays are so refracted that they emerge as parallel rays and no longer can be brought to a point of focus. Note that this condition is the reverse of the one shown in Figure 14.9, where the source is an infinite distance from the lens and the point of focus is one focal length away. Thus, the lens action is reversible. As we bring the source of light to a point that is less than one focal length away from the lens, the emerging rays diverge and cannot be brought to a point of focus, as illustrated in Figure 14.10E.

In the cases illustrated in Figures 14.10D and 14.10E, no real images are formed. In all instances where the source of light is more than one focal length away, however, the convex lens causes light rays to converge or come together at a point to form a real image. Thus, the convex lens is used in cameras, projectors, and so forth to focus images on film or screen.

Now let us examine what happens to light rays when they pass through a lens that is constructed in such fashion that it is thinner in the middle than at the ends. We call such a lens a **concave lens**. If the source of light is so far from the lens that the rays are parallel to one another, as in Figure 14.11A, each ray will be refracted in the same way it was in the convex lens— that is, each ray will be bent toward the thick section of the lens. But since the thick sections now form the ends of the lens, the light rays will be bent *away* from the principal axis. Thus, the rays of light diverge. Because the light rays diverge, no real image of the source can be formed behind the lens.

---

* concave lens
A lens designed to diverge light rays.

If the source of light comes closer to the concave lens, as in Figure 14.11B, the lens will merely cause the diverging rays to diverge more. Thus, a concave lens cannot form real images.

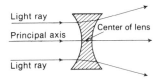

A. Parallel rays of light diverge as they leave the lens.

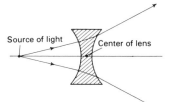

B. Diverging rays of light diverge more as they leave the lens.

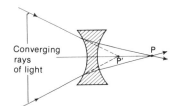

C. Converging rays of light are brought to a point of focus (P) further to the right of the lens than the point of focus without the lens (P').

FIGURE 14.11   Action of a Concave Lens.

Whereas the convex lens converges light rays, the concave lens diverges them. However, if the rays of light are converging as they strike the concave lens (for example, rays of light after emerging from a convex lens), and if this convergence is greater than the divergence caused by the concave lens, the net result is converging rays of light, and a real image is formed, as illustrated in Figure 14.11C. Note that this image appears farther to the right (at point P) than it would appear if the lens were not present. The image would have appeared at point P'.

So far, we have considered only images formed by lenses from point sources of light. Let us now see how lenses form images with light rays reflected from objects. Assume that we have an object, such as the person illustrated in Figure 14.12A, which is more than two focal lengths from a convex lens. We can locate the image by tracing the paths of the light rays from each point of the object.

Starting with the head of the person, consider the path of a light ray (1) that is parallel to the principal axis. As we have learned, this sort of ray is refracted in such a way that it passes through the principal focus (F) of the lens. Somewhere along this ray we shall find the image of the head.

Now take a second ray (2) from the head of the person—this time one that passes through the center of the lens (C). Such a ray is bent a certain amount one way as it enters the lens and is bent the same amount the other way as it leaves the lens. The net result is that any ray passing through point C travels in what is practically a straight line. (Actually, there is a certain amount of sideways displacement, which may be neglected if the lens is thin.) The image of the head lies along this second ray also. Where the two rays intersect, the image appears. In like manner, we may locate the rest of the image.

Note that the image is inverted, that it is smaller than the object, and that it lies between one and two focal lengths away from the lens. Also, as we have seen previously, the image is real and will appear on a screen placed at that point. A convex lens is used in this manner in the camera, where the image of a large object is projected on a small sheet of film.

The magnification of the lens can be determined if the distances from lens to object (D_o) and lens to image (D_i) are known. In Figure 14.12A, notice that the right triangle formed on the left side of the lens by the upper part of the object, the principal axis, and ray 2 is similar to the triangle formed by the upper part of the image (inverted), the principal axis, and ray 2. The triangles are similar because they are both right triangles and the angles at point C are equal. Therefore, the corresponding sides are proportional, and we have:

$$\frac{\text{image height}}{\text{object height}} = \frac{D_i}{D_o}$$

This ratio is called the magnification:

*** MAGNIFICATION
FORMULA**
$$m = \frac{D_i}{D_o}$$

For the particular case shown in Figure 14.12A, the magnification is less than one, because the image is closer to the lens than the object. Let's say the image is 6 inches from the lens and the object is 10 inches from the lens. Then:

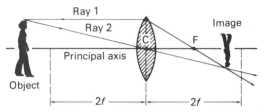

A. An image smaller than the object is formed when the object is more than two focal lengths in front of the lens.

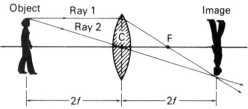

B. An image the same size as the object is formed when the object is two focal lengths in front of the lens.

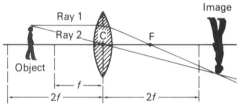

C. An image larger than the object is formed when the object is between one and two focal lengths in front of the lens.

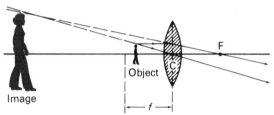

D. A virtual image larger than the object is formed in front of a convex lens when the object is less than one focal length in front of the lens.

FIGURE 14.12 Images Formed by a Convex Lens.

$$m = \frac{D_i}{D_o} = \frac{6 \text{ in.}}{10 \text{ in.}} = 0.6$$

In Figure 14.12B, we see that if the object is placed at a point two focal lengths in front of a convex lens, an image will be formed two focal lengths behind the lens. Note that the image is real, that it is inverted, and that it is the same size as the object ($m = 1.0$). A lens is used in this way in a copying camera, where full-size reproductions of an object are desired.

As the object is moved between one and two focal lengths away from a convex lens (Figure 14.12C), the image, still real and inverted, appears more than two focal lengths behind the lens. Note that the image is larger than the object. A lens is used in this way in the motion-picture projector, in which a picture on a small piece of film is projected to form a large image on a screen. Of course, if the image is to appear right side up on the screen, the picture must be inverted in the projector. When the object is moved to a point that is one focal length in front of a convex lens, the emerging rays are parallel and no image will be formed.

If an object is placed at a distance of less than one focal length in front of a convex lens as in Figure 14.12D, the emerging rays of light diverge and no image can be formed behind the lens. But if the eye is positioned behind the lens (the right side of Figure 14.12D), it will follow the rays back in a straight line and an image in front of the lens—that is, on the same side as the object—will be seen. Such an image is a virtual image. Note that it appears erect and larger than the object. This is the principle used in the magnifying glass.

SAMPLE PROBLEM 14.1

From experimentation, we determined that a particular lens projected an image 250 mm behind the lens when an object was 1.5 m in front of the lens. What is the magnification?

SOLUTION

Wanted

$m = ?$

Given

$D_i = 250$ mm and $D_o = 1.5$ m. Changing millimeters to meters by multiplying by 0.001, we have $D_i = 250$ mm $\times$ 0.001m/mm $= 0.25$ m.

Formula

The formula for magnification is:

$$m = \frac{D_i}{D_o}$$

Calculation

$$m = \frac{0.25 \text{ m}}{1.5 \text{ m}} = 0.17$$

This information is helpful in determining whether or not the lens is suitable for some particular application. It may also be a check to determine if the lens has been made according to the specifications.

We can actually determine graphically and analytically the size and location of the image formed by a lens. We have already gone through the analytical procedure for calculating the size of an image (the magnification). Now we will follow the steps for a graphical solution of image size and distance. Then we will proceed with the analytical solution for image distance.

Figure 14.12A will be redrawn and arrows will be used to represent the object and image, as in Figure 14.13. This figure represents an object 6 inches high placed 15 inches in front of a lens with a 7 inch focal length. We must find the image height and distance.

Steps for Graphically Determining the Size and Location of an Image

Step 1 Draw the principal axis on a sheet of paper and locate the center of the lens (point D in Figure 14.13) about midway along the axis. The drawing may be scaled, but for this demonstration, we will use full scale.

Step 2 Draw the lens about twice the height of the object. Locate the principal focus (F) to the right of the lens.

Step 3 Place the object (AB) on the principal axis to the left of point D and at the proper distance from the lens.

Step 4 Draw two lines, or rays, from the top of the object (point A). One line is drawn parallel to the principal axis and, on reaching the center line of the lens (point C), is drawn through the principal focus. The other line is drawn from the top of the object straight through the center of the lens (point D). The two rays meet at point H, which represents the top of the image.

Step 5 Draw an arrow from H to the principal axis (arrow GH) and measure its length for the image height. Also, measure distance

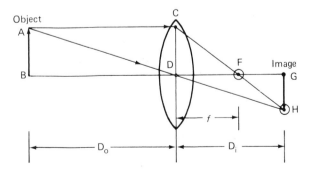

FIGURE 14.13   Graphical Determination of Image Formation by a Convex Lens.

(D_i) for the image distance. If you have constructed your drawing carefully, you should get an image distance of 13.1 inches and an image height of 5.2 inches.

For analytical solutions, we must apply the lens formula shown below. This lens formula relates three distances to each other: the focal length (f), the object distance (D_o), and the image distance (D_i).

---

* LENS FORMULA

$$\frac{1}{f} = \frac{1}{D_o} + \frac{1}{D_i}$$

The lens formula is arranged to give positive answers if the image is real and, therefore, on the opposite side of the lens from the object. However, in Sample Problem 14.3 the image proves to be a virtual image. A virtual image is indicated by the lens formula when the image distance is solved and found to be a negative value.

SAMPLE PROBLEM 14.2

Refer to Figure 14.13. If the object is 30.0 in. from the lens and the focal length is 10.0 in., how far away is the image and what is the magnification?

SOLUTION

Wanted

D_i = ? and m = ?

Given

f = 10.0 in. and D_o = 30.0 in.

Formula

The lens formula:

$$\frac{1}{f} = \frac{1}{D_o} + \frac{1}{D_i}$$

must be rearranged to:

$$\frac{1}{D_i} = \frac{1}{f} - \frac{1}{D_o}$$

Calculation

$$\frac{1}{D_i} = \frac{1}{10.0 \text{ in.}} - \frac{1}{30.0 \text{ in.}} = \frac{1 \times 3}{10 \times 3} - \frac{1}{30} = \frac{2}{30} = \frac{1}{15}$$

Inverting both sides of the equation, we obtain the answer: D_i = 15 in. The magnification formula is:

$$m = \frac{D_i}{D_o} = \frac{15 \text{ in.}}{30 \text{ in.}} = \frac{1}{2}$$

This information is useful, for example, in camera design. The position of the lens in front of the film can be predetermined, as can the size of film required.

SAMPLE PROBLEM 14.3

Refer to Figure 14.12D. The focal length is 8.00 in. and the object is 5.00 in. away from the lens. Find the distance to the image and the magnification.

SOLUTION

Wanted

D_i = ? and m = ?

Given

f = 8.00 in. and D_o = 5.00 in.

Formula

The lens formula:

$$\frac{1}{f} = \frac{1}{D_o} + \frac{1}{D_i}$$

must be rearranged to:

$$\frac{1}{D_i} = \frac{1}{f} - \frac{1}{D_o}$$

Calculation

$$\frac{1}{D_i} = \frac{1 \times 5}{8.00 \text{ in.} \times 5} - \frac{1 \times 8}{5.00 \text{ in.} \times 8} = \frac{5.00}{40.0} - \frac{8.00}{40.0} = -\frac{3.00}{40.0}$$

Inverting both sides of the equation, we have:

$$\frac{D_i}{1} = -\frac{40.0}{3.00} \qquad D_i = -13.3 \text{ in.}$$

The minus sign for the image distance indicates the image is on the same side of the lens as the object. Therefore, the image is virtual, not real. The minus sign is disregarded when using the magnification formula:

$$m = \frac{D_i}{D_o} = \frac{13.3 \text{ in.}}{5.00 \text{ in.}} = 2.66$$

The magnification indicates the image is 2.66 times the object in size. The importance of positioning lenses to provide magnified virtual images will be demonstrated in Chapter 15 on optical instruments.

Summary

In this chapter, we discussed the fundamentals of reflection and refraction, and how lenses make use of refraction to control the path of light rays. Light rays are reflected from a surface at the same angle as they strike the surface. If the surface is smooth, the reflected rays are parallel; if the surface is rough, the reflected rays are scattered, or diffused.

The image we see in a mirror is called a virtual image because the light rays only appear to come from behind the mirror. There are no actual light rays behind the mirror.

When light travels from one transparent medium to another, its speed, and therefore its direction, changes. Its frequency does not change. The bending of light rays is called refraction. When light attempts to go from a slow-speed medium to a high-speed medium, and the angle of incidence is large enough, the light will be totally reflected inside the slow-speed medium. Reflectors on signs, precious gems, and optical glass fibers utilize this characteristic of light.

Lenses are devised to make use of the refractive properties of light. The majority of applications for lenses are to form images of some object. The images may be real or virtual, depending on the type of lens and on the relationship between the object distance and the focal length of the lens.

Key Terms

* angle of incidence
* angle of reflection
* law of reflection
* virtual image
* index of refraction
* refraction
* total internal reflection
* lens

* convex lens
* principal focus
* focal length
* real image
* concave lens
* magnification formula
* lens formula

Questions

Where diagrams are required, be sure that they are neatly drawn and clearly labeled.

1. Explain the difference between light reflected from a smooth surface and light reflected from a rough surface.

2. Why is the image in a mirror called a "virtual" image?

3. How does a prism refract light?

4. Explain how a prism disperses white light to form the visible spectrum.

5. What is meant by the total internal reflection of light? How does it differ from ordinary reflection, as from a mirror? Give an example of how total reflection is used.

6. Explain the action of a convex lens on parallel rays of light; on converging rays; on diverging rays.

7. What is a real image?

8. Explain the action of a concave lens on parallel rays of light; on converging rays; on diverging rays.

9. How does a convex lens form an image of an object that is more than two focal lengths away from the lens? Describe the image.

10. Suppose that you wished to form an enlarged image of an object on a screen. What kind of lens would you use? How would you use this lens to make the enlarged image?

11. Can a concave lens form a real image if the light rays from the object are parallel?

Problems

Check your answers graphically.

1. Do Sample Problem 14.1 if the image is $75\bar{0}$ mm behind the lens.

2. Do Sample Problem 14.2 with the object distance 40.0 in. from the lens.

3. Do Sample Problem 14.3 with the object 4.00 in. from the lens.

4. A particular lens provides a sharp image when the object is 88.0 in. in front of the lens and the image is 11.0 in. behind the lens. What is the focal length of the lens?

5. A camera lens is focused on an object $12\bar{0}$ in. away from the lens. The film is 2.00 in. from the lens (image distance). What is the focal length of the lens?

6. A convex lens has a focal length of 12 in. When an image is formed 18 in. behind the lens, what is the object distance?

7. A slide projector has a lens and object arrangement similar to the one in Figure 14.12C. The slide is 3.0 in. from the lens, and the image is focused on a screen 120 in. away. What is the focal length of the lens?

8. Refer to Figure 14.12A. The object is 72 in. from the lens, and the image is 24 in. from the lens. Specify the focal length and the magnification.

9. A photo enlarger has a lens and setup similar to Figure 14.12C. The object (picture to be enlarged) is 12 in. from the lens, and the focal length is 11 in. Find the image distance and magnification.

10. Refer to Figure 14.12D. The object is 2.00 in. from the lens, and the focal length is 3.00 in.
    a. Find the image distance.
    b. Is the image real or virtual?
    c. What is the magnification?

11. Refer to Figure 14.12B. Let $f = 10.0$ in. and $D_o = 20.0$ in.
    a. Is D_i equal to $2f$, as the figure indicates?
    b. Move the object 100.0 in. from the lens and find D_i.

12. You are given a convex lens with a focal length of 4.00 in. A particular arrangement gives a magnification of 3.00 (the image is 3 times the size of the object and the image is real). Find D_o and D_i.

Computer Program

This program uses the lens formula so that you can find the focal length, the object distance, or the image distance for your lens problems. Also, you may want to experiment by keeping the focal length constant and changing the object distance to see how the image distance changes.

PROGRAM

```
10   REM   CH FOURTEEN PROGRAM
20   PRINT "LENS FORMULA. MEASUREMENTS ARE TO BE IN INCHES."
21   PRINT "TO FIND FOCAL LENGTH (F)-TYPE 1, TO FIND OBJECT."
22   PRINT "DISTANCE (DO)-TYPE 2, AND TO FIND IMAGE DISTANCE (DI)"
23   PRINT "TYPE 3."
30   INPUT A
40   PRINT "YOU HAVE CHOSEN - ";A
50   PRINT
60   IF A = 1 THEN   GOTO 100
70   IF A = 2 THEN   GOTO 200
80   IF A = 3 THEN   GOTO 300
90   END
100   PRINT "TYPE IN THE OBJECT DISTANCE AND THEN THE IMAGE DISTANCE."
110   INPUT DO,DI
120   PRINT "OBJECT DISTANCE = ";DO
130   PRINT "IMAGE DISTANCE = ";DI;" IN."
140   LET F = 1 / ((1 / DO) + (1 / DI))
150   PRINT
160   PRINT "THE FOCAL LENGTH = ";F;" IN."
170   GOTO 90
200   PRINT "TYPE IN THE FOCAL LENGTH AND THEN THE IMAGE DISTANCE."
210   INPUT F,DI
220   PRINT "FOCAL LENGTH = ";F" IN."
230   PRINT "IMAGE DIATANCE = ";DI;" IN."
240   LET DO = 1 / ((1 / F) - (1 / DI))
250   PRINT
260   PRINT "THE OBJECT DISTANCE = ";DO;" IN."
270   GOTO 90
300   PRINT "TYPE IN THE FOCAL LENGTH AND THEN THE OBJECT DISTANCE."
310   INPUT F,DO
320   PRINT "FOCAL LENGTH = ";F;" IN."
330   PRINT "OBJECT DISTANCE =";DO;" IN."
340   LET DI = 1 / ((1 / F) - (1 / DO))
350   PRINT
360   PRINT "THE IMAGE DISTANCE = ";DI;" IN."
370   GOTO 90
```

NOTES

Line 40 confirms your selection.

Lines 60–80 select the correct subroutine for the answer desired.

Lines 120, 130, 220, 230, 320, and 330 display your data input.

Lines 140, 240, and 340 represent the formula variations.

TV cameras use lenses the same way that film cameras do. However, the lens in a TV camera focuses the image on a light-sensitive plate that converts the light image to an electrical signal. (Courtesy of ABC Visual Communication)

Chapter 15

Objectives

When you finish this chapter, you will be able to:

- ☐ Diagram a camera, showing location of lens and film. Explain how focusing is accomplished and discuss the problem of spherical aberration and its correction with a compound lens.
- ☐ Explain what is meant by the *f*-number of a camera.
- ☐ Recognize the *f*-number equation *f*-number $= f/d$.
- ☐ Diagram a projector, noting location of condenser and objective lenses; explain operation.
- ☐ Diagram a compound microscope, noting location of ocular (eyepiece) and objective lenses and optical tube.
- ☐ Diagram a telescope; explain its operation.
- ☐ Diagram binoculars; explain their operation.

Optical Instruments

In this chapter, we will speak of some of the problems and solutions in the construction of optical devices. We will also discuss how two convex lenses can be positioned to increase magnification beyond that possible with a single lens. This information provides a good understanding of lenses and is a preparation for future courses in optics.

15.1 The Camera

In the sixteenth century, an Italian scientist, Giambattista della Porta (1535–1615), performed an interesting experiment. He darkened a room by putting shades over the windows. Then he made a very small hole in one of the shades and noticed that the entering light produced a picture of the outdoor scene on the opposite wall. He called this effect *camera obscura,* which means, literally, "dark room." It is from this expression that we get the modern word "camera."

We can duplicate this experiment by means of the pinhole camera illustrated in Figure 15.1. This device consists of a light-tight box with a pinhole in the middle of the front side. Light

PHOTO 15.1   The principles of optics apply to all kinds of modern cameras. The camera on the left is sensitive to infrared radiation and detects hot areas in operating machinery, electrical systems, and building heating systems. (Courtesy of Westinghouse Electric Corp.) The high-powered camera on the right is being used to photograph fingerprints on a weapon. (Courtesy of the FBI, U.S. Department of Justice)

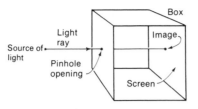

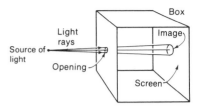

A. A very small opening produces a small, sharp image.

B. A larger opening produces a larger, blurred image.

FIGURE 15.1 The Pinhole Camera.

rays entering through this pinhole form a spot of light on the opposite side, which acts as a screen. If this opposite side or screen is made of some translucent substance, such as oiled paper or ground glass, we can see this spot of light by looking through the back of the box.

If the pinhole is very small, as in Figure 15.1A, the spot of light, too, is small and appears as a point. If the hole is made larger, as in Figure 15.1B, more rays of light enter, and the spot becomes larger. However, if we wish a clear image of the object, the pinhole must be very small (Figure 15.2). If the hole is made larger, light rays from each point of the object will produce a large spot of light on the screen, thus creating a blurry image.

Since light rays travel in straight lines, the image appears inverted. For the same reason, the farther the screen is from the opening, the larger the image appears. Since we have no lens action here, if the pinhole is very small, the image will appear sharp, regardless of the distance between the opening and the screen.

Although the pinhole camera at first was regarded as a scientific toy, artists soon found that they could use it to obtain

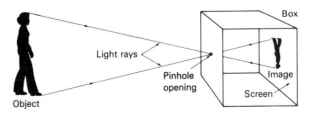

FIGURE 15.2 Illustration Showing How the Image of an Object Is Formed on the Screen of the Pinhole Camera.

a reduced view of a scene that they wished to paint. It was only natural that they should desire some means for preserving permanently the view they saw on the translucent screen.

Of course, today, we can preserve our pictures on film. The pinhole camera is excellent for taking pictures—even the color type—except for one thing: It requires too long an exposure time. We have seen that the opening must be extremely small if clear pictures are to be produced. Thus, very little light can enter, and, even using the fastest modern film, an exposure would take minutes, which is too slow for most photography, where exposures of $\frac{1}{25}$ of a second, or faster, are required to prevent an object's movement from blurring the picture. What we need is to make the opening larger to admit more light in a short time period and yet have the light rays produce a sharply focused image on the screen or film. It is here that we can call upon the convex lens for help.

In Figure 14.9, we saw that the light rays entering the lens are focused to a fine point at the focal point. Here, then, is our answer. All we need do is to replace our pinhole with a suitable convex lens, as in Figure 15.3. Because the lens has a larger diameter than the pinhole, more light will enter. Because of the converging action of the lens, this light will be focused to produce a sharply focused image on the screen. In this way, light from all points of an object in front of the lens will form a sharp, inverted image of the object on the screen.

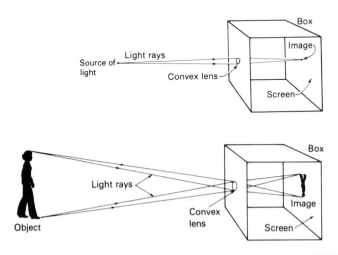

FIGURE 15.3   The Effect of Substituting a Convex Lens for the Pinhole of the Camera.

However, we must pay for this advantage. In the pinhole camera, you will recall, the image is sharp, regardless of how far the object is in front of the pinhole or how far the screen is behind it. When we use a convex lens to produce a sharp image on the screen, however, the screen must be at the point of focus. The point of focus, in turn, will be determined by the focal length of the lens and by the distance the object is in front of the lens.

The focal length of the lens is a fixed value that depends on the curvature of the lens. But as the object moves farther from, or nearer to, the lens, the point of focus, where a clear image will be produced, changes. Thus, we must have some method for varying the distance between the lens and the screen if we wish to obtain clear images of objects at different distances from the lens.

Keep in mind that the chief purpose of the lens is to permit as much light as possible to shine on the film. Since the wider the diameter of the lens is, the more light will be permitted to enter, our first impulse would be to use a very large lens. However, the larger the lens is, the more expensive it is to make. Generally, cameras are not fitted with the largest lenses possible.

We already know that we do not need as much light to illuminate a small area as we need for a large area. Therefore, if you have a small frame of film, you may use a lens of a small diameter. Thus, small cameras that produce small pictures use lenses of smaller diameter than do large cameras that produce large pictures.

Another factor that determines the brightness of the image on the film is the distance between the film and the lens. The shorter this distance is, the brighter is the image. Since the image is formed at the point of focus of the lens, the shorter the focal length of the lens, the nearer is the film to the lens, and consequently the brighter the image. Thus, it would seem that we need a lens with as short a focal length as possible. However, certain complications occur. The shorter the focal length of the lens (and the nearer the point of focus to the lens), the smaller will be the image.

As you see, there is a definite relationship between the brightness of the image, the diameter of the lens, and the focal length of the lens. The greater the diameter of the lens, the brighter is the image; the greater the focal length of the lens, the less bright is the image. Thus, a lens with a large diameter and a longer focal length may produce an image as bright as that formed by a lens of smaller diameter but with a shorter focal length. For this reason, the light-gathering property of a lens is rated in terms of its diameter and its focal length. This light-gathering property is displayed on cameras as an **f-number.**

* *f*-number	A rating that compares the diameter of a lens to its focal length.

For example, a lens is marked *f*/6 if its focal length (f) is six times its diameter (d). Thus, $f = 6 \times d$ can be rearranged to:

$$\frac{f}{6} = d$$

Thus, all *f*/6 lenses, no matter what the diameter, have the same light-gathering properties and will produce images of equal brightness. The smaller the *f*-number of a lens, the brighter the image it can produce; and conversely, the larger the *f*-number, the less bright is the image. The general formula is:

* *f*-NUMBER FORMULA	$$f\text{-number} = \frac{\text{focal length}}{\text{diameter}}$$ $$f\text{-number} = \frac{f}{d}$$

Sample Problem 15.1 examines two lenses with the same light-gathering ability.

SAMPLE PROBLEM 15.1	Figure 15.4 shows two lenses: One lens has a focal length of 12.0 in. and the other has a focal length of 8.00 in. What are the diameters of the lenses when rated *f*/6?
SOLUTION	
Wanted	$d_1 = ?$ $d_2 = ?$
Given	$f_1 = 12.0$ in., $f_2 = 8.00$ in., and *f*-number $= 6$
Formula	Since we are interested in the diameters of the lenses, we rearrange the *f*-number formula:

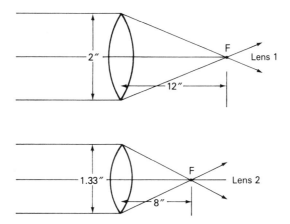

FIGURE 15.4 Two Lenses for Sample Problem 15.1.

Calculation

$$d = \frac{f}{f\text{-number}}$$

$$d_1 = \frac{f_1}{6} = \frac{12.0 \text{ in.}}{6} = 2.00 \text{ in.}$$

$$d_2 = \frac{f_2}{6} = \frac{8.00 \text{ in.}}{6} = 1.33 \text{ in.}$$

This information is helpful in matching the sizes of lenses to camera sizes and in determining the distances from lenses to films.

There is a definite numerical relationship between the f-number and the amount of light that lenses will admit. If we start with the amount of light admitted by $f/11$ lenses as unity, an $f/8$ lens will admit twice as much. An $f/16$ lens will admit half as much. Or, to put it another way, if the exposure time required for an $f/11$ lens is 1 second, then an $f/8$ lens will require an exposure time of 1/2 second and an $f/16$ lens 2 seconds, everything else being equal.

The following table shows the relationship between various f ratings and the relative exposures required:

Relative exposure time (s)	1/10	1/8	1/6	1/4	1/3	1/2	1	2	4	8	16
f-number	3.5	4	4.5	5.6	6.3	8	11	16	22	32	45

Finally, we need to keep in mind that the function of the lens is to permit more light to enter the camera than a pinhole would permit. However, to obtain a clear image, the rays of light from each point of the object must be focused to produce a fine point of light on the film. Should the lens fail to focus each point, the result will be a disk or a circle of light on the film instead of a fine point, and a blurred image will be obtained.

It is here that our simple convex lens fails. Light entering the edges of the lens will be bent more than light passing closer to the middle, as Figure 15.5 shows. Thus, the rays of light passing through the edges will be brought to a point of focus closer to the lens than that produced by the rays through the middle portion, thus causing a blurred image. We call such a fault *spherical aberration,* and it is due to the fact that our manufacturing processes are not perfect. It would be extremely expensive to polish the lens to the exact curvature required. It is less expensive to use a concave lens in combination with the convex lens.

Figure 15.6A shows that the convex lens bends the outer rays more than it does the middle ones. A concave lens (Figure 15.6B), on the other hand, can correct this excessive bending of the outer rays. If we combine the two into a **compound lens**, the spherical aberration will be overcome, and all the light rays will focus at a single point, as Figure 15.6C shows. Of course, spherical aberration cannot be entirely eliminated, but it can be reduced to an acceptable level.

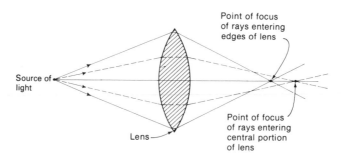

FIGURE 15.5   Spherical Aberration.

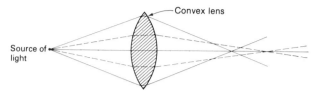

A. Action of convex lens

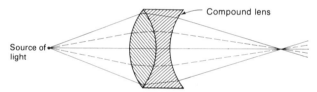

B. Action of concave lens

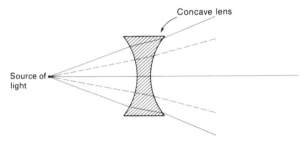

C. Action of compound lens composed of combined convex and concave lenses

FIGURE 15.6   How Spherical Aberration Is Corrected.

* compound lens   A lens made of two or more lenses that are combined to correct for spherical aberration.

15.2 Projectors

In the camera, light from a large, distant object is focused by means of a convex lens upon a nearby screen or sensitized film to form a small, real, inverted image. In the projector, this process is reversed. Light from a small, nearby object is focused by a convex lens to form a large, real, inverted image upon a distant screen.

Figure 15.7 shows light from a powerful source passing through a set of *condenser lenses.* This set consists of two planoconvex lenses mounted with their curved surfaces facing each other. (A planoconvex lens has one plane—that is, flat—surface and one convex surface.) The function of condenser lenses is to concentrate the diverging rays of the source into a slightly converging beam directed upon the object to be projected. This object is usually an image on a glass slide or film transparency. Where the image is opaque, no light will pass through. Where it is clear, all the light will pass through. Different degrees of opacity will permit more or less light to pass.

The light that passes through the slide next strikes the convex *objective lens,* located slightly more than one focal length away from the slide. (The lens is called "objective" because it is relatively near the object.) As a result, a large, inverted image of the picture on the slide is formed upon a screen that is the appropriate distance away from the lens. The image is focused sharply on the screen by moving the objective lens closer to, or farther from, the slide.

Note that the image on the screen is inverted. Thus, if we wish this image to appear right side up, the slide must be placed in the projector upside down.

Now we can use the *f*-number and the magnification properties of a lens to obtain information about a projector. Remember from the previous chapter that the magnification of a lens is the ratio of the image height to the object height, or the image distance to the object distance:

$$m = \frac{\text{image height}}{\text{object height}} = \frac{d_i}{d_o}$$

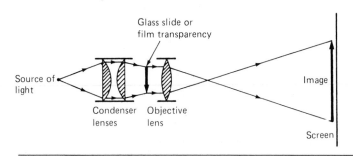

FIGURE 15.7   The Projector.

Of course, the "image height" could also be the "image width," as long as it is compared to the corresponding dimension on the object. Sample Problem 15.2 concerns the f-number and magnification of a projector.

SAMPLE PROBLEM 15.2	An $f/3.5$ projector lens has a focal length of 128 mm. It can project a 0.690 m wide × 1.05 m high image at a distance of 3.96 m. The object is a 35 mm slide that is 23.0 mm wide by 35.0 mm high. A. Determine the lens diameter. B. Find the magnification. C. Determine the distance from lens to slide.
SOLUTION A	
Wanted	$d = ?$
Given	$f = 128$ mm and f-number $= f/3.5$.
Formula	The f-number formula is: $$f\text{-number} = \frac{f}{d}$$
Calculation	We rearrange the formula: $$d = \frac{128}{3.5} = 36.6 \text{ mm}$$
SOLUTION B	
Wanted	$m = ?$
Given	We are given the image and object heights: image height $=$ 1.05 m × 1000 $=$ 1050 mm and object height $=$ 35.0 mm.
Formula	Note that we are not given the object distance, so we cannot use the equation $m = D_i/D_o$. However, we can use the magnification formula in the form: $$m = \frac{\text{image height}}{\text{object height}}$$

Calculation	$$m = \frac{1050 \text{ mm}}{35.0 \text{ mm}} = 30.0$$
SOLUTION C	
Wanted	$D_o = ?$
Given	$D_i = 3.96$ m
Formula	Now that we know the magnification is 30.0, we can use $m = D_i/D_o$ to solve for D_o. We first rearrange the formula to: $$D_o = \frac{D_i}{m}$$
Calculation	$$D_o = \frac{3.96 \text{ m} \times 1000 \text{ mm/m}}{30.0} = 132 \text{ mm}$$
	This information is useful in determining the dimensions of the projector and the size of the room that the projector can be used in.

15.3 Compound Microscopes

When an object is too small to be seen clearly by the eye, we call upon the convex lens for help. A convex lens so used is called a *magnifying glass*. When used in a camera, the convex lens produces a real image of a relatively distant object. But now, as a magnifying glass, the convex lens is used to produce a virtual image of a relatively close object. Therefore, the shorter the focal length of the lens, the greater will be the magnification. However, keep in mind that, conversely, the greater the magnification, the shorter the focal length, and accordingly, the nearer the object must be to the lens. If we wish to greatly magnify a portion of some object, we must hold our lens very close to that portion. We will not be able to see very much of the whole object through the glass. We say that the "size of the field" is reduced as we increase the magnification.

The magnifying glass is sometimes called a *simple microscope*. The highest practical magnification of such an instrument is about 20 times. If greater magnification is needed, a **compound**

PHOTO 15.2   The microscope uses lenses to make large images of small objects, and has created tremendous advances in fields from medicine and biology to crime detection. (Courtesy of the FBI, U.S. Department of Justice)

microscope is employed. In this device, two lenses, or sets of lenses, are used in such a way that one magnifies the magnified image produced by the other.

*** compound
microscope**

A microscope that contains more than one lens.

In the compound microscope shown in Figure 15.8, two convex lenses are mounted at either end of the *optical tube,* which consists of two metal tubes that telescope into each other. This tube is necessary to keep outside light from interfering with the images formed by the lenses. The bottom lens, called the *objective lens,* has a focal length of a fraction of an inch. The object to be examined is placed upon a glass slide a little more than one focal

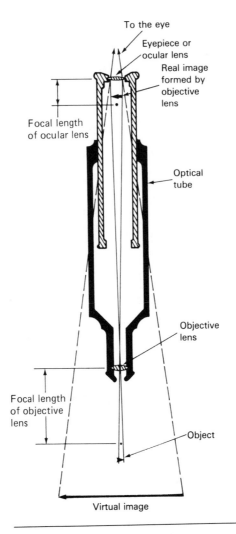

Focal length
of ocular lens

To the eye

Eyepiece or
ocular lens

Real image
formed by
objective
lens

Optical
tube

Objective
lens

Focal length
of objective
lens

Object

Virtual image

FIGURE 15.8 The Compound Microscope.

length below this lens. A mirror (not shown) reflects light up through the slide into the lens.

The objective lens forms a greatly enlarged real image several inches up in the optical tube. This image forms less than one focal length below the upper convex lens, which is called the *eyepiece,* or *ocular lens.* The eye, looking down through the ocular lens, sees a greatly enlarged virtual image of the real image formed by the objective lens. As a result, we get a double magnification of the object.

The two parts of the optical tube are adjustable so that the tube may be made longer or shorter to meet the needs of both lenses. This adjustment is made by a set of knobs that permit very fine control over the distance between the lenses and the distance between the slide and the objective lens. To overcome spherical aberration, each lens is made in several parts. The objective lens may be composed of eight or more combined lenses, and the ocular may be composed of two or more combined lenses.

With some of our modern compound microscopes, objects as small as 0.000 000 2 inch may be seen, and magnifications as great as 2000 times may be obtained. The theoretical limit of such an optical instrument is determined by the wavelength of the light employed. With this in mind, more powerful microscopes have been constructed that operate by ultraviolet rays instead of visible light. Since the wavelengths of these rays are smaller than those of visible light, the limit of magnification possible is extended. Such microscopes may magnify up to about 4000 times.

Since ordinary glass will absorb ultraviolet rays, the lenses of such a microscope must be made of quartz, which transmits ultraviolet rays. Also, because these rays are invisible to the human eye, as well as dangerous to it, the images are formed on photographic plates that are sensitive to ultraviolet light.

15.4 Refracting Telescopes and Binoculars

The **telescope**, illustrated in Figure 15.9, is an instrument used to make faraway objects appear closer. Note its resemblance to the compound microscope. Two lenses are mounted on either end of the tube, or barrel, that is used to keep out interfering light. The convex objective lens forms a real image of the object less

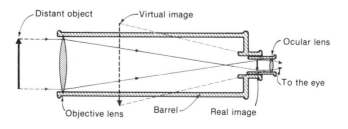

FIGURE 15.9   A Type of Astronomical Telescope.

than one focal length away from the convex ocular lens. As a result, the eye, looking through the ocular lens, sees a large virtual image of the real image.

The objective lens of the telescope differs from that used in the microscope in a number of ways. From our study of the camera, we have learned that when an object is a considerable distance from a convex lens, then the greater the focal length of the lens, the larger will be the image. Since we desire as large an image as possible, the objective lens of the telescope is made with a long focal length, contrasted with the short focal length of the objective lens used in the microscope. Thus, the barrel of the telescope must be long. Also, since the telescope operates on light from distant objects, the objective lens must have as large a diameter as possible to gather in the maximum amount of light. The microscope, however, uses an objective lens of very small diameter.

The ocular lens of the telescope resembles that of the microscope. It is set in a tube that can slide into or out of the barrel in order to adjust the distance between the lenses.

If we examine Figure 15.9 closely, we note that the final virtual image seen by the eye is inverted. An inverted image does not matter if the telescope is used for astronomical purposes. But if we wish to use a telescope to examine a distant landscape, for example, we want the final image to appear right side up. To accomplish this, all we need do is insert a third lens between the objective and ocular lenses of the refracting telescope. Figure 15.10 shows light from a distant object entering the objective lens and forming a real, inverted image in the barrel of the telescope. Light from this image, passing through the erecting lens, forms a second real image farther up the barrel. Note that this second image is re-inverted and thus appears right side up. The ocular lens forms an enlarged virtual image of this second real image,

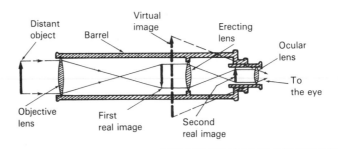

FIGURE 15.10   The Terrestrial Telescope.

as has been described previously. A telescope of this type is called a *terrestrial telescope*.

The terrestrial telescope would be most convenient, except for two details. First of all, it is long—the long focal length of the objective lens requires a long overall size. Second, we are accustomed to seeing things with both eyes. Thus, two telescopes side by side would be required for normal vision.

Binoculars, illustrated in Figure 15.11, overcome both of these objections. To reduce the overall length of the instrument, the light entering through the objective lens is twice reflected, so that the barrel is as effective as that of a telescope three times as long. The reflection is accomplished by two sets of prisms, which act as reflectors.

Now is the time to try out the lens formula from the previous chapter:

$$\frac{1}{f} = \frac{1}{D_o} + \frac{1}{D_i}$$

and the magnification formula:

$$m = \frac{D_i}{D_o}$$

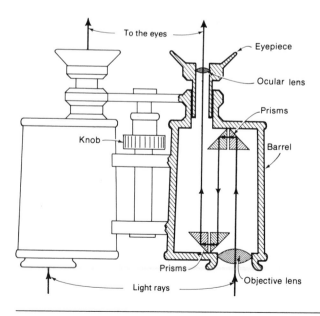

FIGURE 15.11   Binoculars.

on a device with two lenses, such as microscopes, telescopes, and binoculars. Remember, the image of one lens forms the object of the second lens. Sample Problem 15.3 illustrates the method for handling two-lens systems.

SAMPLE PROBLEM 15.3	A special-purpose industrial optical device has two convex lenses set $35\overline{0}$ mm apart. Each has a focal length of $15\overline{0}$ mm. The object is $45\overline{0}$ mm in front of the first lens. A. What is the location of the final image? B. What is the magnification?
SOLUTION A	We begin by sketching the setup, as shown in Figure 15.12. Focal lengths are shown on both sides of the lenses to aid in determining whether images are real or virtual. Our procedure will be to start with the lens closest to the object (lens 1) and find the image distance and magnification. We will then use this image as the object for lens 2 and determine the final image distance.
Wanted (for Lens 1)	D_i = ?

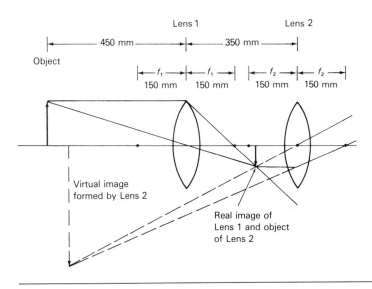

FIGURE 15.12   Sketch of Lens Setup in Sample Problem 15.3.

Given

$f_1 = 15\overline{0}$ mm and $D_o = 45\overline{0}$ mm

Formula

$$\frac{1}{f_1} = \frac{1}{D_o} + \frac{1}{D_i}$$

Calculation

We rearrange to:

$$\frac{1}{D_i} = \frac{1}{f_1} - \frac{1}{D_o} = \frac{1 \times 3}{15\overline{0}\text{ mm} \times 3} - \frac{1}{45\overline{0}\text{ mm}} = \frac{2}{450}$$

Inverting both sides of the equation, we have:

$$\frac{D_i}{1} = \frac{450}{2} \qquad \text{or} \qquad D_i = 225\text{ mm}$$

The image is real as indicated by the positive answer and the position of the image in Figure 15.12. This image is the object of lens 2. As can be seen from the figure, its distance from lens 2 is equal to $D_o = 350$ mm − 225 mm = 125 mm. (This distance is inside f_2; therefore the image of lens 2 will be a virtual image.)

**Wanted
(for Lens 2)**

$D_i = ?$

Given

$f_2 = 15\overline{0}$ mm and $D_o = 125$ m

Formula

$$\frac{1}{f_2} = \frac{1}{D_o} + \frac{1}{D_i}$$

Calculation

We rearrange to:

$$\frac{1}{D_i} = \frac{1}{f_2} - \frac{1}{D_o} = \frac{1 \times 5}{150\text{ mm} \times 5} - \frac{1 \times 6}{125\text{ mm} \times 6} = -\frac{1}{750}$$

Inverting both sides of the equation, we have:

$$\frac{D_i}{1} = -\frac{750}{1} \qquad \text{or} \qquad D_i = -750\text{ mm}$$

The minus sign in the answer indicates that we have a virtual image that is 750 mm in front of lens 2.

We will repeat the calculation for D_i using decimal numbers instead of fractions because in some problems, the fractions may become difficult to work with:

$$\frac{1}{D_i} = \frac{1}{f_2} - \frac{1}{D_o} = \frac{1}{150 \text{ mm}} - \frac{1}{125 \text{ mm}}$$
$$= 0.00667 - 0.00800 = -0.00133$$

Inverting both sides of the equation, we have:

$$\frac{D_i}{1} = \frac{1}{0.00133} = -750 \text{ mm}$$

SOLUTION B

Our procedure will be to find the magnification for each lens and then multiply the two together for the total magnification.

Wanted

$m = ?$

Given

$D_o = 450$ mm and $D_i = 225$ mm for lens 1; $D_o = 125$ mm and $D_i = 750$ mm (the minus sign is ignored) for lens 2

Formula

$$m = \frac{D_i}{D_o}$$

Calculation

First, we find the magnification of lens 1:

$$m = \frac{225 \text{ mm}}{450 \text{ mm}} = 0.5$$

Then, for lens 2:

$$m = \frac{750 \text{ mm}}{125 \text{ mm}} = 6$$

Now we can find the total magnification:

$$m = m_1 \times m_2 = 0.5 \times 6 = 3$$

This answer means the final image is three times the size of the original object. The information is useful in determining whether the lens arrangement is correct and the magnification is sufficient.

Summary

In this chapter, we discussed a few of the applications for lenses and the problems associated with these applications. Generally, when we think of lenses, we think of them being used to magnify our view of an object. However, in a camera, the purpose of a lens is to let sufficient light strike the film in as short a time period as possible. The lens must focus the light correctly and provide a short exposure time that prevents movement of the object or film from blurring the picture.

The f-number of a lens relates the focal length of the lens to its diameter. All lenses with the same f-number, regardless of lens diameter, provide the same amount of light in the same interval of time. An $f/6$ lens means that, regardless of diameter, the focal length (f) is six times the lens diameter (d).

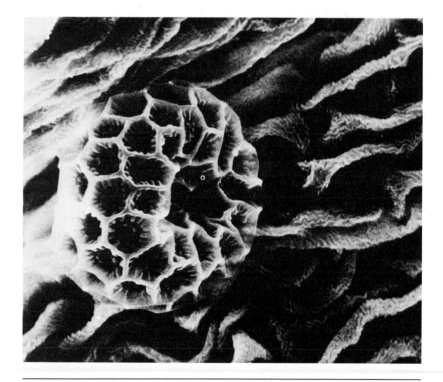

PHOTO 15.3   This picture shows a pollen grain magnified 3000 times. To obtain this great magnification, ultraviolet light must be used, since the wavelength of this light is smaller than that of visible light. (Courtesy of RCA)

Concave lenses may be placed next to convex lenses to correct spherical aberration in the convex lens. Such combinations are called compound lenses.

The objective lens in a projector is used to project a real and enlarged image of a slide (or movie film) onto a screen.

When a microscope uses two or more lenses for the purpose of increasing the magnification of a single lens, the device is referred to as a compound microscope. The lens closest to the object is the objective lens and the lens closest to the eye is the ocular lens. The objective lens is used to focus a real image inside the principal focus point of the ocular lens. The ocular lens then produces an enlarged virtual image for the eye to see.

Telescopes and binoculars work on the same principle as the microscope, except that the objective lens must be designed to view objects at great distances.

Key Terms

* *f*-number
* *f*-number formula
* compound lens

* compound microscope
* telescope
* binoculars

Questions

1. Why is a lens better than a pinhole in a camera?
2. How do we focus a camera to produce sharp images of near and far objects?
3. A lens is rated *f*/3.5. What is the relation of the focal length to the lens diameter?
4. What is spherical aberration?
5. What is the purpose of a compound lens?
6. What is the function of the objective lens in a projector?
7. Draw a simple diagram of a compound microscope and explain the functions of the most important parts. Identify the objective and ocular lenses.
8. Draw a simple diagram of a terrestrial telescope and explain the functions of the most important parts.
9. Why are prisms used to reflect the light in binoculars?

Problems

THE CAMERA

1. Solve Sample Problem 15.1 with the focal lengths changed to $f_1 = 1\overline{0}$ in., $f_2 = 5.0$ in.

2. Refer to Figure 15.4. Find the new f-number for lens 2 if the diameter is doubled but the focal length remains at 8 in.

3. Refer to Sample Problem 15.1. Solve the problem with the f-number changed to 4.

4. A scene requires an exposure of 1/6 second at $f/4.5$ to be photographed. What exposure time is required at $f/16$?

5. If a particular scene requires an exposure time of 1/2 second at $f/8$, what should the time be at $f/5.6$?

6. A camera has an $f/3.5$, $4\overline{0}$ mm diameter lens.
   a. What is the focal length?
   b. What is the distance from lens to film if the object is 4.0 m away? Use the lens formula.

PROJECTORS, MICROSCOPES, AND TELESCOPES

7. A slide projector uses 2.00 in. × 2.00 in. slides. The slides are positioned 5.50 in. behind the convex objective lens, which has a focal length of 5.00 in. How far away should the screen be placed and what is the size of the image?

8. Refer to Sample Problem 15.2.
   a. If the lens is adjusted to within $15\overline{0}$ mm of the slide, at what distance will the image be in focus?
   b. What is the magnification?

9. You are asked to help design a "projecting comparator." (This is an inspecting device that projects an enlarged shadow of a small object on a screen. The shadow, say, of a very fine screw thread, must fit inside certain lines drawn on the screen.) A single $f/3.5$ convex lens will be used. An image distance of 50.0 in. is specified and a magnification of 10.0 is required. Find: (a) the object distance; (b) the focal length; (c) the lens diameter

10. Refer to Figure 15.8. The distance between the two lenses is $8\overline{0}$ mm, the focal length of the objective lens is $1\overline{0}$ mm,

and the focal length of the ocular lens is $4\overline{0}$ mm. If the object is 12 mm from the objective lens, determine the position and magnification of the image.

11. Refer to Figure 15.9. A surveyor's telescope (called a transit) has an objective lens with a focal length of 15.00 in. and an ocular lens with a focal length of 4.820 in. The lenses are 20.00 in. apart. Show the position of the vertical image and determine the magnification when looking at a marker 100.0 ft away.

Computer Program

The computer program in Chapter 14 relating to the lens formula may be used in this chapter.

Part V

Electricity

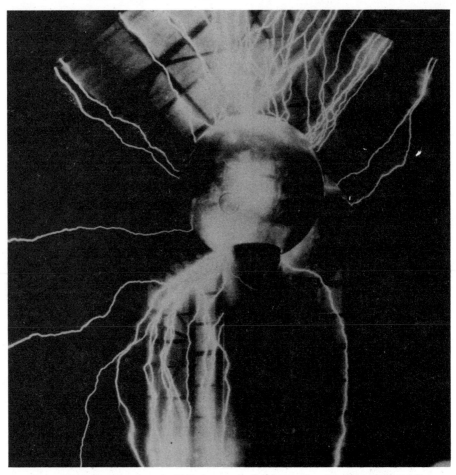

Our understanding of modern electricity is a result of early experiments with static electricity. Here, one of the original Van de Graaff electrostatic generators provides a vivid display of a high voltage electrical discharge. (Courtesy of High Voltage Engineering Corp.)

Chapter 16

Objectives

When you finish this chapter, you will be able to:

☐ Recognize electricity as the accumulation or motion of electrical charge, and the effects of such accumulation or motion.

☐ Create static charge accumulations by using silk and glass or fur and hard rubber. Show how this is applied in fly ash collectors and sandpaper manufacture.

☐ Demonstrate the electrostatic force law that like charges repel and unlike charges attract.

☐ Identify substances as insulators or conductors, and compare their electrical uses.

☐ Diagram an electroscope; explain its use through contact and induction.

☐ Define the strength of an electrical energy source as potential difference, measured in volts.

☐ Diagram a capacitor; show plates, dielectric, and separating distance; explain operation; use the Leyden jar as an example.

☐ Explain what is meant by "grounding."

Static Electricity

Electricity is a rather complicated subject, and scientists are still learning about it. The study of electricity, what it is and what it does, begins with static electricity. In this chapter, we open with a discussion about the beginnings of our knowledge of static electricity, its definition, and its effects. We will then discuss a device called an electroscope, which is used to detect an electrical charge; the measurement of the amount of energy needed to create an electrical charge; the Leyden jar and its successor, the capacitor; and applications of static electricity.

16.1 Static Electricity

The ancient Greeks were aware that amber (a resinous substance), upon rubbing, would attract small objects. About 1600, William Gilbert, physicist and doctor to the English Queen Elizabeth, discovered that other substances, such as sulfur, sealing wax, and glass, had properties similar to that of amber. The

PHOTO 16.1   By touching a source of static electricity, this man has become charged himself. Each strand of his hair repels each other strand because all strands feel a like charge. (Courtesy of the Museum of Science, Boston)

scientists of his time believed that such substances, when rubbed, or *charged,* exuded a sort of fluid that attracted light objects. They called this fluid "electricity," and they said that rubbing the substance "charged, or loaded, it with electricity." (The word "electricity" is derived from the Greek word for amber, *elektron.*)

In 1672, the German scientist Otto von Guericke (gā′rĭ-kə) noticed that if he charged a ball of sulfur by rubbing it, a feather would be attracted to it. But no sooner had the feather touched the sulfur than the feather would be repelled from it. This observation led the French scientist Charles du Fay (1698–1739) to the conclusion that there were two types of "fluids," or electricity. One type he called "vitreous," the electricity present in charged glasslike substances. The other type he called "resinous," the electricity present when objects such as amber, wax, and rubber are charged. Du Fay noted that charged vitreous substances attracted charged resinous substances, but repelled charged vitreous substances. He also noted that charged resinous substances attracted charged vitreous substances, but repelled charged resinous substances.

About 1747, Benjamin Franklin, the American statesman and scientist, came to the conclusion that there was only one type of "fluid," or electricity. The "vitreous" and "resinous" charges, he believed, were only two opposite phases of the same phenomenon. He arbitrarily called the "vitreous" charge *positive* (+) and the "resinous" charge *negative* (−).

Our present theory of the electronic structure of matter enables us to arrive at a plausible explanation of the results observed by du Fay and gives us an insight into what we call **electricity** and **static electricity.** Recall the electron theory in Chapter 2 that indicated all atoms consist of a central nucleus surrounded by shells of planetary electrons. The nucleus contains neutrons, which carry no charge, and protons, each of which carries a single positive charge. In a normal atom, the number of planetary electrons equals the number of protons in the nucleus. Since each such electron carries a negative charge, which is equal and opposite to the positive charge of the proton, the normal atom is neutral—that is, there is no external charge.

---

* **electricity**   Those physical phenomena involving an electric charge and the effects of this charge when at rest and in motion.

* **static electricity**   Those physical phenomena involving an electric charge at rest.

Now let us perform a simple experiment. Suppose, as Figure 16.1A shows, that we suspend two glass rods by pieces of string. The glass rods will neither attract nor repel each other. If we charge the glass rods by rubbing each with a piece of silk, however, they will repel each other, as shown in Figure 16.1B. What has happened? We must recall that the nucleus of the atom holds its planetary electrons in place by the attraction between unlike charges. The electrons of the innermost shells are tightly held, but those of the outermost shells are loosely held and may be detached quite readily. When the glass rods are rubbed with the silk, some of the outer electrons of the atoms that make up the glass are detached by the cloth. Some of the positive charges of the protons in the nucleus are no longer neutralized, and as a result, the atoms acquire external positive charges. Since like charges repel, when the two positively charged glass rods are brought near each other, they swing apart.

The silk, on the other hand, receives an excess of electrons, and thus receives an external negative charge. You can prove this by holding the cloth near either of the two charged rods. The positively charged glass rod will be attracted to the negatively charged cloth, as shown in Figure 16.1C.

If we substitute hard rubber for the glass rod and fur for the silk, we find that when the hard-rubber rod is rubbed by the fur, the fur will lose some of its electrons and acquire a positive charge. The rubber, on the other hand, will receive an excess of electrons from the fur and thus obtain a negative charge. You now can see why du Fay thought that there was "vitreous" and

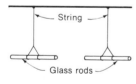

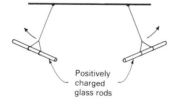

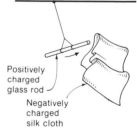

A. Two uncharged glass rods have no attraction or repulsion.

B. After rubbing with a silk cloth, the two positively charged rods repel each other.

C. A positively charged glass rod and a negatively charged silk cloth attract each other.

FIGURE 16.1   How Charged Bodies Interact.

"resinous" electricity. You also can see why Franklin was closer to the truth when he said that these two phenomena were merely two phases of the same thing.

Today, scientists believe that a charged body can exert an attractive force on another nearby uncharged or oppositely charged object. If the object has a similar charge, it will be repelled. This force is somewhat like the attraction of a magnet for steel, or the gravitational attraction of the earth for other objects. The force we are talking about here is called an **electrostatic force.**

---

* electrostatic force

A force exerted by one electrically charged body on another electrically charged body.

We can now explain the action of von Guericke's feather. When he rubbed his sulfur ball, he gave it a negative charge—that is, an excess of electrons. It could then attract the uncharged (neutral) feather. But when the feather touched the charged sulfur, it obtained some of the excess electrons. It, too, became negatively charged. Hence, it was repelled from the similarly charged sulfur ball.

We have seen that as we charge a body by rubbing it, we either add electrons to the body (giving it a negative charge) or subtract electrons from the body (leaving it with a positive charge).

PHOTO 16.2   Modern electronic circuits are often made of conducting material printed on top of an insulating substance. The electricity travels from one part of the circuit to another only along the conducting path.

The nature of the substance—that is, the number and arrangement of its outermost electrons—determines whether we add or take away electrons from a body when we rub it. Some substances, such as glass, have their outermost electrons so arranged that they can lose electrons quite easily, acquiring a positive charge. On the other hand, a piece of hard rubber, for example, has its outermost electrons so arranged that when it is rubbed, it will seize electrons from the material with which it is stroked, and it will accumulate an excess of electrons, acquiring a negative charge.

There is another aspect of electricity to be considered. When a glass or hard-rubber rod is being charged by losing or acquiring an excess of electrons, the action is local. That is, only the portion of the rod being rubbed is affected. If you take away or add electrons to one end of a hard-rubber or glass rod, the atoms of the other end remain neutral, and thus the rod at that end has no charge. The disturbance of the atoms remains local and does not affect other atoms. We call such materials **insulators**. Examples are glass, hard rubber, wax, mica, air, paper, and some plastics.

---

* insulators   Substances that resist the movement of electrons.

On the other hand, there are certain substances, generally metals, whose outer electrons are held very loosely. As a matter of fact, a certain number of these outer electrons are constantly jumping from atom to atom, even without external influence. We call such moving electrons *free electrons,* and we call such substances **conductors**.

---

* conductors   Materials that permit the free movement of electrons

---

If an excess of electrons (negative charge) is placed at one end of a conductor, the repulsion between like charges will cause the loosely held electrons of neighboring atoms to move toward the other end. The movement of these electrons will cause a disturbance among electrons farther away, and as a result, the excess of electrons quickly distributes itself throughout the entire conductor.

Similarly, if electrons are removed from one end of a conductor (causing a positive charge), electrons from neighboring atoms are attracted to compensate for the deficiency. Again, all the atoms are affected, and soon the deficiency is spread throughout the entire conductor. Thus, a charge created at any portion of a conductor quickly spreads itself.

16.2 The Electroscope

The ability of an electrical charge to spread itself over the surface of a conductor led the Rev. Abraham Bennet, an English physicist, to invent the electroscope in 1787. This device can detect an electrical charge.

The electroscope is a simple instrument, as Figure 16.2 shows. A stopper, made of insulating material such as rubber or plastic, is fitted into the mouth of a glass flask. A metal rod is forced through a hole in the stopper so that one end of the rod is in the flask and the other end is outside. A metal ball, or knob, is mounted on the end of the rod extending outside the flask. Two strips of extremely thin gold leaf are attached to the end of the rod within the flask.

Why a metal ball on top? Once a charge is placed on the ball, it will leak off to the surrounding air more slowly from a metal ball than from other shapes. For example, if the rod were pointed on top, the charge would leak off rapidly.

Now suppose we rub a hard-rubber rod with a piece of fur. As we know, the hard rubber will tear away electrons from the fur, and as a result, the rod will have an excess of electrons (negative charge). Now we touch the charged end of the hard-rubber rod to the knob of the electroscope (Figure 16.3A). Some

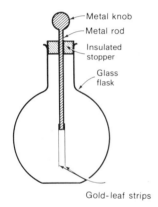

Metal knob
Metal rod
Insulated stopper
Glass flask

Gold-leaf strips

FIGURE 16.2 The Gold-Leaf Electroscope.

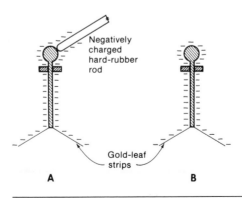

Negatively charged hard-rubber rod

Gold-leaf strips

A   B

FIGURE 16.3 Charging an Electroscope by Contact with a Negatively Charged Body.

of the excess electrons are transferred to the knob and distribute themselves over the surfaces of the knob, the metal rod, and the gold-leaf strips, as indicated by the minus signs (−) in the figure. The insulated stopper prevents any of these electrons from reaching the flask or any neighboring objects.

Note what happens to the gold-leaf strips. Since both strips have negative charges, they repel each other and swing out. The greater the negative charge—that is, the greater the excess of electrons on the strips—the stronger is this repulsion and the farther the strips swing away from each other. We now have a means for measuring roughly the relative strength of a charge.

When, as in Figure 16.3B, the charged hard-rubber rod is taken away from the knob, the excess electrons remain trapped in the electroscope. The gold-leaf strips continue to repel each other. We say the electroscope is charged negatively. It will remain so until the excess electrons leak off into the air or until they are removed by some other means. When this happens, the electroscope loses its charge, the gold-leaf strips lose their mutual repulsion, and they assume their original vertical positions.

We may charge the electroscope positively in a similar manner. When a glass rod is rubbed with silk, the rod loses some of its electrons to the cloth and obtains a positive charge (deficiency of electrons). When the charged glass rod is touched to the knob of the electroscope, some of the electrons on the knob rush over to compensate for the deficiency on the glass rod. As a result, a deficiency of electrons on the knob arises. This deficiency spreads over the knob, the metal rod, and the gold-leaf strips on the electroscope. Since both strips again have similar charges (this time, positive charges), they repel each other once more. When the glass rod is removed, the deficiency of electrons remains. This time, the electroscope is charged positively.

When we charge an electroscope by either of the methods just described, we call it *charging by contact*. To remove a charge from an electroscope (called *discharging*), all we need to do is touch our finger to the metal knob. If the electroscope is charged negatively, the excess electrons will flow out through our body to the ground (earth). If the electroscope is charged positively, a sufficient number of electrons will flow from the ground through our body to neutralize the deficiency on the electroscope. The earth is considered an enormous reservoir for either positive or negative charges. What we have just done is to discharge the electroscope by **grounding**. Grounding is mentioned frequently in this and later chapters.

* grounding	A term used to indicate that a charged object is connected to the earth by a conductor.

There is still another method for charging our electroscope. Suppose you bring a negatively charged body near to, but not touching, the knob of the uncharged electroscope, as shown in Figure 16.4A. The negative charge, as indicated by the minus sign ($-$), repels some of the electrons of the knob down onto the gold-leaf strips. Thus, the knob has an electron deficiency, or positive charge, as indicated by the plus ($+$) signs, and the gold-leaf strips are charged negatively. Since they both have similar charges, the strips repel each other.

Still keeping the negatively charged body near the knob, touch the knob with your finger or connect it to the ground by means of a copper wire, for example, as in Figure 16.4B. Now some of the repelled electrons of the knob are conducted through the wire to the ground instead of to the gold-leaf strips. Because some of these electrons are going to the ground rather than to the strips, the negative charge on the strips is reduced, and their mutual repulsion becomes less. As a result, they come closer together.

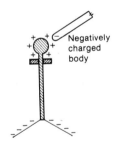

A. A negatively charged body brought near the electroscope knob repels the electrons to the gold-leaf strips.

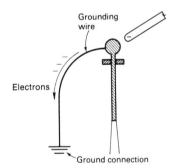

B. If the knob is grounded, the electrons are repelled to the ground connection.

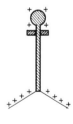

C. When the grounding wire is removed, the remaining positive charges distribute themselves along the metal knob and gold-leaf strips.

FIGURE 16.4   How an Electroscope Is Charged Positively by Induction.

Next, remove the wire, and then remove the charged body. The electroscope has lost a number of its electrons, and since it is no longer connected to the ground, it cannot recover them. Thus, there is a deficiency of electrons, or positive charge, all over the surface of the electroscope, including the gold-leaf strips. Therefore, the electroscope is charged positively, and the strips repel each other, as we see in Figure 16.4C. We call this method **charging by induction.** The electroscope can also be charged inductively by a positive charge.

✴ charging by induction

A method of charging an object without direct contact.

Note that when we charge *by contact,* the electroscope acquires the same type of charge as the charging body. But when we charge *inductively,* the electroscope acquires a charge that is opposite to that of the charging body. As we will see later in this chapter, both methods are used in industry.

We now can see how a charged body can attract an uncharged one. Suppose, as in Figure 16.5, we place a positively charged body near an uncharged one. Because of induction, some of the electrons of the uncharged body are attracted to the side nearest the charged body. This nearer side will then acquire a negative charge, leaving the farther side with a deficiency of electrons, or a positive charge. Because the attraction between the negatively charged nearer side and the positively charged body is greater than the repulsion between the latter and the positively charged farther side, the two bodies are attracted. As we will see later, this attraction is similar to that between a magnet and an unmagnetized piece of iron.

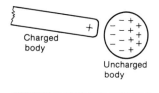

FIGURE 16.5 How a Charged Body Attracts an Uncharged One by Induction.

16.3 Potential Difference and Electrical Pressure

We now know that a force exists between two dissimilarly charged objects. Another way of looking at this same condition is to picture the excess electrons (negative charge) as straining to reach a point where there is a deficiency of electrons (positive charge). If the two charges are connected by a conductor, the excess electrons will have an easy path to the point of deficiency. If the two charges are separated by an insulator, which opposes the passage of electrons, the excess electrons are prevented from reaching the point of deficiency. However, there is no perfect insulator. A certain number of electrons will always leak across to the point of positive charge. For example, if a charged electroscope is permitted to stand for a considerable length of time, it will lose its charge slowly by the leakage of electrons through the air, which is an insulator. The electrons will go to or from the ground depending on whether the electroscope is charged negatively or positively.

Let's perform an experiment. We place two metal spheres on insulated stands side by side, as shown in Figure 16.6. We add electrons to one sphere by repeatedly charging a rubber rod and touching it to the sphere. We subtract electrons from the other sphere by using a glass rod in a similar manner. Bear in mind that we are doing work (expending energy) when we strip electrons from their atoms, and we are doing work when we force electrons from the rubber rod onto the sphere. Electrons must be forced onto the sphere because the electrons already on the sphere try to repel those on the rubber rod. (Recall the glass rods in Figure 16.1.) Once the electrons are on the sphere, the energy expended is stored as potential energy. Of course, work is also required to remove electrons from the other sphere.

By adding electrons to one sphere and subtracting electrons from the other sphere, the difference in potential energy between the spheres (the *potential difference*) is greater than if we had simply charged one sphere and grounded the other. For an anal-

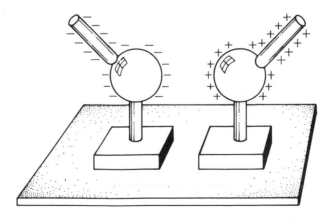

FIGURE 16.6   Potential Energy Stored by the Electric Charges on the Spheres.

ogy of this latter condition—charging one sphere and grounding the other—imagine that you are filling the large tank in Figure 16.7 by carrying buckets of water up the ladder. You are doing work by lifting the buckets of water to the top of the tank. This energy is stored as potential energy by the water and can be recovered by opening a valve at the bottom of the tank and allowing the water to turn a water wheel.

The difference in potential energy between the two charged spheres is measured in volts. For instance, we might say the **potential difference** (the difference in potential energy) between the spheres is 10 000 volts. The *volt* is a basic electrical term and is used extensively in the next few chapters on electricity. As one example, the electrical wall outlet in your home has a potential difference of 120 volts between the two contacts.

＊ potential difference

The difference in the electrical energy stored between two charged bodies.

Let's get back to our spheres. As we continue to pile up electrons on one sphere and remove electrons from the other, the potential difference between the spheres increases. We can think

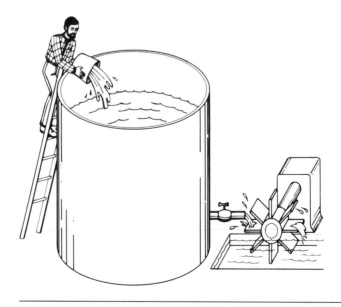

FIGURE 16.7   Potential Energy Stored by the Water in a Tank.

of this increase in potential energy as causing an increase in the *electrical pressure* between the spheres, just as the pressure at the bottom of the water tank increases as the amount of water increases. This pressure is trying to force the electrons to the other sphere. If the electrical pressure becomes great enough, the air between the spheres can no longer restrain the excess electrons, and they rush across to the point of deficiency. This rush of electrons produces light, which we see in the form of a spark. At the same time, the air through which the electrons rush is heated and expands. We hear the inrush of air that follows cooling in the form of a sharp, crackling sound that accompanies the spark.

This rush of electrons, or spark, takes place in a small fraction of a second, and it might seem that the electrons merely jump from the point of excess to the point of deficiency. But closer examination indicates that the electrons jump back and forth between the two points millions of times per second.

We can better understand the behavior of the electrons with an example of a pendulum. Suppose we suspend a small weight from a string tied to a nail. The weight hangs straight down. Now we move the weight several inches to the right. The force of gravity tends to pull the weight back to its original position— that is, straight down. If we release the weight, it swings toward its original position, but keeps right on going and now swings to the left. Again, the force of gravity pulls it back, and once again it swings to the right. This process continues, each swing being a little shorter than before, until the weight comes to rest in its original position.

Similarly, electrons at a point that has a negative charge seek to satisfy the deficiency at the positively charged point. When the electrons rush over, more than enough to satisfy the deficiency move across. The two charges change places. The electrical pressure now is in the opposite direction, and the electrons rush back. This to-and-fro surge of electrons continues until both points are neutral and the electrons are at rest once more.

16.4 The Leyden Jar

In 1746, a Dutch professor at the University of Leyden, Pieter van Musschenbrock, still believing the "fluid" theory of electricity, tried to store this "fluid" in a glass bottle. Let us try to duplicate his experiment. Assume that we have a glass jar partially coated inside and out with a metal foil, as shown in Figure

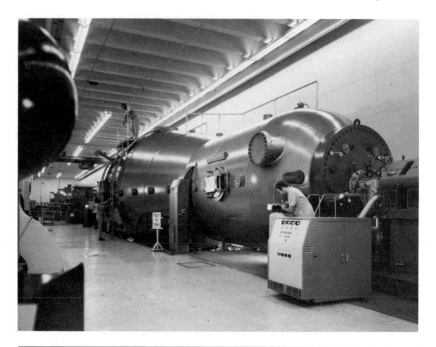

PHOTO 16.3   A common way of storing electrostatic charge is in a Van de Graaff generator, which is very similar in principle to a Leyden jar. These linked Van de Graaff generators are used to accelerate charged particles to high velocities for research. (Courtesy of Brookhaven National Laboratory)

16.8. A stopper, made of rubber or some other insulating material, is fitted into the mouth of the jar. A metal rod runs through this stopper. At its upper end, the metal rod is fitted with a metal knob. A metal chain is attached to the lower end of the rod, and its other end touches the inner foil coating. A copper wire connects the outer foil coating to the ground.

Now suppose we place an excess of electrons (negative charge) on the metal knob. These electrons will travel down the metal rod, through the metal chain, and onto the inner metal foil. Thus, a negative charge is placed upon the metal foil inside the jar. The electrons cannot pass to the outer metal foil because they are stopped by the glass of the jar, which is an insulator. However, because of induction, the electrons on the inner foil will repel electrons on the outer foil. These will flow down the copper wire to the ground. Now we remove the grounding wire. The outer foil has a deficiency of electrons (positive charge), and the inner foil has a surplus of electrons (negative charge).

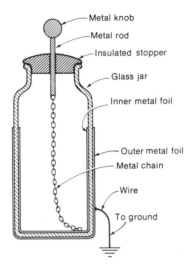

- Metal knob
- Metal rod
- Insulated stopper
- Glass jar
- Inner metal foil
- Outer metal foil
- Metal chain
- Wire
- To ground

FIGURE 16.8   The Leyden Jar.

We have now stored a negative charge in the jar (called a *Leyden jar,* after the city where it was first constructed). This charge will remain in the jar until the excess electrons leak from the inner foil to the outer foil.

Why bother with this arrangement of inner and outer foils? Why not use an electroscope, since it can store a charge also? One reason is that a larger surface area can store a larger charge, similar to the way a larger bucket can hold more water. If the outer foil were not in place, the repelled electrons on the outer surface of the jar would be forced to a location where the repulsion is less, but they would still be on the jar. By removing them with a ground wire to the outer metal foil, the potential difference between the two foils is increased.

You can store a positive charge in a Leyden jar by placing a positive charge (deficiency of electrons) on the knob. The positive charge will be distributed over the inner foil. By induction, electrons will be drawn from the ground, through the wire, and onto the outer foil. When the wire is removed, these excess electrons will be trapped, and hence the outer foil will have a negative charge.

If you connect the two metal foils of a charged Leyden jar by means of a conductor, the excess electrons on one of the foils will rush through to satisfy the deficiency on the other foil. These electrons feel such pressure to get to the point of deficiency that they will not wait till a path has been completed, but will jump across the small gap existing just before the conductor touches the second foil. You see and hear this jump as a spark. As we know, these electrons will surge back and forth between the two foils until the foils become neutral.

The Leyden jar was the original "capacitor." Capacitors are used in many electrical circuits. In the next section, we will learn more about what a capacitor is and how it is constructed.

16.5 Capacitors

For many years, the Leyden jar existed as a scientific toy. Today, though, it is an important electronic component. We have modified its structure somewhat and now call it a **capacitor.** (It used to be called a condenser because it was thought that the Leyden jar stored electrical "fluid" by condensing it.) Instead of using a glass jar for a capacitor, we now use a sheet of some insulating material (called the **dielectric**), such as glass, mica, waxed paper, oil, or even air itself. On both sides of this dielectric, as Figure 16.9 shows, are metal plates, usually made of brass, aluminum, or tin foil. The unit of **capacitance** is the farad (F), named after the English scientist Michael Faraday (1791–1867). The capacitance depends on three factors:

The area of the plates facing each other.
The distance between them.
The nature of the dielectric.

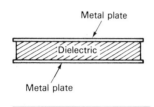

FIGURE 16.9   The Capacitor.

---

* **capacitor**	An electrical device capable of storing an electric charge.
* **dielectric**	An insulating material. The term is used primarily in conjunction with capacitors, which must have a dielectric material between the metallic plates or foils.
* **capacitance**	The ability of a capacitor to store electric charge.

The larger the area of the plates facing each other, the greater the capacitance of the capacitor. The capacitance is also increased as these plates come closer together. Thus, the thinner the dielectric, the greater the capacitance of the capacitor.

The nature of the dielectric will also affect the capacitance. If we have a capacitor of a certain size that uses air as a dielectric

PHOTO 16.4   There are many types of fixed capacitors: paper and plastic film (upper left), electrolytic (upper right), mylar film (center), mica (lower left), mylar film (lower center), and ceramic (lower right).

and we substitute, say, mica for the air, we will find that the capacitance will be increased about seven times (all other factors remaining constant). The number of times a dielectric is as effective as air is called the *dielectric constant* of that substance. Thus, air has a dielectric constant of 1, mica has a dielectric constant of about 7, and waxed paper may have a dielectric constant of from 2 to 3.2.

Capacitors are needed in different sizes depending on the job they must perform. Also, the better dielectrics usually cost more. Therefore, if you need to design a physically small capacitor for an instrument with tight space requirements, you may decide to use the best dielectric available. On the other hand, if you are designing a very large capacitor needed for power circuits, a less expensive dielectric may be the answer.

Mica capacitors generally are constructed of a number of small plates of metal separated by thin sheets of mica. The effective area of the plates is increased by connecting all alternate plates together, and thus the effect is that of two large plates, as shown in Figure 16.10A. The whole device generally is encased in some insulating material, such as Bakelite.

Capacitors that use paper as a dielectric generally are constructed by separating two long ribbons of metal foil by a strip of waxed paper, as shown in Figure 16.10B. The whole is then rolled into a tube to reduce its size and is impregnated with wax to keep out air and moisture.

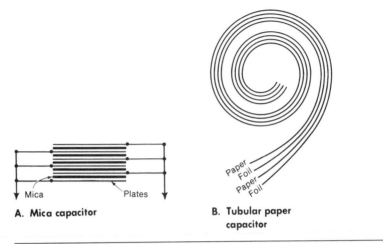

A. **Mica capacitor**   B. **Tubular paper capacitor**

FIGURE 16.10   Two Types of Capacitors.

In the electrolytic capacitor, a sheet of aluminum is placed next to a strip of gauze that is saturated with a borax solution. An extremely thin coating of aluminum oxide and oxygen gas forms on the surface of the aluminum. If we consider the aluminum as one plate of the capacitor and the borax solution as the other plate, the coating of the aluminum oxide and oxygen gas, which is an insulator, becomes the dielectric. The aluminum need not be a straight sheet, but may be folded over many times or loosely rolled to give a greater effective area. Because the "plates" are separated by an extremely thin dielectric, the capacitance of such a capacitor is very great. The whole generally is encased in a cardboard container or an aluminum can.

16.6 Practical Applications

Static electrical charges can be used in a large number of industrial applications. For example, Figure 16.11 illustrates a device used to filter dust and soot that otherwise would go up a factory smokestack, together with the hot gases. It is called an *electrostatic fly ash collector*. Not only is the surrounding air kept purer, but often valuable chemicals are recovered.

The operation of this device is extremely simple. The hot gases, dust, and soot pass through a metal chamber before entering the smokestack. A metal plate is mounted by means of an

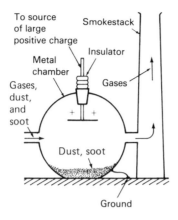

FIGURE 16.11   Electrostatic Fly Ash Collector Used to Remove Soot and Dust from Factory Exhaust Gases.

insulated fitting near the top of this chamber. A large positive charge is placed upon this plate. As the particles of dust and soot pass beneath the charged plate, they are attracted to it. Once they touch the plate, the particles, too, acquire positive charges. As positive charges, the particles are repelled from the charged plate and accumulate at the bottom of the chamber, which is grounded. The hot gases, cleaned of dust and soot, continue up the smokestack.

The electrostatic field is also used in the manufacture of sandpaper and similar abrasives. The abrasive grains are dropped onto a belt, which carries them between two oppositely charged plates. The backing of paper or cloth, coated with adhesive, is passed beneath the upper plate (Figure 16.12). As the abrasive grains enter the area between the two charged plates, they are brought in contact with the lower plate and acquire the same charge as the lower plate. They are then repelled from this plate

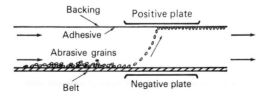

FIGURE 16.12   Electrostatic Field Used in the Manufacture of Sandpaper to Achieve Uniform Spacing of Abrasive Grains.

and fly up to the adhesive backing, where, since all the grains have similar charges, they repel each other, spacing themselves uniformly. When the adhesive hardens, the abrasive grains are found to be standing on end and uniformly spaced. The result is a great improvement in the cutting quality of the abrasive.

A similar method is used in the manufacture of certain fabrics, such as simulated velvet or carpeting. The cloth backing is coated with an adhesive and is passed beneath the upper of two charged plates. A belt carries tiny textile fibers over the lower plate. These fibers are repelled from the lower plate and are shot into the adhesive, where they stand on end and are evenly spaced, often packed in at 250 000 fibers per square inch. Sometimes, instead of spreading the adhesive uniformly, it is applied in the form of a design. The result, then, is a raised design where the tiny fibers stick to the adhesive.

So far, we have dealt with stationary charges. The excess electrons deposited on a body have remained there, except for the brief interval when they have distributed themselves over a conductor or during an electrical discharge. For this reason, we have titled this chapter "Static Electricity." In the following chapters we will study the electrons as they move from point to point under the influence of an electrical pressure. We will then be considering current electricity.

Summary

Our ancestors knew that materials such as amber, sulphur, sealing wax, and glass attracted certain small objects after being rubbed. They said the rubbing charged the material with electricity. We now know that rubbing can dislodge electrons from the atoms of some materials and deposit electrons on others. An object with an excess of electrons is said to have a negative charge. The absence of a sufficient number of electrons creates a positive charge.

Electrons exert a repelling force on other electrons ("like" charges) and exert an attractive force on positively charged particles ("unlike" charges). Positively charged atoms also exert an attractive force on unlike charges and repel like charges. When an object has an excess of electrons (a negative charge), or is positively charged, it will attract an uncharged object or an object of the opposite charge. The reason an uncharged object can be attracted is that its supply of like charges is repelled to the most distant side of the object. This leaves the unlike charges nearest the charged object, and therefore an attractive force occurs. The force exerted by one charged body on another is called an electrostatic force.

The physical phenomena involving electric charges at rest (as opposed to charges continually flowing from one point to another) are referred to as static electricity. These electric charges may be placed on all substances. Materials that allow electrons to move about freely are called conductors. If the material does not allow free movement of electrons, it is called an insulator.

An electroscope is a device that will indicate whether or not an object has a charge. Two movable metal strips in the electroscope separate when they both have the same charge. The electroscope can be charged in two ways. One method is to place the metal knob of the electroscope in contact with the charged object. In this case, the electroscope receives the same charge as the object. The other method of charging is accomplished by bringing the metal knob close to, but not touching, the object. The electroscope is grounded momentarily to allow the repelled charges to escape. This method is called charging by induction. In this case, the electroscope receives a charge opposite to that of the object.

It takes work to overcome electrostatic forces and store a charge on an object, and this stored charge represents potential energy. The difference in the potential energy between a charged object and ground, or between two charged objects, is called the potential difference and is measured in volts. This potential difference creates an electrical pressure tending to force electrons from the body with the greater negative charge.

The Leyden jar is a device for storing electrical charge and was the first of what we now call capacitors. The ability of a capacitor to store electrical charge is called its capacitance. The unit of capacitance is the farad (F). The capacitance depends on the area of the plates facing each other, the distance between the plates, and the type of dielectric used. The larger the plate area, the larger the charge; the closer the plates are to each other, the larger the charge; and the more effective the dielectric, the larger the charge. (The term "dielectric" is used in connection with capacitors and means "insulating material.")

In addition to the use of capacitors for electrical purposes, the effects of static electricity are used in industry to improve the arrangement of grit on sandpaper and the arrangement of fibers on simulated velvet and carpeting. The removal of dust and soot from smokestack gases is also accomplished with the use of static electricity.

Key Terms

* electricity   * charging by induction
* static electricity   * potential difference
* electrostatic force   * capacitor
* insulator   * dielectric
* conductors   * capacitance
* grounding

Questions

1. In terms of the electron theory, what do we mean when we say that an object has a negative charge? A positive charge?

2. What is meant by "static electricity"?

3. What kind of force is exerted by a charged body on other charged bodies?

4. Explain the difference between an insulator and a conductor. Give two examples of each.

5. Describe the gold-leaf electroscope. What are two methods for giving it a negative charge?

6. How can we "ground" a charged electroscope? What happens to the charge on the electroscope?

7. Why can a charged body attract small uncharged bodies, such as scraps of paper or fly ash?

8. What is meant by "potential difference"? What is the unit of measurement?

9. What is a capacitor? What are the factors that determine its capacitance?

10. What dielectric is used in mica capacitors? In paper capacitors?

11. How may static electricity be used to prevent dust and soot from polluting the air near a factory chimney?

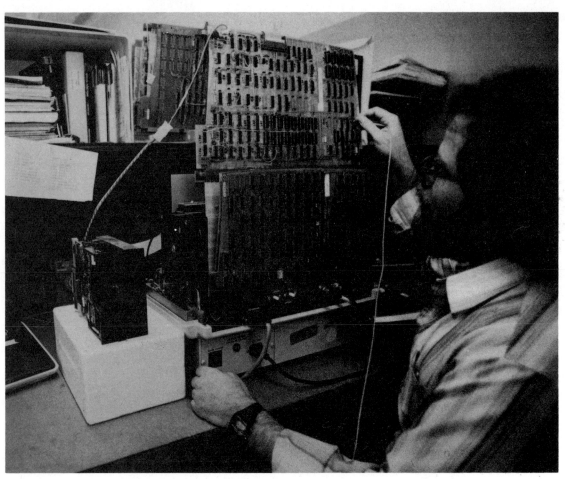

Direct current electricity is utilized in a wide range of electronic circuits. Here, a technician is testing the printed circuit board prior to its installation in a computer. (Courtesy of Wang Labs)

Chapter 17

Objectives

When you finish this chapter, you will be able to:

☐ Recognize electric current as electrical charges (electrons or ions) moving in one direction.

☐ Use standard units of charge, coulombs; use standard units of current, amperes; and use standard units of potential difference, volts.

☐ Diagram a simple series circuit containing an energy source (E), a resistance (R), and connection wires. Show how a current (I) will develop.

☐ Recognize Ohm's law, $I = E/R$, showing how the current (I) generated in a simple circuit equals the potential difference (E) pushing the charges divided by the circuit resistance (R).

☐ Recognize the electric power equations $P = EI = I^2R$.

☐ Diagram series, parallel, and series–parallel circuits of resistors. Calculate equivalent resistances by using the equations $R_t = R_1 + R_2$ and $1/R_t = (1/R_1) + (1/R_2)$.

☐ State the current and potential rules for a series circuit. Do the same for a parallel circuit.

☐ Diagram an electric current and show how a magnetic effect develops around it.

☐ Diagram and compare galvanometers, ammeters, voltmeters, and ohmmeters.

☐ Diagram an electromagnet and explain how the magnetic effect of a current is used in it.

Direct Current Electricity

Most technicians are expected to check the operation of equipment by taking measurements. Quite likely, these measurements are made on electrical components, or electrically operated instruments are used to make mechanical measurements.

In this chapter, we will consider the relationship between the three basic electrical quantities—voltage, current, and resistance. Series and parallel resistance circuits will be analyzed, and the construction and operation of voltmeters, ammeters, and ohmmeters will be discussed. We will use this basic information in later chapters in working with alternating current, electrical prime movers, applications, and wiring practices.

17.1 Three Factors of an Electric Current

ELECTRIC CURRENT—THE AMPERE

If an excess of electrons (negative charge) occurs at one end of a conductor, and a deficiency of electrons (positive charge) arises at the other end, the electrical pressure between the two ends will cause the loosely held electrons from the outer shells of the atoms of the conductor to stream from the point of excess to the point of deficiency. Thus, an **electric current** will move through the conductor from the negative end to the positive end. In the discussion of electricity in this chapter, we will be concerned only with **direct currents** (dc).

---

* electric current   Electrons in motion.

---

* direct current   The flow of electrons in one direction only.

In order to measure current, it is important to know the number of electrons that flow past a given point on a conductor in a certain length of time. Since the charge of one electron is very small, a **coulomb** is used as a unit for measuring the quantity of electric charge. A coulomb is equal to the combined charges of 6 280 000 000 000 000 000 electrons.

---

* coulomb   The unit of electric charge. One coulomb is defined as equal to the combined charges of 6 280 000 000 000 000 000 electrons.

PHOTO 17.1　An electric light bulb lights up when current flows through the filament, heating it up so that it gives off light. This early Edison-type lamp has a sealed tip where air was taken out to create a vacuum in the bulb, allowing the filament to glow without burning.

If 1 coulomb flows past a given point in 1 second, we call this amount 1 **ampere** of electric current. Hence, the unit of electric current is the ampere (A). Aside from the fact that electrons are too small to be seen, we would find it impossible to count them as they flowed by. Fortunately, we have an electrical instrument called the *ammeter* (which will be described later) that indicates directly the amount of current flowing through it.

---

* ampere　　　The unit of current. One ampere is equal to a flow of 1 coulomb per second.

Where the ampere is too large a unit to be used, we may employ the milliampere, which is a thousandth (0.001) of an ampere, or the microampere, which is a millionth (0.000 001) of an ampere. In an electrical formula, the capital letter I stands for current. A statement such as "the current is equal to 30 amperes" could be written as I = 30 A. In actual practice, the full word "ampere" is rarely spoken or written. It is usually shortened to "amp," or in spoken form, "amps."

RESISTANCE—THE OHM

Different substances offer different *resistances* to the flow of electricity. Note that resistance only impedes, or opposes, the flow of electric current—it does not absorb or reflect the current. Metals generally offer little resistance and are good conductors. Silver is the best conductor known, and copper is almost as good. Other substances, such as glass, rubber, and sulfur, offer very high resistance and are good insulators. In fact, because insulators have such a high resistance to electric current, we, for practical situations, consider that they completely block current in the application they are designed for. For example, we have no hesitancy in touching an insulated electrical extension cord in use. However, all substances will permit the passage of some electric current, provided the potential difference creating the current is high enough.

Resistance is affected by the length of the substance in the direction of flow of the current. The longer an object is, the greater its resistance. Another factor is the cross-sectional area of the substance, which is the area of the end exposed if we slice through the substance at right angles to its length. The greater the cross-sectional area, the less the resistance to current. Resistance is

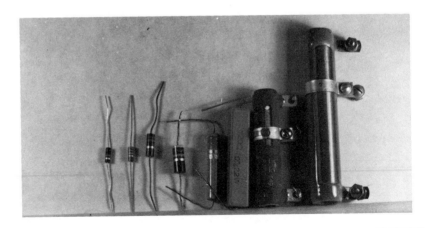

PHOTO 17.2   Typical resistors are made of carbon, such as those on the left, or of tightly wound wire, like those on the right. Resistors of these types can vary from a few ohms to many thousands of ohms.

also affected by the temperature of the substance. Metals generally offer higher resistance at higher temperatures. On the other hand, certain nonmetallic substances, such as carbon, offer lower resistance at higher temperatures.

By international agreement, the **ohm** is the resistance to the flow of electric current offered by a uniform column of mercury, 106.3 centimeters long, having a cross-sectional area of 1 square millimeter, at 0 degree Celsius. In an electrical formula, the capital letter R stands for resistance, and the capital Greek letter omega (Ω) is the symbol for ohm. For example, "the resistance equals 100 ohms" is written as $R = 100\ \Omega$.

---

* ohm   The unit of resistance.

ELECTRIC POTENTIAL—THE VOLT

In Chapter 16, we discussed the fact that a potential difference between two spheres creates an electrical pressure tending to force electrons to jump from one sphere to the other. Potential energy is used up when the spark jumps the gap. However, if we could use this potential difference to cause electrons to flow through some electrical device, the potential energy could do useful work. This condition is exactly what occurs when we have a constant source of electric potential, such as a battery or a generator. Electric generators will be discussed in a later chapter; for now, we will simply discuss the units of measurement.

The unit of potential difference, or electric potential (E), is the **volt** (V). We might say "the electric potential of a battery equals 9 volts," which would be written as $E = 9$ V.

---

* volt   The unit of electric potential. For example, 1 volt will cause 1 ampere of current to flow through a resistance of 1 ohm.

Electrical units are set up so that if the electric potential between two points is 1 volt, then it takes 1 joule of work to move 1 coulomb of charge between the two points. For example, if you have a battery rated at 1 volt, then the battery would do 1 joule of work (about 3/4 foot·pound) to move 1 coulomb of charge through a light bulb and back to the battery. However, we rarely want to know the number of coulombs, so this relationship is not used in most technical applications.

17.2 Relationship between Current, Resistance, and Electric Potential

OHM'S LAW

The relationship between current, resistance, and electric potential was discovered by a German scientist, George Simon Ohm (1784–1854), at the beginning of the nineteenth century. The unit of resistance is named in his honor. This relationship, which is called **Ohm's law**, can be expressed mathematically by the following formula:

*** OHM'S LAW
FORMULA**

current = electric potential ÷ resistance

$$I = \frac{E}{R} \quad \text{or} \quad E = IR$$

where I = current in A
E = electric potential in V
R = resistance in Ω

From this formula, we can see that for a given resistance in a circuit, the greater the electric potential, the larger the current; and for a constant potential, the greater the resistance, the smaller the current.

Now is a good time to point out that industry rarely uses the term "electric potential" and instead uses the term "voltage." For instance, instead of saying that the electric potential of a wall outlet is 120 volts, we say "the voltage at the outlet is 120 volts."

**SAMPLE
PROBLEM
17.1**

How much current will flow through a conductor whose resistance is 10 Ω when the electric potential (voltage) is 100 V?

SOLUTION	
Wanted	$I = ?$
Given	$E = 100$ V and $R = 10 \ \Omega$
Formula	$I = \dfrac{E}{R}$
Calculation	$I = \dfrac{100 \text{ V}}{10 \ \Omega} = 10$ A

This type of problem is useful in demonstrating the relationships indicated by Ohm's law. A little later on, we will see how Ohm's law is used in other practical applications.

ELECTRIC POWER AND ENERGY

When electrons move from the negative to the positive end of a conductor, work is done. The **power** (P) is the rate at which the electrons do work. The unit of power is the **watt** (W), which, for electrical use, is defined as the power used when 1 volt causes a current of 1 ampere to flow through a conductor.

＊ power   The rate at which electrons do work—that is, the amount of work done in a given period of time.

PHOTO 17.3   A watthour meter is actually a form of electric motor that drives the dial hands by a series of gears. It is designed to measure electrical energy in kilowatthours.

The power formula is:

power = voltage × current

$$P = EI$$

Where the watt is too small a unit, we may use the **kilowatt** (1 kW = 1000 W). For example, the power of a motor that draws 5.0 amps and 240 volts is:

$$P = E \times I = 240 \text{ V} \times 5.0 \text{ A} = 1200 \text{ W} \quad \text{(or 1.2 kW)}$$

Note that power is the *rate* at which electrons do work, even though time does not appear in the formula. The time is included in the value for current, since:

1 amp = 1 coulomb per second

Remember from Chapter 6 that power is work divided by time:

$$\text{power} = \frac{\text{work}}{\text{time}}$$

Rearranging the formula, we have:

work = power × time

We can see that the total amount of energy used, or the total amount of work done, is the product of the power (watts) and the time during which it is applied.

The electrical industry has gotten in the habit of measuring energy in wattseconds (1 wattsecond = 1 joule), or watthours, or kilowatthours, instead of foot·pounds and joules. The electric meter on the outside of your house, which shows the total amount of electric energy that has been used, is known as a **watthour meter**.

A meter that measures electrical energy.

We can use these formulas to show the relationship between power (in watts), current (in amps), and resistance (in ohms). Since we know that $P = EI$, we can substitute for E its equivalent from Ohm's law (IR) and get:

---

* POWER FORMULA 2

$$P = (E)I = (IR)I = I^2R \quad \text{or} \quad P = I^2R$$

SAMPLE PROBLEM 17.2

Your manufacturing plant has an electrical heating unit used to seal plastic bags. You need to know how much power it requires. The current flow is $40\overline{0}$ milliamps (mA) through a resistance of 5 kilohms (kΩ).

SOLUTION

Wanted

$P = ?$

Given

$I = 40\overline{0}$ mA $\times$ 0.001A/mA $= 0.4$ A and $R = 5$ kΩ $\times$ 1000 Ω/ kΩ $= 5000$ Ω

Formula

$P = I^2R$

Calculation

$P = (0.4$ A$)^2 \times 5000$ Ω $= 800$ W

This information is useful in later calculations to determine the cost of operation, the voltage required, and the type of power source needed.

17.3 Electrical Circuits

Just as water in a pipe flows from a point of high pressure to a point of low pressure, so the electrons in a conductor flow from a point of high potential (excess of electrons, or negative charge) to a point of low potential (deficiency of electrons, or positive charge). The path or paths followed by the electrons is called an **electrical circuit**. An electrical circuit is said to be *closed* when

PHOTO 17.4   However complicated a circuit may be, it is still a connection of series and parallel circuits. Shown here is the Pennsylvania–New Jersey–Maryland interconnection control room, where the flow of electricity can be adjusted to meet energy needs. (Courtesy of Penn Power & Light)

there is a complete path for electron flow. The circuit is said to be *open* if the flow of electrons is blocked. There are several types of circuits.

* **electrical circuit**   A combination of electrical devices connected together with wires, or other types of conductors, to perform some function.

SERIES CIRCUITS

Figure 17.1 shows a **series circuit**. In this figure, the high-potential (negative) point is indicated by the minus sign $(-)$, and the low-potential (positive) point by the plus sign $(+)$. The direction of electron flow, or current, is indicated by the symbol I.

Although all materials—wires, lights, resistors, and other circuit devices—offer some resistance to current, the resistance of connecting wires is usually ignored since it is very small when compared with other components. Circuit components offer resistance to current at specific locations in the circuit. Each lumped resistance is indicated by a zigzag line in a **schematic diagram** (Figure 17.1A). The circuit and its components can be shown pictorially in a **wiring diagram** (Figure 17.1B).

* series circuit	A circuit in which the electrons can follow only one path from the high-potential point to the low-potential point.
* schematic diagram	A diagram that uses graphic symbols to represent circuit components. It is arranged for easy viewing of the current flow.
* wiring diagram	A diagram that uses pictorial sketches of circuit components, placed in their appropriate locations.

In Figure 17.1, notice that we have two resistors (marked R_1 and R_2) connected together. (The resistors serve no particular function here except to demonstrate a principle.) Since there is only one path, all the electrons must flow through the entire circuit, and the amount of current is the same in all parts of the circuit—that is, if the current in one part of the circuit is measured at, say, 3 amps, the current in any other part of the circuit is also 3 amps. We can use a water analogy to explain why.

The unit "ampere" in electricity is like the unit "gallon per minute" of water. Imagine that you hook up two 25 foot lengths of garden hose together (in series). One hose has a 1 inch inside diameter (low resistance to water flow), and the other hose has a 1/2 inch inside diameter (high resistance). You turn the water

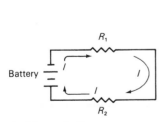

A. Schematic diagram

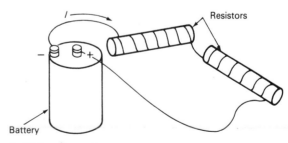

B. Wiring diagram

FIGURE 17.1   The Series Circuit, Showing Two Resistors Connected in Series.

on and find that you can fill a 2 gallon bucket in 1 minute. Therefore, the flow of water past any point in the hose, regardless of size (and resistance), has to be 2 gallons per minute. Thus, the **series circuit current rule** states:

* Series Circuit Current Rule

The current through each resistor or other electrical device in a series circuit must be the same.

In the example of Figure 17.1, the current is supplied by a battery. The symbol for a battery or other dc source is a series of long and short lines. The plus and minus signs are not usually added, but the short line always indicates the negative side of the power source.

The entire voltage supplied by the battery is used to force the electrons from the negative to the positive side of the battery. A portion of this voltage is required to force the electrons through each resistor. The amount of voltage used up for each resistor is called the *voltage drop* across that resistor. If all the voltage drops are added together, they will equal the supply voltage. Thus, the **series circuit voltage rule** can be stated:

* Series Circuit Voltage Rule

The sum of the voltage drops in a series circuit is always equal to the voltage applied.

Expressed mathematically, we have:

* SERIES CIRCUIT VOLTAGE FORMULA

$$E_{\text{applied}} = E_1 + E_2 + \cdots \quad \text{or} \quad E_{\text{applied}} = IR_1 + IR_2 + \cdots$$

The dots $(\cdots)$ in this equation mean that voltage drops can be added to correspond to the number of resistors (or other electrical devices) in the series circuit.

The **series circuit resistance rule** states:

* Series Circuit Resistance Rule

The total resistance in a series circuit is equal to the sum of all the individual resistances.

This rule can be expressed mathematically as:

* **SERIES CIRCUIT RESISTANCE FORMULA**

$$R_t = R_1 + R_2 + \cdots$$

SAMPLE PROBLEM 17.3

Use the information given in Figure 17.2.
- A. Determine the current in the circuit.
- B. Determine the voltage drop across each resistor.

SOLUTION A

Wanted

$I = ?$

Given

$E_a = 10.0$ V, $R_1 = 40.0$ Ω, $R_2 = 30.0$ Ω, and $R_3 = 17$ Ω

Formula

In order to find the current flow, using Ohm's law for the circuit, we must first calculate the total series circuit resistance. Stated another way, we must replace the three resistors with an equivalent single resistor. Thus, we have the formula for total resistance:

$$R_t = R_1 + R_2 + R_3 = 40 + 30 + 17 = 87 \text{ Ω}$$

Ohm's law can then be used to find the current:

$$I = \frac{E_a}{R_t}$$

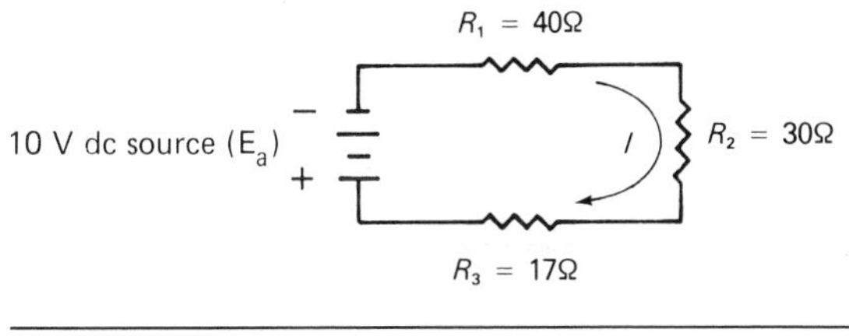

$R_1 = 40\text{Ω}$

10 V dc source (E_a) I $R_2 = 30\text{Ω}$

$R_3 = 17\text{Ω}$

FIGURE 17.2   Schematic Diagram of a Series Circuit with Three Resistors.

Calculation

$$I = \frac{10 \text{ V}}{87 \text{ }\Omega} = 0.1149 \text{ A}$$

This answer is rounded to 0.11 A, or 110 mA.

SOLUTION B

Wanted

$E_1 = ?$, $E_2 = ?$, and $E_3 = ?$

Given

$E_a = 10.0$ V, $I = 0.11$ A, $R_1 = 40.0$ Ω, $R_2 = 30.0$ Ω, and $R_3 = 17$ Ω

Formula

Ohm's law will be used again and rearranged to:

$$E = IR$$

Remember, the current is the same through each resistor in a series circuit.

Calculation

$E_1 = IR_1 = 0.11 \times 40 = 4.4$ V
$E_2 = IR_2 = 0.11 \times 30 = 3.3$ V
$E_3 = IR_3 = 0.11 \times 17 = 1.9$ V

Checking this answer, we find:

$$E_a = E_1 + E_2 + E_3 = 4.4 + 3.3 + 1.9 = 9.6 \text{ V}$$
$$\text{(round to 10 V)}$$

This type of problem has no particular application, but is used to develop your ability to analyze electrical circuits.

PARALLEL CIRCUITS

Figure 17.3 shows a **parallel circuit**. The electron stream flowing from the high-potential (negative) point divides at X. Part flows through resistor R_1 and the rest flows through resistor R_2. Both parts of the electron stream join at Y, and the combined stream then flows on to the low-potential (positive) point.

* parallel circuit

A circuit in which the electrons can follow two or more paths (or branches) simultaneously from the high-potential point to the low-potential point.

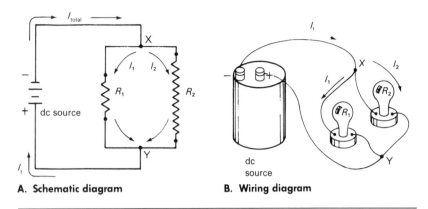

FIGURE 17.3   A Parallel Circuit, Showing Two Resistors Connected in Parallel.

Current will flow more readily through a low-resistance path than through a high-resistance path. Thus, if R_2 has twice the resistance of R_1, twice as much current will flow through R_1 as through R_2. (We disregard here the negligible resistance of the connecting wires.) Thus, if the total current (I_t) is equal to, say, 9 amps, 6 amps will flow through R_1 and 3 amps through R_2. The **parallel circuit current rule** can be stated:

* Parallel Circuit Current Rule	The total current in a parallel circuit is equal to the sum of the currents through each branch of the circuit.

Expressed mathematically, the formula is:

* PARALLEL CIRCUIT CURRENT FORMULA
$$I_t = I_1 + I_2 + \cdots$$

Also, since each parallel resistor has the same difference of potential between its ends, the voltage drop across each resistor is the same. Thus, we have the **parallel circuit voltage rule**:

* Parallel Circuit Voltage Rule	The voltage drop across parallel branches in a circuit is the same.

This statement can be written mathematically as:

$$I_1 R_1 = I_2 R_2 = \cdots$$

When resistors are connected in a parallel circuit (or in *shunt,* as it is often called), we do not add their resistances to obtain the total resistance. Instead, the total resistance for several resistors in parallel is determined by the following formula:

---

* PARALLEL CIRCUIT RESISTANCE FORMULA

$$\frac{1}{R_t} = \frac{1}{R_1} + \frac{1}{R_2} + \cdots$$

Sample Problem 17.4 demonstrates the current, voltage, and resistance relationships in a parallel circuit.

SAMPLE PROBLEM 17.4

Refer to Figure 17.3. Let $R_1 = 15\ \Omega$, $R_2 = 30.0\ \Omega$, and the applied voltage from the battery $E_a = 4\bar{0}$ V.

A. Find the total parallel circuit resistance, or in other words, find a single equivalent resistance that could replace the two resistors.
B. Find the current through each resistor and the total circuit current.

SOLUTION A

Wanted

$R_t = ?$

Given

$R_1 = 15\ \Omega$ and $R_2 = 30.0\ \Omega$

Formula

$$\frac{1}{R_t} = \frac{1}{R_1} + \frac{1}{R_2}$$

Calculation

$$\frac{1}{R_t} = \frac{1 \times 2}{15\ \Omega \times 2} + \frac{1}{30\ \Omega} = \frac{2}{30} + \frac{1}{30} = \frac{3}{30}$$

Inverting both sides of the formula, we have:

$$\frac{R_t}{1} = \frac{30}{3} \qquad R_t = 1\bar{0}\Omega$$

SOLUTION B

Wanted

$I_1 = ?, I_2 = ?,$ and $I_t = ?$

Given

$E_t = 4\bar{0}$ V, $R_t = 1\bar{0}$ Ω, $R_1 = 15$ Ω, and $R_2 = 30.0$ Ω

Formula

The parallel circuit voltage rule says that the voltage drop is the same across each resistor. Therefore, the voltage drops are equal to the applied voltage, as indicated in Figure 17.3:

$$E_a = E_1 = E_2 = 40 \text{ V}$$

With this information, we can apply Ohm's law to each branch of the circuit and determine the individual currents. Then, from the parallel circuit current rule:

$$I_t = I_1 + I_2$$

Calculation

For branch one, we rearrange:

$$E_1 = IR_1 \qquad \text{to} \qquad I_1 = \frac{E_1}{R_1} = \frac{40 \text{ V}}{15 \text{ }\Omega} = 2.67 \text{ A}$$

For branch two, we rearrange:

$$E_2 = IR_2 \qquad \text{to} \qquad I_2 = \frac{E_2}{R_2} = \frac{40 \text{ V}}{30 \text{ }\Omega} = 1.33 \text{ A}$$

$$I_t = 2.67 \text{ A} + 1.33 \text{ A} = 4.0 \text{ A}$$

To check this answer, we can use Ohm's law and the total resistance to find the total current:

$$I_t = \frac{E_a}{R_t} = \frac{40 \text{ V}}{10 \text{ }\Omega} = 4.0 \text{ A}$$

SERIES–PARALLEL CIRCUITS

The **series–parallel circuit** is a combination of the series and parallel circuits, as shown in Figure 17.4. We can determine the entire resistance of this circuit by computing the combined resistance of R_2 and R_3 in parallel and then treating the entire

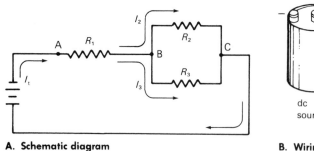

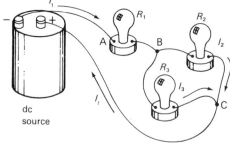

A. Schematic diagram

B. Wiring diagram.

FIGURE 17.4   The Series–Parallel Circuit. Points A, B, and C are labeled for easy comparison of the schematic and wiring diagrams.

circuit as a series circuit consisting of R_1 and the combined resistance of R_2 and R_3. This method is demonstrated in Sample Problem 17.5.

∗ series–parallel circuit   A circuit that is a combination of series and parallel circuits.

SAMPLE PROBLEM 17.5   Refer to Figure 17.4. A battery supplies $6\bar{0}$ V and the resistances $R_1 = 7\bar{0}0$ Ω, $R_2 = 1.5$ kΩ, and $R_3 = 2.0$ kΩ.

A. Find the combined resistance of R_2 and R_3.
B. Find the total circuit resistance.
C. Find the total circuit current.

SOLUTION A

Wanted $R_t = ?$

Given $R_2 = 1.5$ kΩ × 1000 Ω/kΩ = 1500 Ω and R_3
$= 2.0$ kΩ × 1000 Ω/kΩ = 2000 Ω

Formula $\dfrac{1}{R_t} = \dfrac{1}{R_2} + \dfrac{1}{R_3}$

Calculation

$$\frac{1}{R_t} = \frac{1 \times 4}{1500\ \Omega \times 4} + \frac{1 \times 3}{2000\ \Omega \times 3} = \frac{4}{6000} + \frac{3}{6000} = \frac{7}{6000}$$

Inverting both sides of the formula:

$$\frac{R_t}{1} = \frac{6000}{7} = 857\ \Omega \quad \text{(round to 860 } \Omega\text{)}$$

SOLUTION B

Wanted

$R_{circuit} = ?$

Given

$R_t = 860\ \Omega$

Formula

$R_{circuit} = R_1 + R_t$

Calculation

$R_{circuit} = 700 + 860 = 1560\ \Omega \quad \text{(round to 1600 } \Omega\text{)}$

SOLUTION C

Wanted

$I_t = ?$

Given

$E_a = 6\overline{0}\ \text{V and } R_{circuit} = 1600\ \Omega$

Formula

Ohm's law is used for the total circuit:

$$I_t = \frac{E_a}{R_{circuit}}$$

Calculation

$$I_t = \frac{E_a}{R_{circuit}} = \frac{60\ \text{V}}{1600\ \Omega} = 0.0375\ \text{A}$$
(round to 0.038 A or 38 mA)

Why would we want a circuit like the one in Figure 17.4? Well, we wouldn't, really. It is just an example to help you with more complicated circuits later. For instance, it might be part of a larger circuit, and R_1 could be a resistor to reduce the voltage drop across electromagnetic relays R_2 and R_3. However, whatever the actual circuit layout, the method for determining the total circuit values is the same as the steps we have just presented.

17.4   Effects of Electric Current

HEAT EFFECTS

Electricity is a form of energy. As is true of all types of energy, electrical energy may be converted to some other form. For example, as electrons move through a conductor, they knock loose other free electrons. These collisions produce heat. Since a greater force is required to knock loose free electrons from high-resistance materials than from those of low resistance, a given current will generate more heat in materials of high resistance.

The heating effect, then, depends on the current and on the resistance of the conductor through which the current flows. Another way of looking at this concept is to consider the heating effect as being proportional to the electric power dissipated as a current flows through a conductor. You will recall that we found the power to be equal to the square of the current multiplied by the resistance of the conductor ($P = I^2 \times R$). We can therefore use this formula to find the heating effect of a current.

In many instances, this heating effect is not wanted, especially in connecting wires and in such devices as motors, generators, radios, and TV sets. In these devices, steps are taken to reduce the heating effect. However, it is of major importance in many other devices, such as electric resistance furnaces, soldering irons, electric resistance welding equipment, and fuses. We will consider some of these applications further in later chapters. Sample Problem 17.6 illustrates a situation where the heating effect is unwanted.

SAMPLE PROBLEM 17.6	A dc electrical control circuit for electroplating has a total of 800 ft of #24 gage wire, with a total resistance of 24.2 Ω. If the current in the control circuit is 1.5 A, how much power is lost because of wire resistance? (This power is dissipated as heat.)
SOLUTION	
Wanted	$P = ?$
Given	$R = 24.2 \ \Omega$ and $I = 1.5$ A

Formula	$P = I^2R$
Calculation	$P = (1.5)^2 \times 24.2 \ \Omega = 54.5 \ \text{W}$ (round to 54 W)

This power is added to the rest of the power requirement for the operation. The engineer will use this information to determine the size of the power supply apparatus needed.

MAGNETIC EFFECTS

In the thirteenth century, it was discovered that the north pole of one magnet repelled the north pole of another magnet. Similarly, the south pole of one magnet repelled the south pole of another. But if the north pole of one magnet was placed near the south pole of another magnet, the two poles would be attracted.

PHOTO 17.5 Electromagnets can be made with great strength for industrial applications. When current stops flowing through the magnet, it releases the load of metal attracted to it. (Courtesy of Bethlehem Steel Corp.)

From this observation, it was obvious that like poles of magnets repel each other, and unlike poles attract.

It was found that the poles need not touch each other. Even if they are a small distance apart, like poles repel each other, and unlike poles attract each other. Further, even if a nonmagnetic substance is placed between the poles, their attraction or repulsion is unchanged. Thus, if a piece of paper or a sheet of glass is placed between two unlike magnetic poles, they continue to attract each other as though the paper or glass were not there. No wonder that the ancients attributed magical properties to magnets.

To understand these observations, let us try a simple experiment. Place a magnet on a wooden table. Over it, place a sheet of glass. Sprinkle iron filings on the glass and tap the glass lightly. As Figure 17.5 indicates, the iron filings assume a definite pattern on the glass sheet. The iron filings are attracted to the magnet through the glass sheet. Although the glass prevents these iron filings from touching the magnet, nevertheless the filings form a pattern that shows the form of a **magnetic field** that exists between the two opposite poles. Note that the iron filings arrange themselves in a series of closed loops outside the magnet that extend from pole to pole.

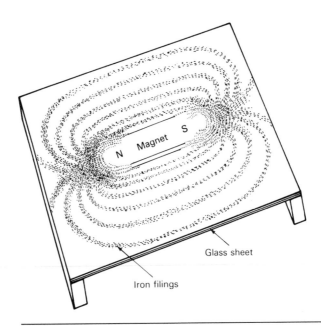

FIGURE 17.5   How Iron Filings Show the Form of the Magnetic Field around a Magnet.

* **magnetic field**	The space occupied by magnetic lines of force.

The pattern formed by the iron filings shows the **magnetic lines of force** surrounding the magnet. Since a force must have a direction in which it acts, we arbitrarily assume that the lines of force leave the magnet at its north (N) pole and re-enter at its south (S) pole. It is as though small north poles were placed on top of the N pole of the magnet. Each small north pole would be repelled and would travel along a line of force until it reached the S pole of the magnet. Note that the lines of force do not cross, but actually appear to repel each other. It would seem that these lines try to follow the shortest distance from pole to pole, at the same time repelling each other.

* **magnetic line of force**	A pattern representing the path along which a magnetic force acts. We assume the force acts in a direction from the north pole to the south pole of a magnet.

Of course, the pattern formed in Figure 17.5 represents the magnetic field only in the horizontal plane. To obtain a true picture, we should consider the magnetic field surrounding the magnet in three dimensions—that is, over and under the magnet as well as on either side of it.

Scientists in the 1700s and early 1800s sought ways to prove or disprove a relationship between electricity and magnetism. In 1819, Hans Christian Oersted (1777–1851), a Danish physicist, brought a small compass near a wire that was carrying an electric current. This compass, as Figure 17.6 shows, consisted of a small magnetized needle pivoted at the center so that it was free to

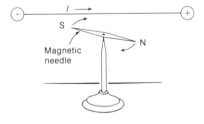

FIGURE 17.6   How the Magnetic Field around a Conductor Carrying an Electric Current Deflects a Compass Needle.

rotate. As he brought the compass near the wire, Oersted noticed that the needle was deflected. This discovery started a chain of events that has helped to shape our industrial civilization.

Let us examine the significance of Oersted's discovery. The compass needle is a small magnet that normally lines up in a north-and-south direction. The deflection of the needle indicates that it is being acted upon by an external magnetic field. Where did this magnetic field come from? Not from the copper wire, which we know is nonmagnetic. Obviously, it could come only from the electric current flowing through the wire, which investigation proved to be true. The needle was deflected only when current flowed through the conductor, and it continued to be deflected only so long as the current continued. When the current stopped, so did the deflection.

Thus, we see that a wire carrying an electric current acts like a temporary magnet. It has been found that the magnetic lines of force exist in the form of concentric circles around the wire and lie in planes perpendicular to the wire (Figure 17.7A). This magnetic field exists only so long as the current continues. When the current is started through the conductor, we may think of the magnetic field as coming into being and sweeping outward from the conductor. When the current stops, the field collapses toward the conductor and disappears.

It was soon discovered that if the wire is formed into a loop, the magnetic field around the wire on each side of the loop reinforces itself in the center of the loop (Figure 17.7B). As more loops, or turns, are added, the resulting magnetic field through the center of the coil becomes stronger, because each turn adds its magnetic field to the fields of the other turns (Figure 17.7C).

A definite relationship exists between the strength of the magnetic field, the number of turns, or loops, of the coil, and the current flowing through the wire. The greater the number of turns, the stronger will be the field. Likewise, the greater the current through the coil, the stronger will be the magnetic field.

The coil, as described, has a core of air. However, if we wind this coil on a core of soft iron, the strength of the magnetic field is increased many times. The reason is that the iron concentrates more of the magnetic field through the center of the coil than does air. Such a coil with its iron core is called an **electromagnet**.

∗ electromagnet   A device that generates a magnetic force only while current flows through it.

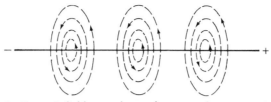

A. Magnetic field around a conductor carrying a current

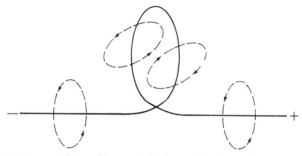

B. Magnetic lines of force reinforcing each other in the center of a loop of wire

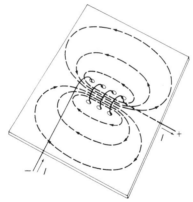

C. Magnetic field around a coil of wire through which a current is flowing (an electromagnet)

FIGURE 17.7   How a Current Carrying Conductor Produces a Magnetic Field.

Electromagnets are used in numerous applications. For example, they are part of electromechanical relays, motors and generators, television picture tubes, and loudspeakers. Some of these applications will be discussed in later chapters.

As long as a current is flowing, the electromagnet behaves as an ordinary magnet. It attracts magnetic materials and has opposite poles. We can determine which end of the electromagnet is a north pole and which end is a south pole by bringing a compass needle near each end of the coil. The north pole of the electromagnet repels the north pole of the compass needle, whereas the south pole of the electromagnet attracts the north pole of the needle.

We have another method for determining the polarity of the electromagnet. If the coil is grasped in the left hand so that the fingers follow around the coil in the direction in which the electrons are flowing, the thumb will point toward the north pole, as Figure 17.8 shows.

17.5 Electrical Measuring Instruments

MEASUREMENT BASED ON THE MAGNETIC EFFECT OF ELECTRIC CURRENT

Most electrical measuring instruments in use today are based on a design invented by the French physicist Arsène d'Arsonval (1851–1940). As Figure 17.9 shows, this instrument consists of an electromagnet pivoted between the two opposite poles of a fixed permanent horseshoe magnet. As current flows through the turns of the electromagnet, it becomes magnetized. If the direction of current is as indicated in Figure 17.9, the left-hand side of the electromagnet becomes a north pole and the other side

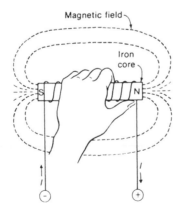

FIGURE 17.8   Method for Determining the Polarity of an Electromagnet.

PHOTO 17.6   This d'Arsonval galvanometer contains two shunts, shown at left, and a multiplier, shown at right, and can function as an ammeter over two different ranges of current values or as a voltmeter. Note the coil in the center.

becomes a south pole. Since like poles are facing each other, they repel. However, the permanent magnet is fixed and cannot move. Accordingly, the electromagnet rotates around its pivot.

The amount of repulsion between the like poles depends on the relative strengths of the magnetic field around the permanent magnet and the magnetic field around the electromagnet. Since the magnetic field around the permanent magnet is a fixed value,

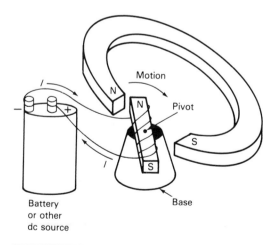

FIGURE 17.9   An Experimental Setup to Demonstrate the d'Arsonval Galvanometer.

the stronger the magnetic field is around the electromagnet, the greater the repulsion and rotating effect. But the strength of the magnetic field around the electromagnet depends on the current flowing through it. Thus, the amount of repulsion and the amount of rotation depend directly on the strength of the electric current flowing through the coil of the electromagnet.

Here, then, is a convenient way to measure the current flowing through the coil of an electromagnet. All we need do is fix a pointer to this electromagnet and measure the amount of rotation on a suitable scale. The **d'Arsonval galvanometer**, shown in Figure 17.10, is a practical instrument using this principle. The electromagnet, called the *armature,* consists of a coil wound on a soft-iron cylinder as a core. The complete armature is delicately pivoted upon jewel bearings (not shown in the illustration) and is mounted between the poles of a permanent horseshoe magnet. Attached to these poles are two soft-iron pole pieces that concentrate the magnetic field. As current flows through the armature coil, a magnetic field is set up around it in such a way that it opposes the field of the permanent magnet. The armature is preset with a slight clockwise displacement. As a result, the opposing magnetic fields cause the armature to rotate in a clockwise direction, carrying the pointer attached to it. A spring opposes the rotation and brings the pointer back to the zero position

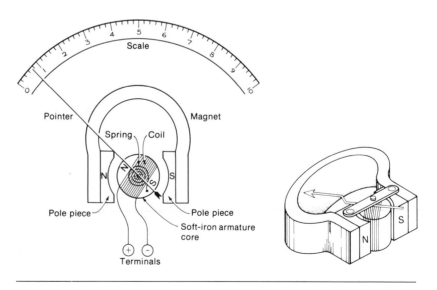

FIGURE 17.10   The d'Arsonval Galvanometer.

when the current stops. The greater the current, the greater is the deflection of the pointer. This deflection is indicated on the scale over which the pointer passes.

---

* d'Arsonval galvanometer An instrument for detecting very small direct currents. The instrument consists of a moving coil and a permanent magnet.

Note that rotation takes place only when the poles of the permanent magnet face like poles on the armature. Thus, the current must flow through the coil of the armature in such a way that this condition prevails. If the current flows in the opposite direction, unlike poles will face and attract each other, and there will be no rotation. Accordingly, the terminals of the instrument are suitably marked plus (+) or minus (–). If the minus terminal is connected to the high-potential side of the circuit being tested and the plus terminal is connected to the low-potential side, current will flow through the instrument in the proper direction.

THE AMMETER

The d'Arsonval galvanometer can also be used as an **ammeter** to measure current flow. However, when used as an ammeter, the galvanometer must be connected in series in the circuit so that all the current will flow through it. Such a connection is shown in Figure 17.11.

---

* ammeter An instrument used to measure the flow of current in a circuit.

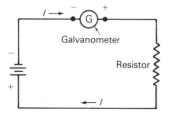

FIGURE 17.11   How a Galvanometer Is Connected to Measure the Current Flowing in a Circuit. Dots indicate terminals.

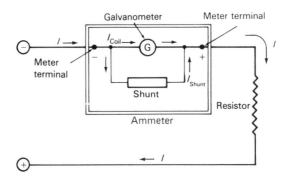

FIGURE 17.12 Circuit Showing How an Ammeter (Galvanometer plus Shunt) Is Used to Measure the Current Through a Resistor.

The moving coil of the d'Arsonval galvanometer must be very light so that it can respond readily to the current flowing through it. Accordingly, it is wound with very fine wire, and if the current through it is too large, this wire may become too hot and melt. These moving coils rarely are wound with wire heavy enough to carry more than 50 milliamps (0.05 amp). This amount is pretty small compared to current drawn by appliance motors, light bulbs, and portable electric tools, which draw on the order of 1/2 to 2 or 3 amps.

Where the current to be measured is greater than that which the moving coil can accept safely, a device called a shunt is used. This shunt consists of a metal wire or ribbon, usually of some alloy such as manganin, which is placed in parallel with the moving coil of the galvanometer, as in Figure 17.12. Thus, the current in the circuit divides, with the greater portion flowing through the coil or the shunt, depending on which has the lower resistance. If the shunt has the same resistance as the coil, half the current will flow through the shunt and half will flow through the coil. If the shunt has half the resistance of the coil, two-thirds of the current will flow through the shunt and one-third through the coil. The scale of the instrument, of course, must be calibrated accordingly. In this way, the ammeter can be used to measure currents that otherwise would burn out the coil. Sample Problem 17.7 illustrates a practical problem in the design of ammeters, putting to use some of the circuit rules discussed in this chapter.

SAMPLE PROBLEM 17.7   You are on the staff of an instrument factory that produces just one type of galvanometer, rated 200.0 mA and 200.0 Ω. (That is, the coil has a resistance of 200.0 Ω and the maximum current that can be put through the coil, giving full scale deflection, is 200.0 mA.) A customer orders an ammeter to measure 5.00 A at full scale deflection. You will supply your galvanometer with a suitable shunt.

    A.   What maximum current must flow through the shunt?

    B.   What is the resistance of the shunt?

SOLUTION A

Wanted

$I_{shunt} = ?$

Given

$I_t = 5.00$ A, $I_{coil} = 200.0$ mA $\times$ 0.001 A/mA $= 0.200$ A

Formula

The parallel circuit current rule states that currents in parallel branches add up to the total current:

$$I_t = I_{coil} + I_{shunt}$$

Calculation

5.00 A $= 0.200$ A $+ I_{shunt}$
$I_{shunt} = 5.00 - 0.200 = 4.80$ A

SOLUTION B

Wanted

$R_{shunt} = ?$

Given

$I_{coil} = 0.200$ A, $R_{coil} = 200.0$ Ω, and $I_{shunt} = 4.80$ A

Formula

The parallel circuit voltage rule states that the voltage drop across parallel branches is the same. This rule can be written:

$$I_{coil} \times R_{coil} = I_{shunt} \times R_{shunt}$$

Rearranged to:

$$R_{shunt} = \frac{I_{coil} \times R_{coil}}{I_{shunt}}$$

Calculation

$$R_{shunt} = \frac{0.200 \text{ A} \times 200 \text{ Ω}}{4.80 \text{ A}} = 8.33 \text{ Ω}$$

This information can now be used to select a suitable shunt and complete the manufacture of the ammeter.

THE VOLTMETER

Suppose, as in Figure 17.13A, we wish to measure the potential difference (voltage drop) between the two ends of a resistor. If we connect a galvanometer across those two points (in parallel with the resistor), the potential difference across the moving coil of the galvanometer will be the same as that across the resistor. This potential difference will cause a certain amount of current to flow through the galvanometer, deflecting the pointer to a certain degree. If the scale is suitably calibrated to read in volts, we can measure the potential drop across the resistor.

Two things are wrong with such a circuit. First of all, the current flowing through the galvanometer may be great enough to melt the coil. Even if this melting did not happen, the comparatively low resistance of the few turns of the galvanometer coil in parallel with the resistance of the resistor would produce a low total resistance for the instrument and resistor, which could result in a lower voltage drop across that part of the circuit than would normally appear there. Consequently, the voltage reading on the instrument could be incorrect, especially when measuring the voltage drop across a resistor that is connected in series with other resistors.

To eliminate these difficulties, a comparatively high resistance, called a multiplier, is placed in series with the galvanometer, as Figure 17.13B clearly shows. This limits the amount of current that can flow through the coil of the galvanometer, thus preventing it from burning up. Also, since the galvanometer and

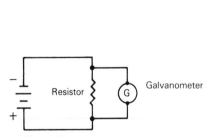

A. Circuit showing how a galvanometer is connected to measure the voltage drop across a resistor.

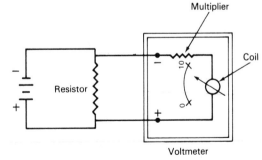

B. Circuit showing how a voltmeter (galvanometer plus multiplier) is connected to measure the voltage drop across a resistor.

FIGURE 17.13   Methods for Measuring Voltage Drop in a Circuit.

its multiplier (the two together constitute a **voltmeter**) offer a high resistance in parallel with the resistor, most of the current flows through the resistor. Thus, a more accurate measurement of the normal voltage drop across the resistor may be obtained. The slight amount of current flowing through the voltmeter may ordinarily be ignored. The scale of the voltmeter, of course, is calibrated in suitable units.

∗ voltmeter	An instrument used to measure the potential difference, or voltage, between two points of an electrical circuit.

Manufacturers take full advantage of the methods we have described for making voltmeters and ammeters. For example, a manufacturer can set up a production line to make just one type of moving coil assembly, perhaps rated at 50 milliamps. Now, one customer wants a 0–100 volt instrument, another wants 0–25 volts, and another wants 0–10 amps. The manufacturer simply assembles different shunts and multipliers to moving coil assemblies in accordance with the orders. If you are familiar with the multimeter test instrument sold by electronic stores, you know that just by turning a switch, you can measure different values of voltage, current, and resistance—all with one moving coil.

SAMPLE PROBLEM 17.8

You are asked to assemble a galvanometer and a multiplier to make a 0–100 V voltmeter. The moving coil of the galvanometer is rated 100.0 Ω and 50.0 mA (for full-scale deflection). What should the resistance of the multiplier be?

SOLUTION

Wanted

$R_{mult} = ?$

Given

The multiplier and coil are connected in series. The voltage drop across the voltmeter is $E = 100.0$ V, $I = 50.0$ mA $\times 0.001$ A/mA $= 0.0500$ A, and $R_{coil} = 100.0$ Ω

Formula

The series circuit resistance rule states that:

$$R_t = R_{coil} + R_{mult}$$

Calculation

Thus, our solution requires two steps because we must first find R_t by rearranging Ohm's law:

$$E = IR \quad \text{to} \quad R_t = \frac{E}{I}$$

$$R_t = \frac{100.0 \text{ V}}{0.0500 \text{ A}} = 2000 \ \Omega$$

Now we can use the resistance rule to find R_{mult}:

$$R_t = R_{\text{coil}} + R_{\text{mult}}$$
$$2000 \ \Omega = 100.0 \ \Omega + R_{\text{mult}}$$
$$R_{\text{mult}} = 2000 - 100 = 1900 \ \Omega$$

With this information, you can proceed to build the 0–100 V voltmeter.

THE OHMMETER

The measurement of resistance is based on the current flowing through the resistor under test and the voltage drop produced by that current across that resistor. If the current and voltage drop are known, the resistance can be calculated from Ohm's law ($R = E/I$). The basic circuit for resistance measurement is shown in Figure 17.14.

Assume that a source of 100 volts is connected in series with an ammeter and with an unknown resistance (R_x). The voltmeter will indicate the voltage across the circuit: 100 volts. If the ammeter reads 10 amperes, then since $R = E/I$, we obtain:

$$R_x = \frac{100 \text{ V}}{10 \text{ A}} = 10 \ \Omega$$

(We have neglected here the slight voltage drop across the ammeter and the slight amount of current flowing through the voltmeter.)

We can readily see that if the voltage remains constant, we may remove the voltmeter and can calibrate the ammeter to read resistance (in ohms) directly. The greater the unknown resistance, the less current flows in the circuit, and the less will be

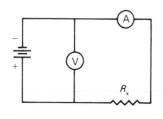

FIGURE 17. 4 The Basic Circuit for Measuring Resistance, Using an Ammeter and a Voltmeter.

the reading on the ammeter. Therefore, when we calibrate the scale to read in ohms, these calibrations go backward: The greater the movement of the pointer to the right, the smaller the number of ohms. We call such a circuit an *ohmmeter circuit* (Figure 17.15), and the calibrated ammeter with its constant voltage supply, an **ohmmeter**.

---

* ohmmeter   An instrument for measuring resistance. It may consist of an ammeter and a constant voltage supply, such as a battery.

Summary

Direct current refers to the situation in which electrons flow in only one direction in a circuit. The electrical potential that causes an electric current is measured in volts; current is measured in amperes; and resistance is measured in ohms. Electrical power is measured in watts; energy in wattseconds or watthours.

There are two basic types of electrical diagrams: a wiring diagram, where the components of a circuit are shown in their appropriate positions, and a schematic diagram, where the components are shown with special symbols and the diagram is arranged for easy viewing of the current path.

In a series resistance circuit, the resistors are connected one after the other in a single branch. That is, the current has only one possible path to follow. The current is the same at all points in a series circuit.

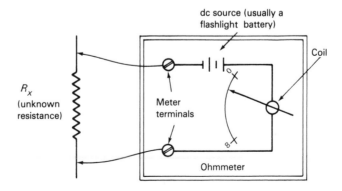

FIGURE 17.15   The Ohmmeter Circuit.

A parallel resistance circuit occurs when the resistors are connected in several intersecting branches. In other words, the current can and does flow through more than one path.

As electrical energy flows through a connecting wire or resistance, heat is produced. This heat loss is unwanted in connecting wires and some electrical components, but is desirable in electric resistance furnaces, electric soldering irons, and so forth.

When current flows in a wire, a magnetic field is formed around the wire. If the wire is wound in a coil, the strength of the magnetic field can be greatly increased. An iron core in the coil will also increase the magnetic field. Such an arrangement is called an electromagnet. The electromagnet can be applied to many components, such as loudspeakers, control relays, motors, and generators.

A large majority of electrical measuring instruments are based on the magnetic effect of an electric current. The most common "meter movement" is the d'Arsonval galvanometer. Just one galvanometer movement can be used in conjunction with the proper multipliers and shunts to measure various voltages and currents. If a battery is added to the instrument, it can be used to measure resistances.

Key Terms

* electric current
* direct current
* coulomb
* ampere
* ohm
* volt
* Ohm's law formula
* power
* watt
* power formula 1
* kilowatt
* watthour meter
* power formula 2
* electrical circuit
* series circuit

* schematic diagram
* wiring diagram
* series circuit current rule
* series circuit voltage rule
* series circuit voltage formula
* series circuit resistance rule
* series circuit resistance formula
* parallel circuit
* parallel circuit current rule
* parallel circuit current formula
* parallel circuit voltage rule

* parallel circuit resistance formula
* series–parallel circuit
* magnetic field
* magnetic line of force
* electromagnet
* d'Arsonval galvanometer
* ammeter
* voltmeter
* ohmmeter

Questions

Where diagrams are required, be sure that they are neatly drawn and carefully labeled.

1. What is meant by electric potential? By current? By resistance?

2. In what units do we measure current? Electric potential? Resistance? Explain the meaning of each unit.

3. What is the relationship among current, voltage, and resistance, as expressed by Ohm's law?

4. What do you buy from the electric company? Identify the units and the product. (Don't answer "electricity.")

5. Draw examples of series, parallel, and series–parallel resistance circuits and explain the differences among them.

6. What is the "current rule" for (a) a series circuit? (b) a parallel circuit?

7. Write the formula for finding the total resistance in (a) a series circuit and (b) a parallel circuit.

8. When is the heating effect of electrical energy desirable? When is it undesirable?

9. What is an electromagnet? Give three factors that determine the strength of its magnetic field.

Problems

ELECTRIC CURRENT, RESISTANCE, AND POTENTIAL

1. A small heating element is connected to a dc outlet. The voltage across the element is 50.0 V, and the resistance is 25.0 Ω. How much current flows through the element?

2. When welding, an arc welder delivers $10\overline{0}$ A at 30.0 V dc. What is the resistance of the arc welding circuit (which is mostly the air gap)?

3. A small electric resistance furnace is connected to $2\overline{0}0$ V dc. If the resistance is 1.5 Ω, how much current does it draw?

4. Solve Sample Problem 17.1 if the resistance of the conductor is 0.5 Ω.

5. A small pilot light in a dc control circuit has a resistance of 3.50 Ω. The voltage drop across the light is 1.75 V. What is the current flow through the light?

6. A current of 0.500 A flows through a resistance of $55\overline{0}$ Ω. What is the voltage drop?

7. A fuse rated at 2.00 A has a resistance of 0.0100 Ω. What is the voltage drop across the fuse at the rated current?

8. How much power is drawn by the arc welder in Problem 2?

9. How much power is drawn by the furnace in Problem 3?

10. An automobile starter motor is rated at 200 A and 12 V. How much power does it draw from the battery when starting an engine?

11. A dc generator for railway coach service is rated 20 kW and 37.5 V. How much current can be drawn when it is operating at its rated power output?

12. Use the alternate power formula to determine the power used by a small industrial furnace that draws a current of 11.5 A and has a resistance of 132 Ω.

ELECTRICAL CIRCUITS

13. A current of 10.0 A flows through four resistors in series. The resistors are each marked 25.0 Ω.
   a. What is the voltage drop across each resistor?
   b. What is the total voltage drop across all the resistors?

14. Three resistors are connected in series: resistor A = 10.0 Ω, resistor B = 20.0 Ω, and resistor C = 30.0 Ω. If a current of 1.50 A flows through the circuit, find (a) the voltage drop across each resistor and (b) the total voltage drop across all the resistors.

15. Refer to Figure 17.2. Suppose $R_1 = 10\overline{0}$ Ω, $R_2 = 50\overline{0}$ Ω, and $R_3 = 30\overline{0}0$ Ω (3 kΩ).
   a. Find the total resistance.
   b. What voltage must the battery supply to provide a flow of $20\overline{0}$ mA?
   c. What is the voltage drop across each resistor?

16. Refer to Figure 17.3. $R_1 = 10.0\ \Omega$, $R_2 = 10\bar{0}\ \Omega$, and the battery supplies 40.0 V.
    a. Find the total parallel circuit resistance.
    b. Find the total circuit current.

17. Refer to Figure 17.16. $R_1 = 100\ \Omega$, $R_2 = 500\ \Omega$, and $R_3 = 3.00$ kΩ. The battery supplies 45.0 V.
    a. Replace the three resistors with an equivalent single resistor. What is its value?
    b. What is the total current in the circuit?
    c. What is the current through each resistor?

18. Refer to Figure 17.4. Suppose $R_1 = 5\bar{0}\ \Omega$, $R_2 = 75\ \Omega$, and $R_3 = 150\ \Omega$. The battery supplies 150 V.
    a. Find the total circuit resistance.
    b. Find the total circuit current.
    c. Find the voltage drop across R_1.
    d. Find the current through R_2 and R_3.

ELECTRICAL MEASURING INSTRUMENTS

19. An ammeter is calibrated so that the pointer at full scale deflection indicates a current in a circuit of $1\bar{0}$ A when the ammeter coil has a real current of 150 mA.
    a. How much current flows through the coil when the circuit current is 7.0 A?
    b. How much current is there in the circuit when there are $5\bar{0}$ mA in the coil?

20. Solve Sample Problem 17.7 if the shunt is changed so the ammeter will read: (a) 20.0 A full scale; (b) $60\bar{0}$ mA full scale.

21. How much voltage will produce full scale deflection in a 50.0 V range voltmeter?

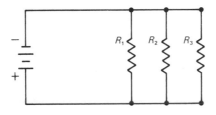

FIGURE 17.16   Sketch for Problem 17. Resistors are in parallel.

22. If the voltmeter in Problem 21 has a $20\overline{0}$ Ω coil that gives full scale deflection at 10.0 mA, what must be the resistance of the multiplier? What is the total resistance of the meter?

Computer Program

This program uses the parallel circuit resistance formula to solve for the total circuit resistance. Use this program for some of the chapter problems and to aid in experiments. Also, you may want to confirm this statement: "The total resistance of a parallel circuit is less than the resistance in any branch." Use two resistors; keep one constant and vary the other to see what happens to the total circuit resistance.

PROGRAM

```
10   REM  CH SEVENTEEN PROGRAM
20   PRINT "THIS PROGRAM WILL CALCULATE THE TOTAL RESISTANCE IN A"
21   PRINT "CIRCUIT FOR 2 OR 3 RESISTORS IN PARALLEL.  IF THE CIRCUIT"
22   PRINT "HAS 2 RESISTORS-TYPE 2, IF THERE ARE 3 RESISTORS-TYPE 3"
30   INPUT A
40   PRINT "YOU HAVE SELECTED -- ";A
50   PRINT
60   IF A = 2 THEN  GOTO 100
70   IF A = 3 THEN  GOTO 200
80   END
100  PRINT "TYPE IN R1 AND THEN R2"
110  INPUT R1,R2
120  PRINT "R1 = ";R1;" OHM."
130  PRINT "R2 = ";R2;" OHM."
140  LET Z = 1 / ((1 / R1) + (1 / R2))
150  PRINT
160  PRINT "TOTAL RESISTANCE = ";Z;" OHM."
170  GOTO 80
200  PRINT "TYPE IN R1,R2,AND THEN R3."
210  INPUT R1,R2,R3
220  PRINT "R1 = ";R1;" OHM."
230  PRINT "R2 = ";R2;" OHM."
240  PRINT "R3 = ";R3;" OHM."
250  LET Z = 1 / ((1 / R1) + (1 / R2) + (1 / R3))
260  PRINT
270  PRINT "TOTAL RESISTANCE = ";Z;" OHM."
280  GOTO 80
```

Line 40 displays the number of resistors you have selected.

Lines 60 and 70 select the proper subroutine for the number of resistors chosen.

Lines 140 and 250 contain the parallel circuit resistance formula.

High frequency alternating current plays an important role in modern electronics. Here, an electronics technician in the U.S. Navy troubleshoots a radar distribution system on board ship. (Courtesy of Naval Photographic Center, Washington, D.C.)

Chapter 18

Objectives

When you finish this chapter, you will be able to:

- [ ] Diagram and explain how induced voltages and currents are produced as a conductor passes through a magnetic field.
- [ ] Use the left-hand induced current rule to establish the direction of the induced current.
- [ ] Explain how induced voltage depends on strength of field, speed of conductor motion, and number of conductors contributing.
- [ ] Sketch and explain a generator's construction and operation. Show how alternating voltages and currents are produced.
- [ ] Sketch and graphically display the sinusoidal alternating voltage and current cycles. Indicate peak and effective voltages.
- [ ] Diagram a circuit containing an ac source and inductor (coil). Explain inductance, back-emf, and reactance: $X_L = 2\pi f L$.
- [ ] Diagram a circuit containing an ac source and a capacitor. Explain back-emf and reactance: $X_C = 1/2\pi f C$.
- [ ] Diagram ac circuits containing both inductance and resistance, or capacitance and resistance. Use the impedance equation to calculate impedance: $Z = \sqrt{R^2 + X_L^2}$ and $Z = \sqrt{R^2 + X_C^2}$
- [ ] Explain how to convert ac voltage and current into dc, or pulsating dc, using rectifiers.

Alternating Current Electricity

Alternating current electricity is supplied to our homes and to industry by electric utilities. It has some effect on our lives every day, regardless of whether or not we are technicians. The electric power is used to supply a very diverse field of electrical equipment, from small light bulbs that require about 1/2 to 1 ampere of current to large electric induction furnaces (for melting metals) that require current of 1000 amps or more.

Alternating current cannot be produced by batteries (batteries produce direct current), but is generated with the aid of a magnetic field. How this is accomplished is explained in the next two sections. The method by which alternating current can "induce" other currents is explained in later sections. Also, a few alternating current circuit components, such as induction coils, capacitors, and rectifiers, are described and their characteristics explained.

* alternating current	A flow of electrons that reverses direction at regular intervals.

18.1   Induced Voltage and Current

Hans Christian Oersted (1777–1851), a Danish physicist, discovered that an electric current flowing through a conductor can create a magnetic field. In 1831, the famous English scientist Michael Faraday (1791–1867) discovered that a magnetic field could create an electric current in a conductor. Strangely enough, at about the same time, Joseph Henry (1797–1878), an American scientist and teacher, working independently, discovered the same thing.

Let us try an experiment to illustrate this point. Assume we have a center-zero galvanometer. A center-zero galvanometer is similar to the type illustrated in Figure 17.10, except that the spring is so adjusted that when no current is flowing through the instrument, the armature (and the pointer attached to it) is positioned so that the pointer is at the center of the scale. This center point is marked zero. If current now flows through the instrument in one direction, the pointer will move from the zero position to the right. If current flows through in the opposite direction, the pointer will move to the left. Hence, this instrument indicates not only how much current is flowing, but also the direction of current flow.

PHOTO 18.1   The Bonneville Dam near Portland, Oregon, uses the power of the Columbia River to drive generators of ac electricity. At present, the dam generates 500 megawatts of power. (Courtesy of the Bonneville Power Administration)

We next connect the ends of this galvanometer to a coil of about 50 turns of wire wound in the shape of a cylinder about 2 inches in diameter. If we plunge the north end of a permanent magnet into the center of the coil, as shown in Figure 18.1A, we will see that the pointer is deflected to the right, showing that an electric current was set moving for a moment in the coil and galvanometer. When the magnet comes to rest inside the coil (Figure 18.1B), the pointer swings back to zero, showing that the current has stopped flowing. Now we remove the magnet from the coil (Figure 18.1C). As we do so, the pointer swings to the left, showing that once more an electric current is set flowing, but this time in the opposite direction. The same effect may be obtained if the magnet is held stationary and the coil is moved.

How can we explain this observation? We know that the permanent magnet is surrounded by magnetic lines of force (also called the magnetic field). As the magnet is moved into or out of the coil, this magnetic field cuts across the wire of the coil. When

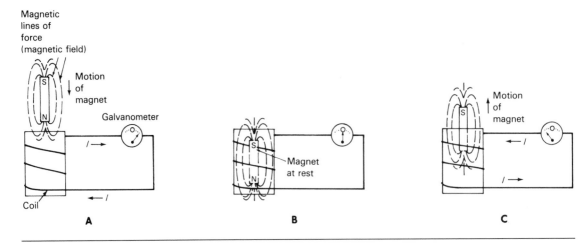

FIGURE 18.1   Demonstration Illustrating How a Current Is Induced in a Conductor As It Cuts Across a Magnetic Field.

a conductor cuts through a magnetic field, a potential difference is set up between the ends of the conductor.

In the case illustrated, the potential difference is set up between the ends of the coil. If the circuit is closed, as in Figure 18.1, a current will flow because of this potential difference. It makes no difference whether the conductor is stationary and the magnetic field is moving across it, or the magnetic field is stationary and the conductor is moving through it. But if both the conductor and the magnetic field are stationary, as in Figure 18.1B, no potential difference will be set up and no current will flow.

The potential difference that is developed by relative motion between a magnetic field and a conductor is said to be caused by an *electromotive force (emf)*. (When speaking, we usually say "ee-em-eff.") This distinction between "potential difference" and "electromotive force" is important to scientists, but for most practical applications, we can think of "electromotive force," "potential difference," and "voltage" as being the same thing. The unit for each is the volt. The terms "voltage" and "emf" are in common use and are frequently used interchangeably.

We call a voltage set up in a conductor in this way an *induced voltage*. The current set flowing as a result is called an *induced current*. The method of producing these voltages and currents is called **electromagnetic induction**.

* electromagnetic induction	The process by which a voltage is induced in a conductor when the conductor cuts or is cut by magnetic lines of force.

Experimentation has evolved a rule to determine in which way an induced current will flow. Figure 18.2 shows a conductor moving across a magnetic field set up between two poles of a horseshoe magnet. Assume that the conductor is moving down between the poles of the magnet. Extend the thumb, the forefinger, and the middle finger of the *left* hand so that they are at right angles to one another. Let the thumb point in the direction in which the conductor is moving (down). Now let the forefinger point in the direction of the magnetic lines of force (we assume they go from the north to the south pole). The middle finger will then indicate the direction in which electrons are set flowing by the induced voltage (away from the observer). We call this procedure the *left-hand rule* of electromagnetic induction.

There is another important principle in connection with induced currents. As the north pole of the magnet enters the coil in Figure 18.1, a current is induced in this coil. We already know that when a current flows through a conductor, it sets up a mag-

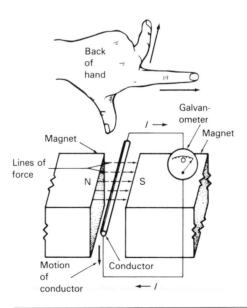

FIGURE 18.2   Left-hand Rule to Determine the Direction of Induced Current Flowing in a Conductor.

netic field around this conductor. Thus, the coil becomes an electromagnet. The induced current in this coil is set flowing in such a direction that the end of the coil facing the north pole of the magnet also becomes a north pole. We can check this statement by referring to Figure 17.8, which illustrates the method for determining the polarity of an electromagnet. Since like poles repel, this arrangement of magnets opposes the insertion of the north pole of the magnet into the coil. Work must be done to overcome the force of repulsion.

When we try to remove the magnet from the coil, the induced current is reversed. The top of the coil becomes a south pole and, by attraction to the north pole of the magnet, tends to prevent us from removing it. Thus, once again, work must be done—this time to overcome the force of attraction. Again, we must perform work to create the induced electric current. Of course, the same holds true if the magnet is stationary and the coil is moved.

These results are summarized in **Lenz's Law**, a law formulated by Heinrich Lenz (1804–1865), a German-Russian scientist who investigated this phenomenon:

* Lenz's Law   An induced current set up by the relative motion of a conductor and a magnetic field always flows in such a direction that it forms a magnetic field opposing the original motion.

Let us return again to the coil and magnet arrangement shown in Figure 18.1. The stronger the field around the permanent magnet, the greater will be the induced current in the coil as it cuts through this magnetic field. Also, the more turns in the coil, the greater the number of conductors cutting the magnetic field. Since each conductor adds its induced voltage to the total, the greater will be the total induced voltage and the greater will be the induced current.

Notice, too, that if we insert the magnet slowly into the coil, the pointer of the galvanometer will be deflected slightly. But if we insert the magnet rapidly, the deflection of the pointer will be much greater. Thus, the faster the conductors cut across the magnetic field, the greater will be the induced current.

We mention current because that is what the galvanometer is measuring. In actual practice, though, current depends on voltage and circuit resistance, so it is voltage that is directly affected by the field strength, the number of turns of the coil, and the speed with which the magnetic field is cut by the conductors.

From all this we may conclude:

1. The stronger the magnetic field, the greater the induced voltage.

2. The greater the number of conductors cutting the magnetic field, the greater the induced voltage.
3. The greater the speed of relative motion between the magnetic field and conductors, the greater the induced voltage.

18.2 The Alternating Current Generator

The discovery of electromagnetic induction was the beginning of our electrical age. We can convert the mechanical energy of a steam engine or water turbine to alternating current for light, heat, and power to operate the marvelous electrical machines that have been invented.

From a mechanical point of view, it is not practical to move our magnet in or out of a stationary coil of wire, or to move the coil over a stationary magnet. The same thing can be accomplished more simply by rotating a loop of wire between the poles of a magnet, thereby inducing a current in the wire as the magnetic field is cut. Figure 18.3 examines such an arrangement.

PHOTO 18.2   The needs of industrialized society can be met only by the creation of modern power-generating plants. This picture of a power plant in Singapore shows four modern steam turbine-generator units. (United Nations photo)

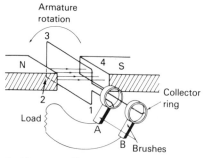

A. No current flow

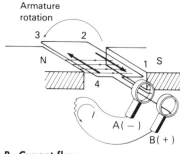

B. Current flow

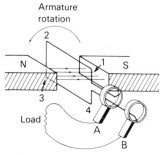

C. No current flow

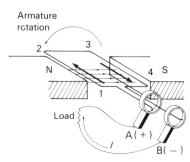

D. Current flow opposite in direction to sketch B

FIGURE 18.3   The Cycle of the ac Generator.

A simple loop of wire (called an *armature coil*) is mounted so that it may be rotated mechanically on a shaft between the north and south poles of a magnet. The two ends of the loop are connected to two brass or copper rings, called *collector rings* (or slip rings), which are insulated from each other and from the shaft on which they are fastened. These collector rings rotate with the loop. Two stationary brushes (A and B) make a wiping contact with these rotating collector rings and lead the current that has been induced in the loop to the external circuit. These brushes usually are made of copper or carbon. This arrangement of loop, magnetic field, collector rings, and brushes constitutes a simple **ac generator**.

* ac generator   A rotating machine that converts mechanical energy into electrical energy in the form of alternating current.

Let us assume that the loop starts from the position shown in Figure 18.3A and rotates at a uniform speed in a counterclockwise direction. In its initial position, no lines of force are being cut, because conductors 1–2 and 3–4 are moving parallel to the lines of force, not across them. As the loop revolves, however, the conductors begin to cut across the lines of force at an increasing rate, and therefore the induced voltage becomes larger and larger.

At the position shown in Figure 18.3B, the loop has the maximum voltage induced in it because conductors 1–2 and 3–4 cut across the maximum number of lines of force per second, since the conductors are moving at right angles to the magnetic field. Notice that the direction of voltage and resulting current produced in each conductor (1–2 and 3–4) add to each other and do not "buck" or cancel.

As the loop rotates to the position of Figure 18.3C, the voltage is still in the same direction, but is diminishing in value until it is zero again. The loop now has made one-half turn, during which the induced voltage increased to a maximum and then gradually fell off to zero. Since conductors 1–2 and 3–4 are now in reversed positions, the induced voltage changes direction in both conductors. The voltage, however, again increases in strength and becomes maximum when the loop is again cutting the lines of force at right angles, as shown in Figure 18.3D.

Finally, the last quarter of rotation brings the loop back to its original position (Figure 18.3A), at which point the voltage is zero again. As the rotation is continued, the cycle is repeated. While Figure 18.3 shows how a generator operates, practical generators are not constructed as simply as the one illustrated. We will discuss practical generators further in Chapter 19.

18.3 The Alternating Current Cycle

The term "cycle" really means circle—that is, a circle or series of events that recur in the same order. A complete turn of the loop in an ac generator is a cycle, and so is the series of changes in the induced voltage and the current set flowing by it. As the loop of the generator makes one complete revolution, every point in the two conductors of the loop describes a circle. Since a circle has 360 degrees (360°), a quarter-turn is equal to 90 degrees (90°); a half-turn, 180 degrees (180°); a three-quarter-turn, 270 degrees (270°); and a full turn, 360 degrees. The number of degrees, measured from the starting point, is called the *angle of rotation*.

Thus, Figure 18.3A represents the starting point, or zero degree (0°) position; Figure 18.3B, the 90 degree position; Figure 18.3C, the 180 degree position; Figure 18.3D, the 270 degree position; and Figure 18.3A again (after a complete revolution), the 360 degree position. Of course, positions between these points may be designated by the corresponding degrees. However, it is customary to use the quadrants—that is, the four quarters of a circle—as the angles of rotation for reference.

We are now ready to examine more closely the induced voltage in the loop of the generator as it goes through a complete cycle or revolution. Note that during half the cycle, the direction of the induced voltage is such as to cause electrons to move onto brush A. Recall from the previous chapter on static electricity that an excess of electrons is indicated by a minus sign (−). Brush A has this excess of electrons and therefore is marked (−) in Figure 18.3B. During the next half-cycle, the direction of the induced voltage is reversed so that the electrons move onto brush B. The current through the load is now reversed and flows from brush B to brush A.

Let us assume that the armature loop makes a complete revolution (360°) in 1 second. Then at 1/4 second, the loop will be at the 90 degree position; at 1/2 second, the loop will be at the 180 degree position; and so on. Assume, too, that the maximum voltage generated by this machine is 10 volts. Now we are able to make a table showing the voltage being generated during each angle of rotation:

Time in seconds	0	1/4	1/2	3/4	1
Angle of rotation	0°	90°	180°	270°	360°
Induced voltage	0	+ 10	0	− 10	0

You will note that in one complete revolution of the loop, there are two positions (Figures 18.3A and C) at which there is no induced voltage and, therefore, no current flowing to the brushes (or to the external circuit connected to the brushes). There are also two positions (Figures 18.3B and D) at which the induced voltage is at its maximum value, although in opposite directions. At intermediate positions, the voltage has intermediate values.

Also note that as the loop rotates, two factors are continuously changing: the position of the loop, and the value of the induced voltage. A graph, like the one shown in Figure 18.4, is a useful device to show instantaneous relationships between two such changing factors.

First we draw the horizontal line of zero voltage and divide

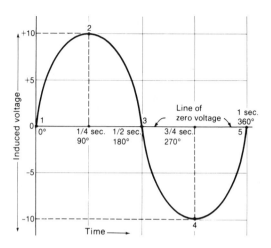

FIGURE 18.4   Graph Showing the Sinusoidal Waveform of the Alternating-Voltage Output of the Generator.

it into four equal sections of a quarter-second each. Since we assume that the loop makes one revolution (360°) in 1 second, each quarter-second corresponds to 90 degrees. Accordingly, these sections may be marked in degrees as well. Next, we draw the vertical line showing the induced voltage. You will recall that this voltage is in one direction for half the cycle and then reverses and is in the other direction during the next half-cycle. We will indicate the voltage in one direction by a plus sign (+) and the voltage in the other direction by a minus sign (−). All the plus values of voltage will appear above the line of zero voltage, and all the minus values will appear below it. Accordingly, all the numbers indicating induced voltage that appear above the line of zero voltage bear plus signs, and all those below bear minus signs.

Now we are ready to transpose the values of our table to the graph. At 0 degrees rotation, the table shows the time to be zero and the induced voltage to be zero as well. Point 1 is located on the graph where the horizontal line meets the vertical line. At the 90 degree or 1/4 second mark, the induced voltage has risen to a value of + 10 volts. We find this point on the graph by drawing a vertical line up from the 1/4 second (90°) mark and a horizontal line to the right from the + 10 volts mark. Point 2 is where these two lines intersect. At the 1/2 second (180°) mark, the voltage has fallen to zero again, and point 3 on the graph lies on the line of zero voltage. At the 3/4 second (270°) mark, the voltage has risen once more to 10 volts, but this time it is in the opposite (−) direction. We find point 4 on the graph by dropping a vertical line down from the 3/4 second mark and drawing a horizontal line to the right from the −10 volt mark. The point lies on the intersection of these two lines. At the 1 second (360°) mark, the voltage again has become zero. Accordingly, point 5 lies on the line of zero voltage.

Having thus established the points on the graph corresponding to the quadrants of the circle, how do we go about filling in the rest of the curve? Well, if we measured the voltages at every 10 degrees of rotation, and plotted them just as we have done for the quadrants, the points would fall on the curve drawn in Figure 18.4. The measurements have been eliminated to shorten our discussion.

Thus, we obtain a graph that illustrates the voltage generated by the armature coil as it makes one complete revolution. This graph also illustrates the flow of current in the external circuit connected to the brushes, resulting from this electromotive force. You must not get the impression that the current is flowing in this scenic-railway type of path; actually, the current is flowing back and forth—alternating—through the external circuit. What this curve does show, however, is the strength of

PHOTO 18.3   This oscilloscope has been adjusted to display one cycle of alternating current. The result is a typical sine wave pattern, like the one shown in Figure 18.4.

the induced voltage (and the resulting current flow) and its direction (+ or −) at any instant during one revolution. This curve is typical for alternating currents. Known as the *sine curve,* it is said to be the wave form of the alternating current produced by the generator.

Each cycle of alternating current represents one complete revolution of the armature coil of the generator. The number of cycles per second, known as the *frequency,* depends on the number of revolutions per second. Recall from our previous discussions on sound waves and light waves that the unit of frequency is the hertz (Hz). This unit applies to alternating current also. In Figure 18.4, the alternating current has a frequency of 1 cycle per second, and since 1 cycle per second is equal to 1 hertz, the frequency is 1 hertz. Alternating current supplied by electric utilities in the United States has a standard frequency of 60 hertz.

Figure 18.4 shows that alternating current and voltage are changing constantly in magnitude—that is, the *instantaneous* values are changing. What do we mean by "instantaneous"? At 60 hertz, the current completes 1 cycle (Figure 18.4) in 1/60 of a second, and that is instantaneous to most of us. But in our present discussion, instantaneous has to be an even shorter time interval. For our purposes here, instantaneous has to be such a small fraction of a second that we can show it only as a dot on the graph in Figure 18.4.

From the sine curve, we can see that there are two maximum values, or instantaneous peak values, for each cycle: a positive maximum and a negative maximum. We call the magnitude of these peak values—that is, the values represented by the distance of these peaks from the zero line in the graph—the amplitude. Thus, in our illustration, the amplitude of the generated voltage is 10 volts.

In practice, we rarely use the instantaneous values of alternating voltage or current in our calculations. Look at Figure 18.4 again. The voltage is near 10 volts for just a short time in each cycle, so it seems reasonable that we can't expect this output to do the same work as a 10 volt dc power source. For comparison, we can run an alternating current through a heating element in a small pot of water. The temperature rise of the water is measured in, say, 5 minutes. Then, we hook the heating element to a dc power source and find what voltage will heat the water to the same temperature in the same time. Suppose we get a value slightly over 7 volts. Since the ac has the same effect as 7 volts dc, we say the effective voltage is 7 volts. The effective voltage of an alternating current is 0.707 of the peak value. Thus, the formula for effective voltage is:

effective voltage = peak voltage × 0.707

Since the peak value is seldom used, alternating current and voltage instruments (except oscilloscopes) measure the effective values. For example, the voltage at the wall outlet in your home is specified as 120 volts, which is the effective voltage. To find the peak voltage, we rearrange the formula. Thus:

$$\frac{120}{0.707} = 170 \text{ V (peak)}$$

In general, the formulas in Chapter 17 for direct current apply to ac circuits as well, especially when the circuit components offer only resistance to the flow of current. However, reactive components, such as inductors and capacitors, must be handled differently. The following sections cover some of the differences.

18.4   Inductance

We have seen how a resistor in a circuit opposes the flow of current through the circuit. But what happens if we place a coil of wire in the circuit?

Assume that we have such a coil connected to a direct current source. The moment the circuit is completed, electrons will start flowing toward the coil. In the coil itself, the electron current will encounter a certain amount of resistance (that offered by the wire of the coil), which will remain constant all during the current flow.

We know that a conductor carrying a current is surrounded by a magnetic field. Thus, the moment a current starts flowing through the coil, a magnetic field starts growing around the coil. As this field expands, it cuts across the conductors or turns of the coil itself, inducing a second current in them. Heinrich Lenz discovered that the direction of this induced current is such that it opposes the original current. In other words, the direction of the induced current is such that it tends to reduce the original current, and thereby tends to oppose the *expansion* of the mag-

PHOTO 18.4   Typical air-core inductors are shown on the left. The larger units are from radio transmitters, and the small coils are from radio receivers. The setup on the right illustrates how a pulse of battery current sent through one coil can induce a voltage and current in a second coil.

netic field. When the original current reaches a steady level, the magnetic field becomes stationary and no longer cuts across the turns of the coil. Thus, there is no longer an induced current in the coil.

Now let us see what happens when the original current begins to decrease. The magnetic field around the coil starts to collapse. In so doing, it cuts across the turns of the coil, and once again a second current is induced in the coil. The direction of the induced current again opposes the change of the original current, which is decreasing. Thus, the induced current tends to keep current flowing in the coil for a time after the original current has ceased. The induced current, therefore, tends to oppose the *collapse* of the magnetic field. Because the induced voltage (which produces the induced current) always acts to oppose changes in the original voltage, the former is often referred to as *counter-voltage,* or *back-electromotive force,* or more briefly, back-emf.

We can look upon the coil as a device for storing energy in the form of a magnetic field. Extra energy is needed to build up the field because the field itself opposes an increase in the applied current. When the applied current reaches a maximum, the magnetic field reaches a maximum. As the applied current starts to drop in value, the magnetic field begins to return its energy to the circuit. We see this as an effort to prevent the reduction of the applied current.

That property by which a coil opposes any changes in the current flowing in it is called *inductance.* Since this opposition is caused by voltages induced in the coil itself by the changing magnetic field, anything that affects this magnetic field must also affect the inductance. When the effect of inductance is such as to cause an induced voltage in the same circuit in which the changing current is flowing, the term **self-inductance** is applied to the phenomenon.

---

* self-inductance   The ability of a coil to oppose any change of current in a circuit.

Except when currents of extremely high frequencies are flowing through them, the inductance of straight wires can be neglected. But the self-inductance of coils (sometimes called inductors), especially when wound on cores of magnetic materials, can be very great, the amount being determined by the number of turns; the size, shape, and type of windings; and other physical factors. Recall from Chapter 17 that a coil of wire produces a stronger magnetic field if it is wound on an iron bar (iron core) than if the coil encloses only air (air core). Electronic circuits use both air-core and iron-core inductors.

Inductors are used in radio, TV, and other electronic circuits. They help in selecting a particular frequency and allowing that frequency to pass through to the rest of the circuit. The induction principle is also used in transformers, which will be discussed in Chapter 19.

The **henry** (H), named for the American scientist Joseph Henry, can be defined as the inductance present when a current change of 1 amp per second produces an induced voltage of 1 volt. In electronics, it is often convenient to employ the **millihenry** (mH), which is 1/1,000 of a henry (0.001 H), and the **microhenry** (μH), which is 1/1 000 000 of a henry (0.000 001 H). In electrical formulas, the symbol used for inductance is L.

---

* henry   The unit used to measure the inductance of a circuit.

When a steady direct current flows through a circuit containing an inductor, the effect of inductance comes into play only when the current is changing—that is, only when the current starts flowing and when it stops. In between, when the current is at a steady value, the magnetic field around the inductor is constant, the countervoltage (back-emf) is zero, and the only opposition to the current is due to the resistance of the wire in the inductor. Remember, an induced voltage can be caused only by a changing magnetic field.

If an alternating current is sent through the circuit, the current is constantly changing, and as a result, a back-emf is constantly generated by the inductor. Since this back-emf opposes the applied voltage, and therefore restricts the applied current, it has the same effect as a resistor. A similar effect occurs when we hook up two 1.5 volt batteries as shown in Figure 18.5. The current can't flow because the voltages from the batteries are against each other. In this instance, each battery appears to the other as an infinite resistance.

Since an inductor "reacts" differently to alternating current than it does to direct current, the opposition to current in an inductor is called *reactance* and is designated by the letter X. More specifically, it is called **inductive reactance** to distinguish it from capacitive reactance, discussed in the next section, and the subscript L is added (X_L).

---

* inductive reactance   The opposition of an inductor to alternating current.

Since inductive reactance has the same effect as that of resistance (namely, opposition to current), we use the same unit for it: the ohm. The formula relating inductive reactance to inductance is:

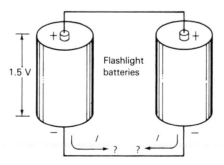

FIGURE 18.5   Opposing Voltages, with Each Appearing As Resistance to the Other Voltage Source.

* INDUCTIVE REACTANCE FORMULA

$$X_L = 2\pi f L$$

where X_L = the inductive reactance in Ω
   f = the frequency in Hz
   L = the inductance in H

The formula shows that the larger the inductance, the larger the reactance. (The inductance can be made larger by adding, for example, more turns of wire to a coil.) Also, the higher the frequency, the larger the reactance. Note that when the frequency is zero (direct current), the reactance is zero.

An inductor in an ac circuit opposes the current in two ways: one due to the inductance of the coil, and the other due to the resistance of the wire used in making the coil. However, in order to avoid complications in this introductory text, we will deal only with those inductors whose resistance values are so small that they may be considered zero. Therefore, when applying Ohm's law to an inductor, the reactance X_L may be used in place of resistance R. Thus, $I = E/R$ is changed to:

$$I = \frac{E}{X_L}$$

SAMPLE PROBLEM 18.1

A $20\bar{0}$ mH coil is placed in a 60.0 Hz ac circuit. The voltage drop across the coil is 15.0 V.
  A. Find the reactance.
  B. Find the current through the coil.

SOLUTION A

Wanted $X_L = ?$

Given $L = 20\bar{0}$ mH $\times$ 0.001H/mH $= 0.200$ H and $f = 60.0$ Hz

Formula $X_L = 2\pi f L$

Calculation $X_L = 2\pi f L = 2 \times 3.14 \times 60.0$ Hz $\times 0.200$ H $= 75.4$ Ω

SOLUTION B

Wanted

$I = ?$

Given

$E = 15.0$ V and $X_L = 75.4$ Ω

Formula

Ohm's law can be used, except that we replace the resistance R with the reactance X_L, since both are measured in ohms:

$$I = \frac{E}{X_L}$$

Calculation

$$I = \frac{15.0 \text{ V}}{75.4 \text{ Ω}} = 0.199 \text{ A } (199 \text{ mA})$$

This information is important when designing or troubleshooting a circuit with inductance.

18.5 Capacitance

Capacitors are useful for storing an electrical charge. However, the amount of charge is generally very small and cannot be compared to the electrical energy of a storage battery. In photography, capacitors are used to store electricity at high voltage to operate flash bulbs. They also have a number of uses in electronic circuits. We will mention two: (1) They can block the flow of direct current and let alternating current pass, and (2) they are used in oscillator circuits with inductors to tune radio and TV sets.

So far, we have investigated the effects of resistance and inductance on the flow of current in a circuit. Now what about capacitance? What happens when there is a capacitor in the circuit?

Figure 18.6 shows a capacitor connected to a direct current source. The moment the circuit is completed, electrons flow from the negative electrode of the source and pile upon plate A of the capacitor. These electrons cannot flow through the dielectric, which is an insulator, but their presence on plate A repels electrons from plate B, and these electrons flow to the positive electrode of the source. So, at the moment the circuit is completed, there is a flow of current through all portions of the circuit (except through the dielectric of the capacitor).

PHOTO 18.5 A television technician must be able to adjust video and audio signals in the control room before they go to the transmitters. This work usually involves adjusting values of capacitive reactance to change the signals as desired. (Courtesy of KVAL-TV)

Note that as electrons are being piled up on plate A of the capacitor, an electrical pressure is being created that tends to make the electrons flow back from that plate to the negative electrode of the source. This flow is a back- or counterelectromotive force. As long as the back-emf is less than the voltage of the source, electrons will flow onto plate A. But when the back-emf becomes equal to the voltage of the source, they will neu-

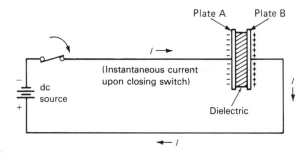

FIGURE 18.6 The Capacitor in a dc Circuit.

tralize each other and no more electrons (no more current) will flow in the circuit, and we say the capacitor is fully charged.

Thus, when a series circuit containing a capacitor is connected to a direct current source, current will flow for an instant until the capacitor is charged to the point where its back-emf is equal to the voltage of the source. At this point, the current will stop flowing, although the capacitor retains its charge. If the voltage of the source drops to zero or to a value less than the back-emf, current will flow from the capacitor to the source.

The time interval it takes to charge a capacitor can be varied by placing a resistor in series with the capacitor. The larger the value of resistance, the smaller the flow of electrons and the longer the time for charging. Remember $E = I \times R$? If resistance is increased and the applied voltage of the source is held constant, the flow of electrons is decreased.

Thus, we see that the effect of the capacitor is to build up a back-emf, which is always opposed to the voltage of the source. Note that current flows in the circuit only when the applied voltage is changing. When the applied voltage reaches its steady value, the back-emf is equal and opposite to it, and no current flows.

Recall that in Chapter 16, the unit of capacitance was defined as the farad (F). For practical use, the farad is too large a value to be handled conveniently. Consequently, we use the **microfarad** (μF), which is one-millionth of a farad, or the **picofarad** (pF), which is one-millionth of a millionth of a farad. In electrical formulas, the symbol used for capacitance is C.

So far we have considered the effect of capacitance on a direct current. What happens when an alternating current, whose electromotive force is constantly changing, replaces the direct current?

Figure 18.7 shows an ac generator connected in series with a capacitor and a lamp. The lamp is used merely to indicate by its light when current is flowing through the circuit. Assume that the current is in the direction indicated in Figure 18.7A. Electrons flow from the negative brush of the generator, through the lamp, and pile up on plate A of the capacitor. At the same time, electrons flow from plate B to the positive brush of the generator.

At the beginning of the next half-cycle (Figure 18.7B), the electromotive force of the generator has decreased to zero and then has reversed its direction. Now the back-emf of the electrons piled up on plate A of the capacitor sends current flowing to the generator. But now the current resulting from this back-emf is in the same direction as that produced by the generator. As this half-cycle continues, current flows as indicated by Figure 18.7C,

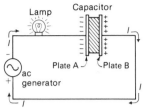
A. Current flow during the first half-cyle

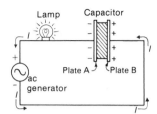
B. Current flow at the beginning of the second half-cycle

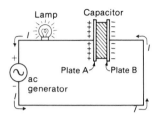
C. Current flow toward the end of the second half- cycle

FIGURE 18.7   The Capacitor in an ac Circuit.

piling electrons onto plate B of the capacitor. Then the cycle repeats itself.

Note that current is continually flowing through the circuit (except through the dielectric of the capacitor), first in one direction and then in the other, in step with the alternations of the generator. The lamp continues to glow, showing that current is flowing through it. Hence, alternating current can flow through a circuit containing a capacitor.

As in a circuit containing inductance, there is opposition to the flow of alternating current in a circuit containing capacitance. This opposition is due to the back-emf arising from the electrons piled up on the plates of the capacitor. We call this type of opposition **capacitive reactance**, and use the symbol X_C in electrical formulas. Its unit of measurement is the ohm.

---

* capacitive reactance

The opposition of a capacitor to alternating current.

The formula relating capacitive reactance to capacitance is:

---

* CAPACITIVE REACTANCE FORMULA

$$X_C = \frac{1}{2\pi f C}$$

where X_C = the capacitive reactance in Ω
f = the frequency in Hz
C = the capacitance in F

Note that as the frequency and/or the capacitance is increased, the reactance is *decreased,* and vice versa. If the frequency is zero (direct current), the reactance approaches infinity and no current exists.

It is usual practice to consider that capacitors offer no resistance to alternating current, only reactance. Ohm's law may be used with capacitors, as with inductors. In this case, the capacitive reactance X_C replaces the resistance R. Thus, $I = E/R$ is changed to:

$$I = \frac{E}{X_C}$$

SAMPLE PROBLEM 18.2	A 5.00 µF capacitor is placed in a 60.0 Hz circuit. A. What is the capacitive reactance? B. At what frequency will the capacitive reactance equal 20.0 Ω?
SOLUTION A	
Wanted	$X_C = ?$
Given	$C = 5.00 \text{ µF} \times 0.000\ 001 \text{ F/µF} = 0.000\ 005\ 00 \text{ F}$ and $f = 60.0$ Hz
Formula	$X_C = \dfrac{1}{2\pi f C}$
Calculation	$X_C = \dfrac{1}{2 \times 3.14 \times 60.0 \text{ Hz} \times 0.000\ 005\ 00 \text{ F}} = 531 \ \Omega$
SOLUTION B	
Wanted	$f = ?$
Given	$C = 0.000\ 005\ 00 \text{ F}$ and $X_C = 20.0 \ \Omega$
Formula	We rearrange the capacitive reactance formula to: $f = \dfrac{1}{2\pi X_C C}$

Calculation

$$f = \frac{1}{2\pi X_C C} = \frac{1}{2 \times 3.14 \times 20.0\Omega \times 0.000\ 005\ 00\ \text{F}}$$
$$= 1590\ \text{Hz}$$

This information can be used to help select a compatible inductor for use with the capacitor in an oscillator circuit (discussed in Chapter 19).

18.6 Impedance

When we deal with circuits in which the current is steady and direct, the opposition to current is essentially due to the resistance offered by the various portions of the circuit. But when we deal with ac circuits, we encounter inductive and capacitive reactances, as well as resistance. The total opposition to alternating current in a circuit is called the **impedance**. The symbol used to represent impedance is Z, and the unit of impedance is the same as that used for any opposition to current (namely, the ohm).

∗ impedance

The total opposition to current in a circuit that has resistance and inductance and/or capacitance.

Resistance, inductive reactance, and capacitive reactance are always present in an ac circuit, even though the inductive reactance may be due to the inductance of a straight connecting wire and the capacitive reactance may be due to the capacitance between two such wires. However, the inductive and capacitive reactances of such wires may be disregarded at low frequencies, although they may assume considerable importance at the high frequencies encountered in radio, TV, and other electronic communication equipment.

As mentioned earlier, the magnitude of the current and voltage at any given time in an ac circuit varies according to the sine wave, as illustrated in Figure 18.4. In a pure resistance circuit, the current and voltage peak at the same instant. However, if an inductor is placed in the circuit, the current is delayed by the induced back-emf, so that the current peaks at a later time than the voltage does. We say the current lags the voltage. A capacitor does just the reverse: It causes the current to lead the voltage.

Recall that when two or more resistors are in series in a circuit, the total resistance is the sum of all the individual resistances. However, when a series circuit has resistance and inductive reactance, we cannot obtain the impedance by simple addition, even though they all use the same unit, the ohm.

We will not go into an explanation of how the following formulas are derived, but will show the formulas that can be used to find impedance in series circuits.

For series circuits, the impedance formula that includes resistance and inductive reactance is:

* **IMPEDANCE FORMULA 1**

$$Z = \sqrt{R^2 + X_L^2}$$

The impedance Z, resistance R, and inductive reactance X_L are all expressed in ohms. Similarly, the impedance of a circuit containing resistance and capacitive reactance may be obtained from the following formula:

* **IMPEDANCE FORMULA 2**

$$Z = \sqrt{R^2 + X_C^2}$$

Again, all are in units of ohms.

If we replace resistance R in the direct current Ohm's law by impedance Z, the law will apply to alternating current circuits. Thus, Ohm's law for alternating current can be written:

$$I = \frac{E}{Z} \quad \text{or} \quad E = I \times Z \quad \text{or} \quad Z = \frac{E}{I}$$

where Z = the impedance in Ω
E = the voltage
I = the current in A

SAMPLE PROBLEM 18.3

A 34.5 mH coil is placed in series with a 17.0 Ω resistor. The circuit has a frequency of 60.0 Hz. What is the impedance of the coil and resistor?

SOLUTION

Wanted

$Z = ?$

Given

$L = 34.5 \text{ mH} \times 0.001 \text{ H/mH} = 0.0345 \text{ H}, f = 60.0 \text{ Hz, and}$
$R = 17.0 \ \Omega$

Formula

We must solve this problem in two parts. First, we must determine the inductive reactance of the coil:

$$X_L = 2\pi f L$$

Then we can use impedance formula 1:

$$Z = \sqrt{R^2 + X_L^2}$$

Calculation

$$X_L = 2\pi f L = 2 \times 3.14 \times 60.0 \text{ Hz} \times 0.0345 \text{ H} = 13 \ \Omega$$

We can now use X_L in the impedance formula:

$$Z = \sqrt{R^2 + X_L^2} = \sqrt{(17 \ \Omega)^2 + (13 \ \Omega)^2} = \sqrt{289 \ \Omega + 169 \ \Omega}$$
$$= 21.4 \ \Omega$$

This information will be helpful in determining the load on the power supply.

Note two points. First, induction coils may have some pure resistance value that might make a difference in the impedance. For this text, we will consider that inductors have only reactance and no resistance. When dealing with capacitors, it is normal procedure to consider that they have only reactance and no resistance. Second, our discussion covers only series circuits. Parallel circuits are handled a bit differently, and so we will leave such circuit explanations to texts on electronics.

18.7 The Rectifier

The extreme sensitivity and accuracy of the d'Arsonval galvanometer (Figure 17.10) make it a desirable instrument to use. Unfortunately, the current must flow through the armature coil in one direction only, and hence it is unsuitable for ac operation.

PHOTO 18.6   Diodes can be large or small. Those at the left are older vacuum-tube diodes from military equipment and TV sets; the right photo shows modern solid-state diodes, with a match shown for size comparison.

Accordingly, a number of methods have been devised to adapt it for alternating currents. One such method is to change the alternating current to direct current (we call this *rectification*) and then measure the resulting direct current with the d'Arsonval galvanometer.

Certain substances, such as galena (which is a crystal of lead sulfide), selenium, and germanium, have a peculiar property that permits current to flow readily through them in one direction, but offers a very high resistance to current flowing through them in the opposite direction. We call such substances **rectifiers**. (The term *diode* is also used.)

＊ rectifier   An electrical device that changes alternating current to direct current.

We can understand the action of the rectifier if we examine the graphs shown in Figure 18.8. In Figure 18.8A, see the graph of an alternating current. Note that the current flows first in one direction (+) and then in the other (−). In Figure 18.8B, we see the graph of this current after it has passed through a rectifier. Note that the rectifier has permitted the current to pass through it in one direction (+), but has blocked its passage in the other direction (−).

When current is blocked, it doesn't flow through the circuit. The circuit appears to be "open" to the power source. Because the resulting current flows only in one direction, it is a direct

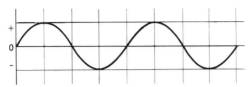

A. Input to rectifier

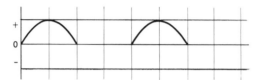

B. Output of rectifier

FIGURE 18.8   Graphs Showing How a Rectifier Changes Alternating Current into a Pulsating Direct Current.

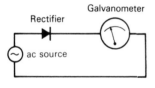

FIGURE 18.9   Circuit Showing How a Rectifier and a d'Arsonval Galvanometer May Be Used to Measure Alternating Current.

current. Note, however, that it is not a steady direct current. We call such a current a pulsating direct current.

If we connect such a rectifier in series with a d'Arsonval galvanometer, as in Figure 18.9, we can use it to measure alternating current. Of course, as each pulse of rectified current starts from zero, rises to its maximum value, and falls to zero again, the pointer of the instrument is deflected back and forth, and we would not be able to obtain a steady reading. However, if the pulses come fast enough (in the case of 60 hertz alternating current, there are 60 such pulses every second), the pointer will be unable to follow their variations and will remain stationary at a point that approximates the average value of such pulses.

The rectifier, or diode, can be used with electrical components and equipment other than galvanometers, such as computers, battery chargers, and electroplating equipment.

Summary

Alternating current (ac) is generated by moving a conductor across a magnetic field. This motion induces a potential difference (voltage) in the wire, with one end accumulating electrons and the other end losing them. If both ends of the wire are connected to

PHOTO 18.7   This large piece of equipment is a type of rectifier called a mercury arc valve that converts alternating current to pulsating direct current in a power station. The output circuit leads to a transformer. (Courtesy of the Bonneville Power Administration)

another wire, current will flow, thereby completing the circuit. The left-hand rule for induced current can indicate the direction of the current.

A conductor carrying a current creates a magnetic field of its own. If this conductor is moving in a magnetic field, then the conductor's magnetic field tends to oppose the motion that created the current in it. This effect is stated by Lenz's Law: An induced current set up by the relative motion of a conductor and a magnetic field always flows in such a direction that it forms a magnetic field opposing the original motion.

Instead of moving a wire back and forth in a magnetic field with a reciprocating motion, it is easier to rotate a loop of wire to produce an alternating voltage (and therefore, if the circuit is complete, an alternating current) in the shape of a sine wave. Instruments normally measure the effective voltage and current in an ac circuit, and not the maximum, or peak, value. The effective value has the same heating effect as direct current of the same value. This effective value is 0.707 of the peak ac value.

A coil of wire in a dc circuit has a certain amount of resistance to the flow of current because of the length and size of wire in the coil. The same coil in an ac circuit will have a greater resistance to current, consisting of two components. One component is the "pure" resistance it had in the dc circuit. The other component, called inductive reactance (X_L) and measured in ohms, is due to the changing magnetic field of the coil, inducing a back-emf in the coil that opposes the applied voltage. It has the effect of opposing the current flow and acting as a resistance. The higher the ac frequency, the higher the reactance, and therefore the greater the opposition to current flow.

Magnetic coils, with either air cores or iron cores, are called inductors and are used in electronic circuits. In conjunction with other components, inductors can aid in tuning a radio receiver to a particular frequency—that is, tuning to a station—or it can pass a direct current.

In some respects, the capacitor is the opposite of the inductor. The capacitor can block direct current and pass alternating current—the higher the frequency, the lower the reactance. (Thus, as the reactance is reduced, the opposition to current is reduced.) The capacitor's main function is to store a small charge of electricity and release this charge when desired. One example is the mechanism on cameras to fire flash bulbs.

When resistors, capacitors, and/or inductors are combined in a circuit, the total opposition to current flow is called impedance. Special formulas can be used to determine impedance.

In situations where an ac supply must be converted to dc, a rectifier, or diode, is used. Basically, a diode blocks the flow of current in one direction so that the current through the circuit flows in only one direction.

Key Terms

* alternating current
* electromagnetic induction
* Lenz's Law
* ac generator
* effective voltage formula
* self-inductance
* henry
* millihenry
* microhenry
* inductance reactance

* inductive reactance formula
* microfarad
* picofarad
* capacitive reactance
* capacitive reactance formula
* impedance
* impedance formula 1
* impedance formula 2
* rectifier

Questions

Where diagrams are required, be sure that they are neatly drawn and clearly labeled.

1. What is the difference between direct current and alternating current?
2. How may a voltage be induced in a conductor? When does an induced voltage cause an induced current to flow?
3. Explain Lenz's Law.
4. State three factors that affect the strength of an induced voltage.
5. Describe how a generator produces an alternating voltage. What is the function of the collector rings? Of the brushes?
6. What is meant by an ac cycle?
7. What is meant by the frequency of alternating current? The amplitude?
8. What is meant by inductance? By inductive reactance?
9. What is meant by capacitance? By capacitive reactance?
10. What is meant by impedance? What is its unit of measurement?
11. What is the function of a rectifier?
12. How might a d'Arsonval galvanometer be used to measure alternating current?

Problems

ALTERNATING CURRENT AND INDUCTANCE

1. An ac voltage has a peak value of $56\overline{0}$ V. What is its effective value?

2. A particular industrial circuit has an effective ac voltage of $100\overline{0}$ V. What is its peak value?

3. Solve Sample Problem 18.1 with the coil in a radio circuit receiving radio waves having a frequency of $55\overline{0}$ kHz.

4. Find the inductive reactance of a $70\overline{0}$ μH coil: (a) in a circuit operated at $6\overline{0}$ Hz; (b) at $6\overline{0}$ kHz.

5. A $60\overline{0}$ mH coil has an inductive reactance of 226 Ω.
   a. What is the frequency of the circuit?
   b. If the voltage drop across the coil is 45 V, find the current through the coil.

CAPACITANCE AND IMPEDANCE

6. Solve Sample Problem 18.2A if the frequency is $80\overline{0}$ Hz.

7. Find the capacitive reactance of a $30\overline{0}$ pF capacitor at a frequency of $1\overline{0}$ mHz.

8. a. Find the capacitance (in farads) of a capacitor that has a capacitive reactance of 375 Ω at a frequency of $20\overline{0}$ Hz.
   b. What is the current flow if the voltage drop across the capacitor is 10.0 V?

9. Solve Sample Problem 18.3 if the coil is rated $20\overline{0}$ mH.

10. The resistance of a coil is considered to be zero, and its inductive reactance is $10\overline{0}$ Ω. It is placed in series with a $10\overline{0}$ Ω resistor.
    a. What is the impedance of the circuit?
    b. If the voltage drop across the circuit is $12\overline{0}$ V, what is the current flow?
    c. Using the current flow in (b), find the voltage drop across each component.

11. A 4 H coil is placed in series with a 100 Ω resistor. Frequency is 60 Hz. Find the impedance of the circuit.

12. A capacitor (with 0 Ω resistance) has a capacitive reactance of $2\overline{0}$ Ω. It is placed in series with a $1\overline{0}0$ Ω resistor.
   a. What is the impedance of the circuit?
   b. What is the current flow if the voltage drop for the circuit is 9.7 V?

Computer Program

This program offers a menu of three formulas. You can find: (1) inductive reactance, (2) capacitive reactance, and (3) impedance of a series circuit with resistance and reactance (either inductive or capacitive). The program will help you plot graphs of inductive reactance versus inductance, or capacitive reactance versus capacitance, or reactance versus frequency. From these graphs, you can determine the answers to such questions as "What happens to X_C if C is doubled?"

PROGRAM

```
10    REM  CH EIGHTEEN PROGRAM
20    PRINT "AC CIRCUIT FORMULAS.  TO FIND INDUCTIVE REACTANCE (XL)-"
21    PRINT "TYPE 1, TO FIND CAPACITIVE REACTANCE (XC)-TYPE 2, AND"
22    PRINT "TO FIND IMPEDANCE (Z)-TYPE 3."
30    INPUT A
40    PRINT "YOU HAVE SELECTED -";A
50    PRINT
60    IF A = 1 THEN  GOTO 100
70    IF A = 2 THEN  GOTO 200
80    IF A = 3 THEN  GOTO 300
90    END
100   PRINT "TYPE IN INDUCTANCE IN HENRIES, THEN FREQUENCY IN HERTZ."
110   INPUT L,F
120   PRINT "INDUCTANCE = ";L;" H."
130   PRINT "FREQUENCY = ";F;" HZ."
140   PRINT
150   LET XL = 2 * 3.1416 * F * L
160   PRINT "INDUCTIVE REACTANCE = ";XL;" OHM."
170   GOTO 90
200   PRINT "TYPE IN CAPACITANCE IN FARADS,THEN FREQUENCY IN HERTZ."
210   INPUT C,F
220   PRINT "CAPACITANCE = ";C;" F."
230   PRINT "FREQUENCY = ";F;" HZ."
240   LET XC = 1 / (2 * 3.1416 * F * C)
250   PRINT
260   PRINT "CAPACITIVE REACTANCE = ";XC;" OHM."
```

```
270   GOTO 90
300   PRINT "TYPE IN RESISTANCE IN OHMS THEN EITHER INDUCTIVE OR "
301   PRINT "CAPACITIVE REACTANCE IN OHMS."
310   INPUT R,X
320   PRINT "RESISTANCE = ";R;" OHM."
330   PRINT "REACTANCE = ";X;" OHM."
340   PRINT
350   LET Z =  SQR ((R ^ 2) + (X ^ 2))
360   PRINT "IMPEDANCE = ";Z;" OHM."
370   GOTO 90
```

NOTES

Line 40 confirms your selection.

Lines 60–80 select the correct subroutine for the answer desired.

Line 150 is the formula for inductive reactance: $X_L = 2\pi fL$.

Line 240 is the formula for capacitive reactance: $X_C = 1/2\pi fC$.

Line 350 is the formula for impedance: $Z = \sqrt{R^2 + X_L^2}$.

This hydroelectric power plant is 1000 feet underground in a cavern hewn from solid granite. The three hydroelectric generators in view can supply 1.2 million kilowatts, which is enough to meet the needs of a city of 1 million people. Hydraulic turbines that drive the generators are below floor level. (Courtesy of Pacific Gas & Electric Company)

Chapter 19

Objectives

When you finish this chapter, you will be able to:

- [ ] Define a prime mover.
- [ ] Describe the differences and similarities between dc and ac generators.
- [ ] Explain the difference between a shunt generator and a series generator.
- [ ] Diagram and discuss the transformer.
- [ ] Use the transformer voltage–turns formula: $E_p/E_s = N_p/N_s$.
- [ ] Describe the dc motor and the induction motor.
- [ ] Apply the efficiency formula $e(\%) = $ power out/power in $\times$ 100 to motors and generators.

Electrical Prime Movers

A **prime mover** is a device that converts energy into useful work. As examples of mechanical prime movers—that is, devices that operate by mechanical energy—we can list such machines as the windmill, waterwheel, spring motor, steam turbine, and the gasoline and diesel engines.

Today, so many industrial machines are operated by electrical devices that all technicians should have some knowledge in this area. This chapter covers electrical prime movers, such as ac and dc generators, ac and dc motors, and transformers. Some typical application problems involving efficiency and power are included.

19.1 Generators of Electrical Energy

ALTERNATING CURRENT AND DIRECT CURRENT GENERATORS

We have already discussed the ac generator (Chapter 18), which converts mechanical energy into electrical energy. This device, you will recall, consists of a loop of wire mounted in a **magnetic field**. As the loop is rotated by mechanical means, it cuts across the magnetic lines of force and, as a result, an alternating voltage is induced in the loop. This voltage produces an alternating current through the collector rings (sometimes called *slip rings*) and brushes and through the external circuit connected to the brushes.

＊ magnetic field　　　　The space occupied by the lines of force present between the poles of a magnet.

Such a simple generator cannot generate very much voltage because the armature consists of a single loop and the magnetic field produced by the permanent magnet is not very strong. If we substitute a coil of many turns for the single loop, a great many more conductors cut the magnetic lines of force during the revolutions, and as a result, a greater voltage is induced. Further, if we provide a soft-iron core for the coil, the induced voltage is increased even more. The induced voltage can be increased still more if we strengthen the magnetic field cut by the conductors. We can do so by substituting an electromagnet for the permanent magnet, as in Figure 19.1, thus creating, in simplified form, a practical ac generator.

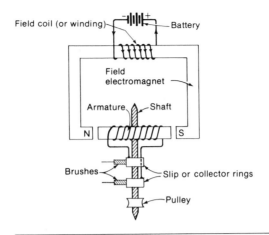

FIGURE 19.1 The ac Generator.

The armature consists of a soft-iron core on which many turns of wire are wound. Although this core is shown as a rod in the illustration, it actually consists of an iron cylinder containing slots in which the wire is laid. This armature is mounted on a shaft supported by bearings (not shown here) at each end. The slip rings are mounted on this shaft, as is a pulley, which is attached to the source of mechanical power and by which the shaft, slip rings, and armature are rotated.

The field electromagnet is a U-shaped yoke of soft iron upon which many turns of wire are wound. The turns of wire are called the **field coil**, or field winding, and it produces the magnetic lines of force in which the armature turns. The field coil is attached to a source of direct current, which may be a battery (as shown here), or a direct current generator (which will be described later). This dc generator (called an *exciter*) may be a separate device or may be built right into the ac generator. The exciter generally is rotated by the same mechanical-energy source that rotates the armature of the ac generator. Most generators use an exciter rather than batteries.

∗ field coil

Turns of wire that produce the magnetic field in which the armature of a generator turns.

The generator pictured in Figure 19.1 has two poles for the field electromagnet. Thus, it is called a *two-pole generator*. Since it produces one cycle of alternating current for each complete revolution, if we wish to obtain 60 hertz alternating current (the usual frequency in this country), the armature must make 60 revolutions every second, or 3600 revolutions per minute.

Large utilities with high-speed steam turbines use two-pole generators. Of course, the generator must have at least two field poles; however, we could add another field magnet perpendicular to the one illustrated in Figure 19.1. Then we would have a *four-pole generator*. To reduce the speed of rotation (say, for a slow-speed diesel-driven generator), we generally use *multipolar* generators.

Where very little current is needed, we may use a type of ac generator called a **magneto**. In this device, the armature coil consists of a great many turns, but the field is produced by several permanent magnets, eliminating the need for a separate exciter or batteries. Magnetos are used on power lawn mowers, small snow blowers, and some motorcycles.

---

* magneto

A small generator of alternating current that uses permanent magnets to produce the magnetic field in which the armature turns.

The **dc generator** resembles the ac type except for a device called the **commutator**. This device mechanically converts the generated alternating current into direct current delivered to the brushes. Looking at the ac generator in Figure 18.3, we see that during one-half a revolution, electrons are streaming out of one brush, and during the next half-revolution, out of the other brush. If, after a half-revolution, the positions of the brushes in relation to the ends of the armature coil were reversed, the electrons would always stream out of the same brush. One brush would always be negative and the other always positive. The output of the generator would then be a direct current.

---

* dc generator

A rotating machine that converts mechanical energy into electrical energy in the form of direct current.

* commutator

A device that mechanically converts the generated alternating current into direct current by making contact with stationary brushes and transferring electrical energy between the armature and the external circuit.

In Figure 19.2, we see a split ring fastened to the shaft by means of some insulating material. This ring is not only insulated from the shaft, but each half of the ring is insulated from the other. The ends of the armature coil are attached to each half of the ring. The brushes are stationary, as before, and now each makes contact with one segment of the ring during one half-revolution and with the other segment during the next half-revolution. There are no slip rings.

As the armature makes a half-revolution (and the current flow in the armature coil is reversed), the positions of the brushes, in relation to the ends of the coil, are reversed, too. Thus, one brush is always positive and the other always negative.

Figure 19.3 is a repeat of Figure 18.3 except that a commutator replaces the slip rings. This figure shows that although current is alternating in the armature, it flows in only one direction through the load.

Since the output of the dc generator is a direct current, we need no external source to excite the field winding (field coil). All we must do is divert some of the direct current output to the field coil. There are several ways of doing so. We may connect the field coil in parallel, or shunt, with the armature and with the direct current output of the generator, as shown in Figure 19.4A. The field winding consists of many turns of fine wire, has a high resistance, and draws only a small portion of the output current. However, there is a drawback. If the load draws more current and the current increases, the generator voltage drops a bit, which reduces the current through the field coil. Thus, the magnetic field is reduced, and the voltage of such a generator drops further. One of the prime reasons this voltage drop happens is that the generated voltage has to push current through its own armature resistance as well as through the load resistance and the field resistance. We can think of a **shunt generator** as consisting of a battery plus the armature resistance. The voltage drop through the armature resistance increases as the current increases, and this leaves a smaller voltage at the terminals to supply the load and the field coil.

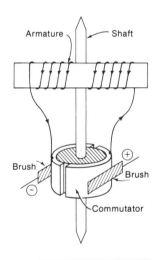

FIGURE 19.2   The Commutator of a dc Generator.

---

* shunt generator   A generator that has the field coil connected in parallel with the output and the armature.

As Figure 19.4B indicates, another way of connecting the field coil is in series with the armature and with the direct current output of the generator. In a **series generator**, the entire output

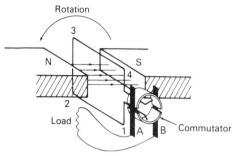

A. No current flow

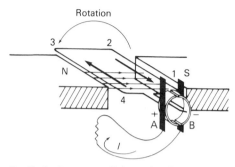

**B. Clockwise current flow with side 1-2 of the
armature in contact with brush B**

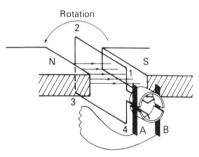

C. No current flow

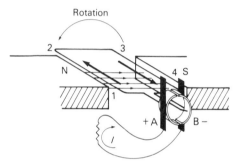

**D. Clockwise current flow with side 3-4 of the
armature in contact with brush B**

FIGURE 19.3   A dc Generator.

current flows through the coil, which must contain relatively few turns of heavy wire. In this case, the greater the current drawn by the external circuit, the more current flows through the coil, and the stronger the magnetic field. Hence the voltage tends to rise as more current is drawn.

* series generator   A generator in which the field coil is connected in series with the output and the armature.

However, we generally want the voltage to remain constant, regardless of how much current is drawn from the generator. For this reason, the shunt and series generators are seldom used.

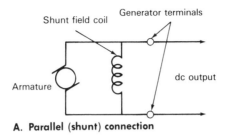

A. Parallel (shunt) connection

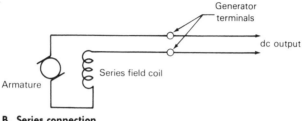

B. Series connection

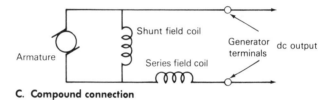

C. Compound connection

FIGURE 19.4   How the Field and Armature Coils of a dc Generator May Be Connected.

Instead, we usually employ a combination type, called the **compound generator**. This type, as Figure 19.4C shows, contains two field windings. One of the field windings has many turns of fine wire and is in shunt. The other field winding has few turns of heavy wire and is in series. As current is drawn, the drop in voltage caused by the shunt winding is offset by the rise in voltage caused by the series winding. Hence, the voltage tends to remain fairly constant.

* **compound generator**

A generator that has two field coils. One coil is connected in shunt and the other coil is connected in series with the output and the armature.

We have seen that the induced voltage depends on the strength of the magnetic field, the number of conductors cutting it, and the speed at which they cut the magnetic field (see Chapter 18). In both ac and dc generators, the number of conductors is constant and cannot be changed readily. If we wish to vary the voltage, we must vary either the speed of rotation or the strength of the magnetic field. However, in the case of the ac generator, varying the speed of rotation also varies the frequency of the current, which we usually desire to keep constant. Therefore, the voltage of an ac generator is varied by changing the strength of the magnetic field. This effect is accomplished by means of a variable resistor placed in series with one of the field windings. The greater the resistance, the less current flows through the winding, the weaker the magnetic field, and the lower the voltage of the generator. In the dc generator, where frequency is not a problem, we may vary the voltage by changing the speed of rotation or by varying the strength of the magnetic field, just as in the ac type.

19.2   The Transformer

We now have an understanding of the electric generator as a prime mover. However, in a more basic sense, any device that generates a voltage and causes electrons to move in a circuit is considered an electrical prime mover. In our previous discussions, the generator has been connected directly to the circuit in which the current flows. However, there is a method for causing an alternating current to flow in a circuit, even though a generator is not connected directly to that circuit. The device by which this effect is accomplished is called the **transformer**. The transformer is a very useful device, and we will discuss some of its applications after explaining how it operates.

---

* transformer

An electrical device for transferring electrical energy (in the form of alternating current) from one circuit to another circuit.

We know that when an alternating current flows through the turns of a coil of wire, the fluctuating magnetic field around the coil continuously cuts across the turns of wire, inducing a back-emf in the coil. This phenomenon is called *self-induction*.

PHOTO 19.1   This step-down transformer in a switching yard converts high-voltage current from transmission lines to lower voltage for use in the city's power distribution system. The "radiators" on the right of the transformer help keep it cool by circulating oil around the electric coils. (Courtesy of Eastern Edison Company)

Now, what will happen if we place another coil near the first, but in no way connected to it, as shown in Figure 19.5?

As the fluctuating magnetic field surrounds the original coil, called the **primary coil**, the field not only cuts across the turns of that coil, but also cuts across the turns of the second coil, called the **secondary coil**. As a result of the magnetic field cutting across its turns, a voltage is induced in the secondary coil, and if it has a circuit connected to it, current will flow in the circuit. When two coils are linked in this way by a magnetic field, we call it a magnetic coupling, or **inductive coupling**.

* **primary coil**   A coil connected to the source of the alternating current. The current through its winding induces alternating voltage in the secondary winding.

* **secondary coil**   A coil connected to the load circuit. The alternating voltage and current in its winding are induced.

PHOTO 19.2   Transformers are needed whenever current is changed from high voltage in distribution lines to more usable voltage in industry. This transformer is in Lake Maracaibo, Venezuela, where about 4500 oil rigs drill for oil. (United Nations photo)

---

* inductive coupling

A coupling that exists between two coils if the alternating magnetic lines of force of one coil induce a voltage in the second coil.

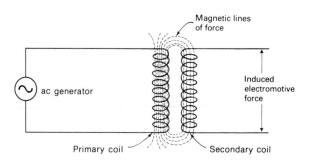

FIGURE 19.5   Magnetic Coupling between the Primary and Secondary Coils of a Transformer.

The combination of a primary and secondary coil so coupled makes up the transformer. We can see that the transformer transfers electrical energy from one circuit to another, even though these circuits have no direct electrical contact with each other. We can see, too, that the transformer is an ac device. If a steady direct current were to flow through the primary coil, the resulting magnetic field would be constant, and as a result, no electromotive force would be induced in the secondary coil.

Notice in Figure 19.5 that some of the magnetic lines of force escape into the surrounding air and do not cut across the turns of the secondary coil. But if we mount both coils on an iron ring (an idea credited to Michael Faraday) as in Figure 19.6, almost all the lines of force will travel through this ring, and since the secondary coil is affected by a stronger magnetic field, the induced voltage will be greater.

In the ideal transformer, all the electrical energy in the primary circuit would be transferred to the secondary. However, this condition can never be achieved because of the presence of certain energy losses. First of all, some of the magnetic lines of force are lost to the surrounding air, although this loss can be reduced to a minimum by the use of a proper iron core. Then, there are certain copper and iron losses in a transformer. The copper loss is due to the resistance of the wire making up the turns of the windings, which dissipates electrical energy as heat energy.

The iron losses may be divided into two parts. Since the core is in the magnetic field, it is magnetized. But the alternating current causes the iron core to change the polarity of its poles in step with the frequency of the current. A certain amount of energy is required to make this change. This energy comes from the electrical source, and therefore is a loss. This iron loss, which we call **hysteresis loss,** may be partially reduced by using cores

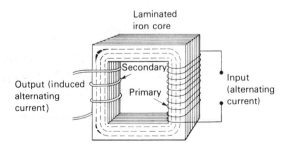

FIGURE 19.6 A Step-Down Transformer.

of silicon steel or certain other alloys that are much more permeable than iron—that is, that are easier to magnetize and demagnetize.

The other iron loss is due to the electric current induced in the iron core by the changing magnetic fields of the coils wound upon it. This induced current is called the *eddy current*. Since the eddy current must come from the electrical source, it too is a loss. To reduce eddy current losses, transformer cores are not made of solid metal, but are built up of very thin strips, called laminations. Each lamination is coated with an insulation of oxide or varnish so that the eddy currents cannot circulate through the core.

In addition to its function of transferring electrical energy from one circuit to another, there is another important use for the transformer. Assume that we have an ideal transformer with no losses and with 100 percent efficiency (large transformers operate at 98–99% efficiency). Further, assume that the primary winding contains 100 turns and that it is connected to an ac generator producing 100 volts. Since there are no losses, the back-emf generated by the primary winding must be equal to the applied voltage (namely, 100 volts). Thus, the magnetic field must be sufficient to generate 1 volt for each turn of the primary winding. This condition must hold true also for the turns of the secondary winding. Therefore, each turn of the secondary winding will generate 1 volt.

If the secondary coil of this transformer has 1000 turns, 1000 volts will be induced in it. The transformer has increased, or stepped up, the input of 100 volts to an output of 1000 volts because the secondary coil has ten times as many turns as the primary coil. Such a transformer is called a **step-up transformer**. However, if the secondary coil has only ten turns, only 10 volts will be induced in it. Such a transformer is called a **step-down transformer** (Figure 19.6). Therefore, the transformer can be used to step up or to step down alternating voltage.

---

* step-up transformer

A transformer in which the voltage is increased from the primary to the secondary winding.

* step-down transformer

A transformer in which the voltage is decreased from the primary to the secondary winding.

From the preceding discussion, we can see that the ratio between the voltage across the primary winding and the voltage across the secondary winding is equal to the ratio between the number of turns of the primary winding and the number of turns of the secondary winding. This relationship may be expressed in a formula as follows:

* VOLTAGE–TURNS FORMULA

$$\frac{E_p}{E_s} = \frac{N_p}{N_s}$$

where E_p = the voltage across the primary coil
 E_s = the voltage across the secondary winding
 N_p = the number of turns in the primary winding
 N_s = the number of turns in the secondary winding

SAMPLE PROBLEM 19.1

A transformer is required to deliver $33\overline{0}$ V of alternating current across the secondary winding. Assume a primary winding of $100\overline{0}$ turns connected across a $11\overline{0}$ V ac line. How many turns must we have in the secondary winding?

SOLUTION

Wanted

$N_s = ?$

Given

$E_s = 33\overline{0}$ V, $E_p = 11\overline{0}$ V, and $N_p = 100\overline{0}$ turns

Formula

The voltage–turns formula is:

$$\frac{E_p}{E_s} = \frac{N_p}{N_s}$$

We rearrange to:

$$N_s = \frac{N_p \times E_s}{E_p}$$

Calculation

$$N_s = \frac{1000 \text{ turns} \times 330 \text{ V}}{110 \text{ V}} = 300\overline{0} \text{ turns}$$

This information can be helpful in designing transformers and can be used in conjunction with experimental data to determine the efficiency of the transformer.

What is happening to the current while the voltage is being stepped up or down? In our ideal transformer, we assume no losses. Therefore, the power ($E \times I$) of the secondary circuit must be equal to the power of the primary circuit:

* POWER RELATION FORMULA

power out = power in

$$E_s I_s = E_p I_p$$

Therefore, in Sample Problem 19.1, since the voltage in the secondary winding is three times that of the primary, the current set flowing in the secondary winding must be 1/3 of the current in the primary. As the voltage is stepped up, the current is proportionately reduced, and vice versa.

SAMPLE PROBLEM 19.2

If the secondary winding in Sample Problem 19.1 supplies a load drawing $33\overline{0}$ V and 5.0 A, what is the current in the primary?

SOLUTION

Wanted

$I_p = ?$

Given

$E_p = 110$ V, $E_s = 330$ V, and $I_s = 5$ A

Formula

The power relation formula is:

$$E_s I_s = E_p I_p$$

We rearrange to:

$$I_p = \frac{E_s I_s}{E_p}$$

Calculation

$$I_p = \frac{330 \text{ V} \times 5 \text{ A}}{110 \text{ V}} = 15 \text{ A}$$

This information is helpful when establishing the ratings for circuit breakers in the primary and secondary circuits.

We have stated that in the ideal transformer, the power in the primary circuit is equal to the power in the secondary circuit. If the primary winding of a transformer is connected to a source of alternating current, and the secondary circuit is left open (no power is drawn), there should be, theoretically, no current flowing in the primary circuit. However, in practical transformers, a certain amount of current will flow in the primary circuit as a result of the various losses present. If the transformer is properly designed, the back-emf induced in the primary winding will keep this current down to a very small value.

If the secondary circuit is completed, it will draw a certain amount of power. Since this power must come from the primary circuit, the current flowing through the primary winding must increase accordingly. The more power consumed by the secondary circuit, the greater will be the current flowing in the primary winding.

The ability to use the transformer for step-up or step-down purposes is one of the chief advantages of alternating over direct current. For example, let us consider the problem of transmission of electrical power from a source, such as the hydroelectric plant at Hoover Dam, to Los Angeles, several hundred miles away. At the plant, the generators produce the power at about 11 000 volts. This voltage is the maximum that can be produced because of the necessity for heavy insulation of the wires of the field coil and armatures of the generators. At higher voltages, the insulation would become so bulky as to render the generators impractical.

Since the power $(E \times I)$ that must be transmitted is tremendous, if it were sent at 11 000 volts, the current would be enormous. As the current flows through the wires, it heats them because of the resistance encountered. The loss of energy due to the heating effect $(I^2 \times R)$ would, accordingly, become so large as to render the transmission of power impractical, except for comparatively short distances. Even then, the wire would have to be of very large diameter in order to have low resistance.

It is here that the transformer comes to the rescue. As Figure 19.7 shows, after the power is generated at 11 000 volts, it is sent into transformers that step up the voltage 25 times to 275 000 volts. Because the voltage has been multiplied 25 times, we obtain the same power with 1/25 the current. Therefore, the power is transmitted at 275 000 volts and, since the current is only 1/25 its original value, the loss due to heat (I^2R) is very small.

At Los Angeles, the voltage is stepped down by other transformers to the 110–120 volt range supplied to the house mains. There is no power loss (except for the slight amount in the transformers) because the current is stepped up in equal degree.

19.3 The Motor

THE DIRECT CURRENT MOTOR

Let us look at the illustration of the d'Arsonval galvanometer shown in Figure 17.9. As direct current flows in the proper direction through the armature coil, the north pole of the armature is repelled from the north pole of the field magnet. As a result, the armature makes a half-revolution until its north pole is opposite the south pole of the magnet. Since unlike poles attract, the armature then stops rotating.

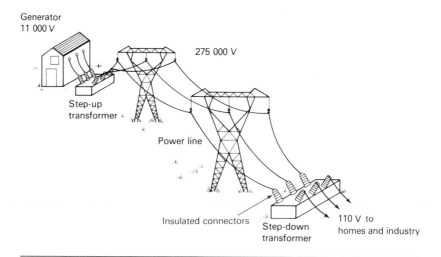

FIGURE 19.7 How Transformers Are Used to Transmit Electrical Power over Long Distances.

PHOTO 19.3   The type of electric motor exemplified by this model can produce from 1 to 200 horsepower at speeds of 1200, 1800, or 3600 rpm.

Now suppose that at the instant the north pole of the armature comes opposite the south pole of the magnet, the direction of current through the armature coil is reversed. Then the north pole of the armature becomes a south pole. Since it is opposite the south pole of the magnet, the armature is repelled again and is forced to make another half-revolution.

We can see that if we could change the direction of current flow through the armature coil every time the armature makes a half-revolution, the armature would revolve continuously. If a pulley were fastened to the revolving armature, we could use its power to turn machinery.

Do we have such a device for changing the direction of current flow every half-revolution? Figure 19.2 shows that the commutator (which we used to convert the alternating current generated in the armature of the generator into the direct current supplied to the brushes) is just such a device. Therefore, we may use the dc generator as a motor if we supply a direct current to the brushes.

As a matter of fact, in 1873, Zénobe T. Gramme, a Belgian engineer, had two identical dc generators standing side by side. One of these generators was being rotated by a steam engine. When the current output of this generator was accidentally fed into the second generator, the latter started to rotate as a motor.

Believe it or not, the dc motor is used in some applications on ac circuits! Small ac motors for vacuum cleaners, portable tools, and so on are among those types of ac motors designed in a manner similar to dc motors. For this reason, they can start with a load already on their shafts and their speeds can be varied better than in most other ac motors. This motor is called a **universal motor** when designed for ac circuits. In fact, some electric locomotives are designed with these motors for use on either ac or dc power.

---

* universal motor | A motor designed with a commutator and usable on either ac or dc circuits.

THE ALTERNATING CURRENT MOTOR

A great many types of motors are designed expressly for alternating current use. Space prevents discussion of all these types. Instead, we will consider the most common type of ac motor, the **induction motor**. The induction motor gets its name because the current in the armature is an induced current. There are no electrical connections to the armature, as in the motors previously described.

---

* induction motor | A motor that operates on ac only and gets its name from the fact that current is induced in the armature instead of being fed into the armature through a commutator or slip rings.

To understand how an induction motor operates, look at Figure 19.8. The figure shows the armature winding to be a closed loop, or coil. The field windings are referred to as stator (sta-

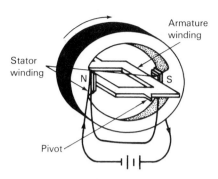

FIGURE 19.8 Diagram Illustrating the Basic Principle of the Induction Motor.

tionary) windings. Industry is in the habit of referring to the windings attached to the frame of ac generators and ac motors as stator windings. This has come about primarily because the large ac generators used by electric utilities have their field windings placed on the rotor and the armature windings placed on the frame. This arrangement allows a smaller and lighter rotor to be used, because the voltage and current applied to the field coil are quite a bit less than the voltage and current output of the armature. Therefore, the rotor windings can be made from smaller wire and the insulation can be thinner and lighter. But let's get back to our induction motor.

Imagine a source of direct current connected to the stator windings. North and south poles appear at opposite poles of the stator, and magnetic lines of force are produced between them. (The lines of force are not shown in Figure 19.8.) The armature winding is a closed loop. If a current can be induced in the loop, the armature will have its own magnetic field, which will interact with the magnetic field of the stator and produce motion.

Now imagine that our motor is stopped with the armature in the position shown and we switch on the dc power. Nothing happens. There is no turning effect on the armature because no current has been induced in its coil. To understand the reason, recall our discussion of how voltage and current are induced in a conductor if there is relative motion between the magnetic field and the conductor. In this case, no relative motion is present. However, let's suppose we can physically rotate the stator clockwise. Now we have relative motion between the magnetic lines of force (the magnetic field) of the stator and the coil. A current is induced in the coil, which, in turn, creates its own electromagnet with north and south poles. These magnetic poles interact with the rotating stator poles, and we get rotation of the armature in the same direction as the stator.

If we could make the stator poles rotate, or, what amounts to the same thing, if we could make the magnetic field rotate, then the rotor loop (armature) would rotate with it. The problem, of course, is to obtain the rotating magnetic field.

Assume that we have two identical ac generators, one starting up a quarter-revolution (90°) behind the other. The graph in Figure 19.9 shows the current output from each armature winding. We call currents that bear this relationship to each other **two-phase alternating current**. In practice, it is not uncommon for these currents to "peak" at different times, thus providing the graph in Figure 19.9.

---

* two-phase alternating current   A combination of two currents, each with its own wave form.

Now suppose that we have four poles set in a stator ring, as in Figure 19.10. Assume that one winding, connected to the first generator (1), is wound on two opposite poles, as shown. Another winding, which is connected to the second generator (2), is wound on the other two poles.

At point A on the graph in Figure 19.9, the output from generator 1 is at its positive maximum. The current flowing through winding 1 produces N and S poles, as indicated in Figure 19.10A. At that instant, there is no current from generator 2, and consequently, there are no magnetic poles for winding 2.

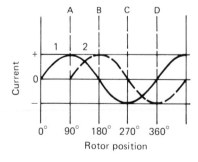

FIGURE 19.9   *Graph Showing What Is Meant by a Two-Phase Alternating Current.*

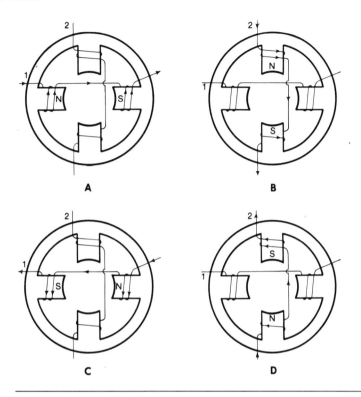

FIGURE 19.10 Diagrams Showing How a Two-Phase Alternating Current Is Used to Obtain a Rotating Magnetic Field.

At point B of the graph (90° later), the current from generator 1 has fallen to zero and there are no magnetic poles for winding 1 (Figure 19.10B). However, the current output from generator 2 has risen to its positive maximum, and winding 2 now has N and S poles as indicated.

At point C of the graph, current 2 has fallen to zero and current 1 is at its negative maximum. Accordingly, winding 2 has no magnetic poles (Figure 19.10C) and winding 1 has N and S poles again. However, because the current is flowing in the opposite direction now, the poles are in reverse position from those of Figure 19.10A.

At point D of the graph, current 1 has fallen to zero and current 2 is at its negative maximum. Winding 1 now has no magnetic poles (Figure 19.10D), and the poles of winding 2 are reversed from Figure 19.10B.

Look closely at the four illustrations in Figure 19.10. Note that the north and south poles have, in effect, revolved in a clockwise direction. Or, what is the same thing, the magnetic field has revolved as the currents proceeded through their alternations. This is the principle of the induction motor.

This, then, is how we obtain a revolving magnetic field. We wind the stator as illustrated in Figure 19.10 and connect its windings to a source of two-phase alternating current. The rotor (armature) consists of a single loop, and as the magnetic field of the stator revolves, so does the loop.

In most commercial installations, *three-phase alternating current* is used. The effect is the same as if three generators were employed, operating 120 degrees apart. Accordingly, the motor must be three-phase, with the rotating field being obtained by three stator windings wound on three sets of poles, 120 degrees apart.

In commercial motors of the type described here, the rotor generally consists of heavy copper bars arranged in a drum and short-circuited at the ends. The result is many loops. The rotor looks like a "squirrel cage" and is known by that name.

19.4 Efficiency and Power Related to Motors and Generators

Recall from Chapter 11 on heat engines that the efficiency (*e*) of a machine is the ratio of the work out to the work in: $e(\%) = $ work out/work in $\times$ 100. We can also get the same ratio by comparing the power out to the power in:

---

* EFFICIENCY FORMULA

$$e(\%) = \frac{\text{power out}}{\text{power in}} \times 100$$

This formula is useful when dealing with electrical machines because power measurements are more easily obtained than work measurements.

Most generators and motors operate in the 75–95 percent efficiency range. An exception is the fractional horsepower motor, which may have an efficiency of about 20 percent.

We have already discussed mechanical and electrical units of power and work (see Chapters 6 and 17). Remember that in

PHOTO 19.4 A switching yard is where the changes in current and voltage take place for transmission of electric power. (Courtesy of the Bonneville Power Administration)

electrical measurements, energy is given in units of watthours:

energy in watthours = power in watts × time in hours

The following list gives the multiplying factors for unit conversion:

To change the number of watthours to joules, multiply by 3600.
To change horsepower to watts, multiply by 746.
To change watts to horsepower, multiply by 0.001 34.

Eventually, electrical and mechanical energy (and heat energy) will be measured in joules, and power will be measured in watts. However, industry still uses the other units mentioned. For example, a few paragraphs above, "fractional horsepower" motors were mentioned, a common term used in industry. The common unit of energy used in the electrical industry is the watthour. Electric utilities need to know how much energy a customer has used in order to make out the proper billing. Your home has a watthour meter, as do commercial and industrial customers of the utility. Utilities charge residential users about 6¢ to 10¢ for each kilowatthour used. This charge varies depending on locality and type of use.

SAMPLE PROBLEM 19.3	An electric utility plant can generate 70.0 MW of power. The generator operates at 95.0% efficiency. A. What is the output in horsepower? B. How much power is applied to the generator?
SOLUTION A	
Wanted	P_{out} = ?
Given	70.0 MW × 1 000 000 W/MW = 70 000 000 W
Formula	To find horsepower, we will multiply watts by 0.001 34.
Calculation	P_{out} = 70 000 000 W × 0.001 34 = 93 800 HP
SOLUTION B	
Wanted	P_{in} = ?
Given	P_{out} = 70.0 MW and e = 95.0%
Formula	$$e(\%) = \frac{P_{out}}{P_{in}} \times 100$$ We rearrange to: $$P_{in} = \frac{P_{out} \times 100}{e(\%)}$$
Calculation	$$P_{in} = \frac{70.0 \text{ MW} \times 100}{95\%} = 73.7 \text{ MW}$$
	This information is useful when making comparisons with other types of power systems.

**SAMPLE
PROBLEM
19.4**

An industrial plant has a large motor that delivers 35.0 HP for 8.00 hr at 85.0% efficiency.
  A. Find the power in to the motor in kW.
  B. Find the total energy input for 8.00 hr in kW·hr.
  C. What is the 8 hr cost of the electricity if the charge is $0.05 per kW·hr?

SOLUTION A

Wanted

$P_{in} = ?$

Given

$e = 85\%$ and $P_{out} = 35.0$ HP $\times$ 746 $= 26\ 100$ W

Formula

$$e(\%) = \frac{P_{out}}{P_{in}} \times 100$$

We rearrange to:

$$P_{in} = \frac{P_{out} \times 100}{e(\%)}$$

Calculation

$$P_{in} = \frac{26\ 100\ \text{W} \times 100}{85\%} = 30\ 700\ \text{W}$$

P_{in} is read "30 thousand 700 watts"; therefore, we can replace the word "thousand" with the prefix "kilo":

  power in $= 30.7$ kilowatts

SOLUTION B

Wanted

energy input $= ?$

Given

$P_{in} = 30.7$ kW and $t = 8.00$ hr

Formula

We simply multiply the power by time, because

  energy $=$ power $\times$ time

Calculation

energy in $= 30.7$ kW $\times$ 8.00 hr $= 246$ kW·hr

SOLUTION C

Wanted

total cost for 8 hr $= ?$

Given	energy consumption = 246 kW·hr and cost = \$0.05 per kW·hr
Formula	We will multiply the number of kW·hr by the cost for each kW·hr.
Calculation	cost = 246 kW·hr × 0.05 = \$12.30

This information is useful in determining power requirements and the cost of manufacturing operations.

Summary

Electrical prime movers convert electrical energy from and to other forms of energy. In this chapter, we discussed prime movers that converted electrical energy to and from mechanical energy. We also discussed the transformer, which converts electrical energy in one circuit into electrical energy in another circuit.

The principle of operation of an ac generator, with a single turn of wire for the armature, had already been covered in Chapter 18. In this chapter, we showed that the armature winding actually is a coil with many turns wound on an iron core. In the ac generator, one end of the armature coil is always in contact with the same brush. Therefore, as current is induced in different directions in the armature, depending on its position, current in the load circuit alternated also.

A dc generator is similar to an ac generator, but current is supplied to a load circuit through a commutator instead of slip rings. Therefore, as the position of the armature changes, the end of the armature coil contacts an alternate brush, thus maintaining a constant direction for the current in the load circuit.

The transformer is primarily a device for changing the voltage in electrical circuits. It can increase the voltage supplied by some power source (a step-up transformer) or it can decrease the voltage (a step-down transformer). The transformer works on the principle of induction. The changing magnetic field produced by alternating current in one coil (the primary coil) induces a current in an adjacent coil (the secondary coil). The voltage induced depends on the ratio of the number of turns in each coil. Also, the power supplied by the primary circuit is transferred to the secondary circuit.

PHOTO 19.5   High-voltage power lines must be carefully secured and insulated before becoming functional. Although the current flowing through them is lower than the original current generated at the power station, the voltage is correspondingly higher. (United Nations photo)

When the d'Arsonval galvanometer was discussed in Chapter 17, electrical energy was used to produce mechanical energy (turning the armature). A motor operates similarly, except full 360 degree rotation of the armature is obtained. The dc generator will act as a motor if a current is supplied to the armature winding. This type of motor will also operate if alternating current is supplied to the armature winding and the field winding. When designed for ac circuits, it is called a universal motor, and is used in such machines as vacuum cleaners and portable tools.

The induction motor for ac circuits differs from the universal motor in that current is not supplied to the armature winding. The current is induced in the winding from the changing magnetic field of the motor. If the magnetic field can be rotated, or at least appear to be rotating, the armature will try to follow. This effect is due to the interaction of the magnetic fields, one set up by the field windings, and the other set up by the induced current in the armature winding.

The efficiency of machines can be obtained from the ratio of power output to power input:

$$e(\%) = \frac{\text{power out}}{\text{power in}} \times 100$$

The joule is the standard unit of energy and the watt is the standard unit of power in the SI metric system. However, industry still uses the watthour for energy units, and horsepower, along with the watt, for power units.

Key Terms

* prime mover
* magnetic field
* field coil
* magneto
* dc generator
* commutator
* shunt generator
* series generator
* compound generator
* transformer
* primary coil

* secondary coil
* inductive coupling
* step-up transformer
* step-down transformer
* voltage–turns formula
* power relation formula
* universal motor
* induction motor
* two-phase alternating current
* efficiency formula

Questions

1. How does a practical ac generator differ from the simple type illustrated in Figure 18.3?
2. In a dc generator, how does the commutator convert the alternating voltage generated in the armature into direct voltage at the brushes?
3. In the dc generator, what is meant by a shunt field connection? A series field connection? A compound field connection? What are the characteristics of each type?
4. Explain the principle of the transformer. What is a step-up transformer? What is a step-down transformer?
5. Explain the principle of the dc motor.

6. How is the dc motor related to the d'Arsonval galvanometer?
7. How does the ac induction motor operate?
8. How is the efficiency of a machine determined?
9. What is the standard unit for energy in the SI metric system? What unit does industry usually use?
10. Name two units for power.

Problems

TRANSFORMERS

1. Do Sample Problem 19.1 with the primary winding containing 1700 turns.
2. A transformer has $20\overline{0}$ turns for its primary winding and is designed to step down voltage from 120 V to 9.0 V. How many turns are required in the secondary winding?
3. A transformer is to step down voltage from 67 000 V to 15 000 V. What is the primary-to-secondary turns ratio?
4. A transformer is needed to step up the voltage from 120 V to 1500 V. If there are 750 turns in the secondary coil, how many turns are in the primary coil?
5. A home is wired to receive a maximum of 50.0 A at $12\overline{0}$ V. The transformer supplying the neighborhood must step down the primary voltage from $30\overline{0}0$ V to $12\overline{0}$ V. When 50 A flow in the secondary coil, what current is in the primary coil?
6. A transformer is used to reduce current so a low current instrument can be used to measure the primary current. The voltage drop across the primary coil is 3.00 V and the primary current is $10\overline{0}$ A. The secondary current through the instrument is 5.00 A. What is the secondary voltage?

EFFICIENCY AND POWER

7. A dc motor is rated (output) at 3.50 HP.
   a. What is the rating in kilowatts?
   b. How much work is done (in joules and watthours) if the motor runs for 6.25 hr?
8. If the motor in Problem 7 is 70.0% efficient, what is the kilowatt input?
9. A motor draws 5.00 A from a $12\overline{0}$ V line. It is rated at 0.500 HP output. What is its efficiency?

10. A motor rated at 1.0 HP output is $8\overline{0}\%$ efficient. How many amps does it draw from a 240 V line?

11. A mobile diesel-electric generating set can deliver $40\overline{0}$ kW of power. What size of diesel engine is needed if the generator is 87.0% efficient? Give the answer in horsepower.

12. An industrial furnace has three 2100 W heating units supplied by a small diesel generator set. If all three units are used for 6.0 hr a day for $3\overline{0}$ days, find: (a) the total kilowatthours of energy used, and (b) the total cost if 1 kW·hr costs 4.5¢.

13. A $1\overline{0}$ HP motor, at 75% efficiency, drives a sawmill for 8.0 hr.
    a. How much energy must be supplied by the power lines (answer in joules and watthours)?
    b. If the electricity costs 6.0¢ per kW·hr, how much does it cost to run the sawmill for 8.0 hr?

14. Refer to Sample Problem 19.3. How much energy is generated in 1.0 hr? Give the answer in joules.

15. Refer to Sample Problem 19.4. If the owner of the industrial plant buys a new motor with an efficiency of 95%, find: (a) the power input in kilowatts, and (b) the total energy input in 8.0 hr in kilowatthours.

Computer Program

This program uses the efficiency formula:

$$e(\%) = \frac{\text{power out}}{\text{power in}} \times 100$$

to solve for efficiency, power out, or power in. Units must be the same for power out and power in. The units can be in horsepower or watts. Use this program to check your answers to some of the chapter problems.

PROGRAM

```
10  REM  CH NINTEEN PROGRAM
20  PRINT "EFFICIENCY FORMULA.  THE UNITS FOR POWER IN AND OUT "
21  PRINT "MUST BE THE SAME.  IF HORSEPOWER-TYPE HP, IF WATTS-"
22  PRINT "TYPE W, IF KILOWATTS-TYPE KW.
23  INPUT A$
```

```
30    PRINT "YOU HAVE SELECTED ";A$
40    PRINT
50    PRINT "TO FIND EFFICIENCY-TYPE 1, TO FIND POWER OUT-TYPE 2,"
51    PRINT "AND TO FIND POWER IN-TYPE 3 "
60    INPUT A
70    PRINT "YOU HAVE SELECTED - ";A
80    PRINT
90    IF A = 1 THEN   GOTO 200
100   IF A = 2 THEN   GOTO 300
110   IF A = 3 THEN   GOTO 400
120   END
200   PRINT "TYPE IN POWER OUT AND THEN POWER IN."
210   INPUT PO,PI
220   PRINT "POWER OUT = ";PO;A$
230   PRINT "POWER IN = ";PI;A$
240   LET E = (PO / PI) * 100
250   PRINT
260   PRINT "EFFICIENCY = ";E;" %."
270   GOTO 120
300   PRINT "TYPE IN EFFICIENCY THEN POWER IN."
310   INPUT E,PI
320   PRINT "EFFICIENCY = ";E;" %."
330   PRINT "POWER IN = ";PI;" A$
340   LET PO = (E * PI) / 100
350   PRINT
360   PRINT "POWER OUT = ";PO;A$
370   GOTO 120
400   PRINT "TYPE IN EFFICIENCY THEN POWER OUT."
410   INPUT E,PO
420   PRINT "EFFICIENCY = ";E;" %."
430   PRINT "POWER OUT = ";PO;A$
440   LET PI = (PO * 100) / E
450   PRINT
460   PRINT "POWER IN = ";PI;A$
470   GOTO 120
```

NOTES

Line 23 keeps track of your units.

Lines 90–110 select the correct subroutine for the answer desired.

Lines 240, 340, and 440 represent the formula variations.

Appendix

TABLE A.1 Numbers, Prefixes, and Symbols

Number		Prefix	Symbol
1 000 000 000 000	(one trillion)	tera	T
1 000 000 000	(one billion)	giga	G
1 000 000	(one million)	mega	M
1 000	(one thousand)	kilo	k
0.001	(one-thousandth of)	milli	m
0.000 001	(one-millionth of)	micro	μ
0.000 000 001	(one-billionth of)	nano	n
0.000 000 000 001	(one-trillionth of)	pico	p

TABLE A.2   Mechanical Properties of Materials

Material	Ultimate Tensile Strength (Stress)		Modulus of Elasticity (E)	
	lb/in.²	MPa	lb/in.²	GPa
Class 30 cast iron	30 000	207	15 000 000	103
Typical structural steel	64 000	441	30 000 000	207
Typical stainless steel	90 000	621	28 000 000	193
Ultrastrength steel	220 000	1517	30 000 000	207
Copper (annealed)	32 000	221	15 600 000	108
Typical aluminum alloy for structural shapes	45 000	310	10 000 000	69
Douglas fir (tension in bending)	12 200	84.1	1 760 000	12.1

Note: All numbers are accurate to at least 2 significant figures.

TABLE A.3   Miscellaneous Heat Values

Heat given off in combustion
  Wood/coal: 14 000 Btu/lb
  Spent coffee: 10 000 Btu/lb
  Gasoline: 20 000 Btu/lb
  Fuel oil: 19 000 Btu/lb
  Propane: 21 600 Btu/lb

Heat given off by average person (inactive):
  about 400 Btu/hr
Heat gained in a typical solar house: 90 Btu/hr·ft²
Heat gained by south-facing window: 1500 Btu/ft²·day

Answers to Selected Problems

Chapter 2
1. 110 N
3. 1140 kg
5. (a) 5.75 slugs
   (b) 84.0 kg
7. 13.2 in.
9. 1.60 m
11. 35.3 ft^3
13. 6.56 l
15. 0–150 000 g
17. 300 nanometers (nm)
19. 100 microseconds (μs)

Chapter 3
1. 61 000 lb
3. 30 800 lb
5. 2700 lb/in.^{2}
7. (a) 0.000 808 in./in.
   (b) 0.0121 in.
9. (a) 45 000 psi
   (b) 0.001 50 in./in.
   (c) 0.150 in.
11. (a) 21 900 psi
    (b) 0.000 730 in./in.
    (c) 0.003 65 in.
    (d) 11.7 in.^{2}

Chapter 4
1. 250 psi
3. 6.00 in.^{2}
5. (a) 6.40 MPa
   (b) 928 lb/in.^{2}
7. F_{pump} = 260 lb, F_{lift} = 2800 lb
9. 2360 psi
11. 0.11 lb/in.^{3}
13. 62.4 lb
15. 78 lb
17. 3.1
19. Mercury
21. p_{gage} = 483 kPa
23. 103 kPa
25. (a) 0.367
    (b) 15.1
    (c) 2530 Pa (2.53 kPa)
27. 9120 in.^{3}
29. 220 kPa
31. (a) 516 in.
    (b) 43.0 ft

Chapter 5
1. Figures 5.11 A and D.
3. (a) 14.1 lb
   (b) 20.0 lb

5. 216 lb
7. (a) 106 lb in each wire
   (b) No
9. 11.5 lb in each member
11. F_{boom} = 500 lb, F_{wire} = 707 lb
13. 103 ft/s, 31.4 m/s
15. (a) 243 in./s
    (b) 20.3 ft/s
17. 270 ft
19. (a) 4 ft/s
    (b) 3.01 ft
21. (a) 5.51 ft/s^2
    (b) 55 100 lb
23. (a) 54 100 mi/hr^2
    (b) 22.1 ft/s^2
25. (a) -14.0 ft/s^2
    (b) 1390 lb
27. (a) -11 000 ft/s^2
    (b) 34 200 lb
29. 80 ft/s

Chapter 6
1. (a) 500 ft·lb
   (b) 7500 ft·lb
3. 45 000 ft· lb
5. Hoisting 200 lb 60 ft

578

7. 5170 ft·lb
9. 6670 J
11. 147 GJ
13. 7300 ft·lb
15. (a) 120 000 ft·lb
    (b) 22 HP
17. (a) 4.4 HP
    (b) 3.3 kW
19. 5.5 HP

Chapter 7

1. (a) 11 500 ft/min
   (b) 131 mi/hr
3. (a) 81 ft/min
   (b) 27 ft/min
5. 6.5 in.
7. 1220 ft/min
9. 106 rpm
11. 3.8 m/s
13. 400 rpm
15. 637 rpm
17. 6.66 lb
19. 95 lb·in.
21. 95 N
23. 930 lb·in.

Chapter 8

1. $F_{applied}$ = 60 lb down,
   $F_{fulcrum}$ = 120 lb up
3. 95.8 lb
5. 751 N up
7. 1300 N up
9. 75.0 lb up
11. R_L = 357 lb up,
    R_R = 143 lb up
13. R_L = 3490 N up,
    R_R = 3370 N up
15. R_L = 200 N down,
    R_R = 70 N up
17. 2100 rpm
19. 663 rpm
21. 2100 rpm
23. The 22–28 teeth pair
25. 31 in.
27. 2 592 000 : 1

Chapter 9

1. 0°, 22°, 38°, −29°,
   −273°C

3. −40°
5. (a) 1220°F
   (b) 931 K
7. (a) 21 000 cal
   (b) 83 Btu
   (c) 88 000 J (88 kJ)
9. (a) 24 000 Btu
   (b) 6.0 Mcal
   (c) 25 MJ
11. Aluminum
13. 6.6 lb
15. 1.3 min
17. (a) 4800 Btu
    (b) 1.2 Mcal
    (c) 5 MJ
19. 7.6 lb
21. See Figure A.1.
23. 43 kcal
25. (a) 440 Btu
    (b) 2920 Btu
27. 5200 Btu
29. 39.8 in.
31. 0.056 in.
33. 250°F
35. $Dia._{cyl}$ = 99.844 mm,
    $Dia._{piston}$ = 99.701 mm
37. 688°F
39. 40 m³
41. 300 kPa
43. 233°C

Chapter 10

1. 8.02 CFM, 60 GPM
3. 0.0984 CFM, 0.736
   GPM

5. 7.3 in.
7. 49 000 Btu/hr
9. 3070 Btu/hr
11. 52 000 Btu/hr
13. 0.065
15. 3 in.

Chapter 11

1. 48.9%
3. 56.5%
5. 61.4%
7. (a) 115 000 Btu/hr
   (b) 45.2 HP
   (c) 37.3%
9. (a) 130 000 Btu/hr
   (b) 52 HP
11. 27.4%
13. 41.2%

Chapter 12

1. 0.183 m
3. 0.055 ft (0.66 in.)
5. 0.40 ft (4.8 in.)
7. 14.6 m
9. Steel
11. 0.0073 ft (0.088 in.)
13. 11 200 ft/s
15. (a) 7190 ft
    (b) 2190 m
17. 2.8 ft
19. (a) 0.84 s
    (b) 2.0 s

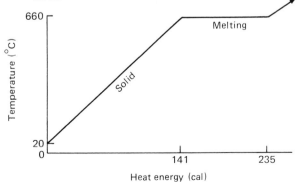

FIGURE A.1

21. (a) 188 Hz
    (b) 375 Hz
23. 1.0 ft
25. 200 Hz
27. (a) 9590 ft
    (b) 35.2 kHz

Chapter 14
1. 0.5
3. 2
5. 1.97 in.
7. 2.93 in.
9. 11
11. (a) Yes
    (b) 11.1 in.

Chapter 15
1. (a) 1.7 in.
   (b) 0.83 in.
3. $D_1 = 3.0$ in.,
   $D_2 = 2.0$ in.
5. 1/4 s
7. $D_i = 55.0$ in.,
   Image size = 20 in. ×
   20 in.

9. (a) 5.00 in.
   (b) 4.55 in.
   (c) 1.30 in.
11. Total magnification =
    6.10

Chapter 17
1. 2.00 A
3. 130 Ω
5. 0.500 A (500 mA)
7. 0.0200 V (20.0 mV)
9. 26 kW
11. 530 A
13. (a) 100 V
    (b) 400 V
15. (a) 3.60 kΩ
    (b) 720 V
    (c) $E_1 = 20.0$ V,
        $E_2 = 100$ V,
        $E_3 = 600$ V
17. (a) 81.1 Ω
    (b) 0.555 A
    (c) $I_1 = 450$ mA,
        $I_2 = 90.0$ mA,
        $I_3 = 15$ mA

19. (a) 110 mA
    (b) 3.3 A
21. 50.0 V

Chapter 18
1. 396 V
3. (a) 691 kΩ
   (b) 21.7 μA
5. (a) 60.0 Hz
   (b) 199 mA
7. 53.0 Ω
9. 77.3 Ω
11. 1510 Ω

Chapter 19
1. 5100 turns
3. 4.47:1
5. 2.00 A
7. (a) 2.61 kW
   (b) 58.7 MJ (16.3
       kW·hr)
9. 62.2%
11. 616 HP
13. (a) 290 MJ (80 kW·hr)
    (b) $4.80
15. (a) 28 kW
    (b) 220 kW·hr

Index

Key Formulas